Functional Electrical Rehabilitation

Chandler Allen Phillips

Functional Electrical Rehabilitation

Technological Restoration After Spinal Cord Injury

With 151 Illustrations

Springer-Verlag
New York Berlin Heidelberg London
Paris Tokyo Hong Kong Barcelona

Chandler Allen Phillips, M.D., P.E., Ph.D. (Hon.)
Department of Biomedical and Human Factors Engineering
College of Engineering and Computer Science/School of Medicine
Wright State University
Dayton, OH 45435, USA

Library of Congress Cataloging-in-Publication Data
Phillips, Chandler A.
Functional electrical rehabilitation : technological restoration after spinal cord injury / Chandler Allen Phillips.
p. cm.
Includes bibliographical references and index.
ISBN-13:978-1-4612-7796-5

1. Spinal cord—Wounds and injuries—Patients—Rehabilitation.
2. Exercise therapy. 3. Electric stimulation. I. Title.
[DNLM: 1. Electric Stimulation Therapy. 2. Spinal Cord Injuries—rehabilitation. WL 400 P558f]
RD594.3.P47 1991
617.4'820645—dc20
DNLM/DLC
for Library of Congress 90-10416
CIP

Printed on acid-free paper.

Softcover reprint of the hardcover 1st edition 1991

Typeset by Publishers Service of Montana, Bozeman, MT.

9 8 7 6 5 4 3 2 1

ISBN-13:978-1-4612-7796-5 e-ISBN-13:978-1-4612-3096-0
DOI: 10.1007/978-1-4612-3096-0

In memory of my father and brother,
Chandler and Robert;

In appreciation of my mother and sister,
Ann and Marilyn;

And thanks to my wife and the girls,
Janie and Nutmeg and Holly.

Preface

On one of my returns to California, I attended the "Disabilities Expo 88" at the Los Angeles Convention Center. Among the various marvels of technology for the wheelchair disabled were stair-climbing wheelchairs, self-raising and lowering kitchen cabinetry, and even a completely accessible "dude ranch" experience. At the same time, as a guest of the Southern California Chapter of the National Spinal Cord Injury Association, I was part of a small booth (among the more than two hundred exhibitors) in which we had spinal cord injured people up and walking with a lower-extremity bracing system (the reciprocating gait orthosis) used at the PEERS Spinal Injury Program in Los Angeles. I had a young man, a C6/7 level quadriplegic, walking with electrical muscle stimulation and lower-extremity bracing. The system is reviewed in Chapter 8 of this book.

As these "disabled" persons walked erect and upright among their wheelchair bound colleagues and took long, confident strides past exhibits extolling the latest technological virtues of yet another "new" wheelchair (Fig. 1), I reflected on the paradox of it all. What a majority of these paralyzed people were really looking for was an alteration of their disability so that they could more normally function (in an unaltered environment). What the great majority of the exhibitors were offering was an alteration of the environment so that they could more normally function (with an unaltered disability).

One association of spinal cord individuals has tersely condensed these observations to simply "cure not care." But they also miss the mark! For in their zeal to modify their disability, they go to the extreme by focusing on complete removal of their disability. The quest for "neural regeneration" of the injured spinal cord is laudatory. But it will not happen tomorrow, and more important, it is not here today, certainly not at the human species level. For these spinal cord injured individuals to continue to wait expectantly for their "cure" (and ignore modern prosthetic advances!) is equally as foolish as the amputee who rejects the artificial arm, hand, leg, foot, or other body part while waiting for "limb regeneration."

What is here today are modern prosthetic devices ranging from internal devices (ligaments, bones, knees, hips, etc.) to external devices (the Seattle foot, the Utah arm, etc.) and most recently, neuroprostheses (cochlear implants, functional electrical stimulation, etc.). The decade of the 1980s

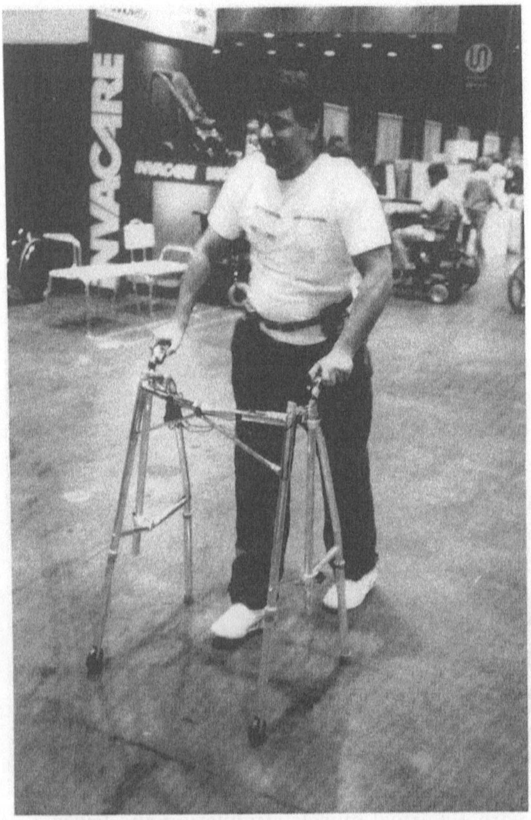

FIGURE 1. Quadriplegic subject walking with the EMS-RGO system at the Los Angeles Convention Center (June, 1988).

has seen tremendous development in these areas, and in one area especially–functional electrical rehabilitation.

Functional electrical rehabilitation is defined as the use of paralyzed muscle as an actuator that actively participates in the physical reconditioning of a previously deconditioned individual. It differs from conventional rehabilitation therapy, in which the paralyzed muscle is passive and is manipulated by various external influences. Functional electrical rehabilitation requires that paralyzed muscle be electrically stimulated in a manner that is functional (FES), i.e., the muscle is made to perform useful external work.

This book has been written for a broad and diversified audience. First and foremost, it has been written for physiatrists, orthopedists, and neurosurgeons who may wish to prescribe this emerging treatment modality for their patients. It is equally intended for the allied health professionals (such as physical therapists and rehabilitation nurses), who will be actively applying this technology for the benefit of spinal cord injured persons.

Other professionals should find this volume useful. For example, exercise physiologists should be particularly interested in the response of spinal cord injured individuals to these modalities. Also, biomedical engineers may find instructive the principles of theory and design that result in these clinically useful systems. Finally, this book should be of particular interest to any individual who desires to obtain the best rehabilitation that state-of-the-art technology can provide for spinal cord injury.

I have organized this book as a progressive narrative. Having worked in the field for more than a decade, I have synthesized systems, procedures, results, and conclusions. The original journal publications are extensively referenced for the interested reader. In the course of this unfolding narrative, it is hoped that new insights will be developed by the reader. It is certainly expected that the clinical application of functional electrical rehabilitation will be facilitated.

The individual chapters are organized to reflect the historical progression and chronological development of functional electrical rehabilitation. The theory and background of feedback control of paralyzed muscle movement is first introduced (Chapter 1). This general theme is continued with the application of motor feedback control to extremity prostheses (Chapter 2). The theory and background section is then concluded with the application of sensory feedback control to upper and lower extremity prostheses (Chapter 3).

The individual chapters then concentrate on clinical application. Stationary exercise rehabilitation is first introduced and overviewed (Chapter 4). There is then a description of the acute effects of the leg exercise system (Chapter 5). This is followed by a discussion of the acute exercise effects of the exercise bicycle system (Chapter 6). These two chapters reflect the physiological process. I conclude with a review of the chronic exercise effects that constitute the therapeutic outcome (Chapter 7).

Proceeding to ambulatory exercise rehabilitation, I introduce and overview the ambulatory exercise systems and effects (Chapter 8). Next, the ambulatory exercise systems are described specifically for both the paraplegic individual and the quadriplegic individual (Chapter 9). The clinical application of these systems is concluded with the ambulatory exercise procedures and resultant exercise effects (Chapter 10).

The final chapters are devoted to the prescription of functional electrical rehabilitation. Initially, I elucidate systematically the medical criteria by which the physician can be guided in the prescription of functional electrical rehabilitation (Chapter 11). Finally, the patient evaluation and prescription program is presented in detail (Chapter 12).

If this book is successful, it will expose the reader to new horizons in the clinical rehabilitation of the spinal cord paralyzed patient. It will unfold a panorama in which the mathematical, physical, and engineering sciences are assisting the medical sciences in controlling the movement of paralyzed muscle so that improved physical rehabilitation may be realized. The end result is high-technology, state-of-the-art rehabilitation modalities that result in a positive therapeutic outcome for the spinal cord injured individual.

The material presented in the following pages is my interpretation of the current state of the art. As a scientist who can express only cautious optimism, I would request that my readers regard this publication as simply a first edition: My eyes will continue to gaze toward the future; my feet (of necessity) are planted in the present; and in my heart I acknowledge our debt to the past.

Chandler Allen Phillips

Acknowledgment

During the course of this book, I will often refer to the "Wright State group" or use the plural personal pronoun for it. Numerous individuals (in various capacities and at various periods from 1978 through 1990) were and are responsible for the work reported in this book. Twenty-four individuals, however, met the following criteria: (a) authored a published article that is referenced in this book; and (b) were simultaneously affiliated with Wright State University.

The twenty-four individuals of the "Wright State group" include (in alphabetical order): Jose Almeyda, Roderick Briggs, Carol Brunsman, John Buhrman, William Colby, William Couch, Damian Danopulos, Paul Danset, Roger Glaser, David Hanpeter, Denise Heard, Harry Heaton, Debra Hendershot, Paul Kezdi, Richard Koubek, Loretta Meyer, Mary Moore, Jerrold Petrofsky, Chandler Phillips, Michael Sawka, Ralph Stacy, Donald Stafford, Edwin Stanley, and Cynthia Weber.

I hereby acknowledge the contributions of the "Wright State group," with the following thought:

"The hurried scribbles in our laboratory notebooks are intelligible only to ourselves, and the seminar and lecture have only a temporary and narrow influence. The published record, however, is permanent; it is there for all time as a source of pride or shame as the case may be."*

*From H.B. Vickery, To Rudolph J. Anderson, © 1958 American Society of Biological Chemists. Reproduced with permission from *Journal of Biological Chemistry*, 233:1249–1250, 1958.

Contents

Part 4 Prescription of Functional Electrical Rehabilitation

Part 1
Theory and Background

1
The Theory and Background of Functional Electrical Rehabilitation

Currently, a spinal cord injury patient is confined to life in a wheelchair, with occupational and vocational rehabilitation being geared to the limitations thus imposed. On the other hand, the possibility of electrical stimulation of paralyzed muscles to control (for example) bladder and limb function would considerably enhance the quality of life of a paraplegic. To control complex limb movements, a microprocessor can be used to control and stimulate muscles isometrically (i.e., to exert tension while muscle length remains constant), to stabilize the joints, as well as permit isotonic control (i.e., to enable muscle to shorten while exerting a constant force) while lifting a load.

Our early work involved a computer-controlled system (for isometrically stimulating muscles) to exert proportionate control over tension and to minimize muscle fatigue by means of an electrode array around the peripheral motor nerve. We then developed a computer-controlled system (to stimulate isotonic contractions). The system first stimulates a known number of motor units, measures the target velocity, computes the appropriate number of motor units that must be recruited, and applies appropriate increases in stimulation voltage to the electrode to enable the muscle to move at the proper velocity.

Computer-controlled stimulation of muscle groups to induce movement of paralyzed muscles requires control of voltage level to ensure stimulation of the appropriate number of agonist motor units to achieve the corresponding target tension, as well as stimulation of some antagonist activity to achieve the required target velocity. This chapter overviews the logic and principles involved in the development of the various systems employed and also the experimental studies directed toward the realization of functional electrical rehabilitation.

I. Introduction and the Basic Animal Studies

A. Introduction

During voluntary movement, nerve impulses leave the brain and travel down the spinal cord. These impulses travel to the nerves that go to skeletal muscle. Cell bodies of these nerves, or alpha motor neurons, are found in the various sections of the spinal cord (Fig. 1.1). The impulses are conducted between the brain and these alpha motor neurons by a second set of neurons called interneurons. These interneurons are similar in many ways to the large cables that conduct impulses throughout a computer. The spinal cord is constructed of millions of interneurons, which connect signals from the periphery to the brain and from the brain back to the peripheral organs, glands, and skeletal muscle. In the normal healthy individual, thousands of nerve impulses travel in both directions up and down the spinal cord, enabling the muscles of the body to move in a coordinated fashion. This requires both impulses traveling from the brain and a very fine system of sensors (which respond to the tension generated in muscle, the length of muscle, and the position of the joints). The sensory information is called afferent nerve traffic and travels up the spinal cord toward the brain. However, if for any reason the nerves become damaged, this

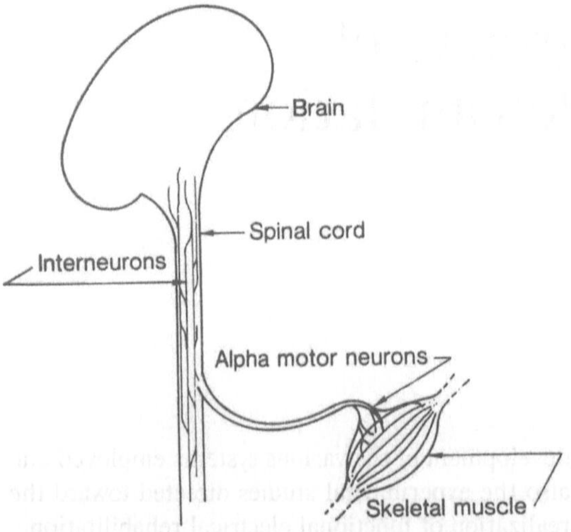

FIGURE 1.1. Neuromuscular system organization.

system of communications between the brain and skeletal muscle is lost. If the motor nerves that go to the muscle itself are not damaged, however, then the muscle remains healthy, although it becomes much smaller as a result of disuse atrophy (Guttman, 1976). Although such a muscle is still fully functional, information from the brain is never transmitted to it; therefore, the muscle remains paralyzed for the patient's entire life. This type of disability can occur because of war-related injuries to the spinal cord, or as a result of stab wounds, birth defects, stroke, tumors, and compression injuries from sports and auto accidents, as well as from a variety of other sources.

Until World War II, this area of injury health care was largely ignored because, before this time, most people had died within a few weeks of any spinal injury. There were two causes of death: spinal shock and urinary tract infection. Normally the bladder is under voluntary control, but since it is impossible to control bladder emptying with some types of spinal injury, urine must be emptied via an indwelling catheter. This causes a high incidence of bladder infections in paralyzed patients. Once an infection has developed, it spreads to the kidneys and can be fatal. The appropriate therapy for renal and bladder infections is an antibiotic (i.e., the sulfa drugs). However, these drugs were not discovered until the 1930s.

With the advent of the sulfa drugs and appropriate treatment for spinal shock, a person sustaining a spinal injury can live a normal lifespan. However, the average age of those who sustain spinal injuries is 25 to 30 years, and victims can anticipate living until age 70–80. The cost of health care of maintaining a spinal patient over this period of time is enormous. As of 1980, it was estimated that the cost of maintaining a spinal patient for conventional medical care alone was about $2 million per patient per lifetime.

The current therapy is to place a spinal patient in a wheelchair. Although this allows mobility, it causes both psychological and physical problems for the patients. For example, there are barriers preventing such patients from moving freely in, around, and out of many buildings. Furthermore, there are the psychological problems due to feelings of dependence induced by being continuously in a wheelchair. Finally, there is a great deal of cardiovascular deconditioning associated with life in a wheelchair.

One alternative to existence in a wheelchair would be to electrically stimulate muscle. Because of Galvani's initial experiments with electricity, it has been known since 1791 that muscles will contract when an electric current flows through them. For this reason, a number of people in the 1900s tried using a variety of electrical stimulation techniques to cause muscles such as the bladder or diaphragm to move (Fig. 1.2). In both these cases, the muscles either completely contract or entirely relax. It is, therefore, simply a matter of placing electrical wires into the tissue and stimulating the area with an appropriate voltage source to cause contractions. However, complex movements such as standing and walking require proportional control of movement and coordination between the movements of various muscles.

If electrical stimulation of paralyzed muscles is to be employed to control complex movements, muscle must be controlled under two types of condition. To stabilize the joints, it is necessary to have muscles contract isometrically (i.e., the muscle exerts tension but does not change length). Two criteria must be satisfied during these isometric contractions. First, the contractions must be controlled to allow for proportional control of tension. Second, this control must be achieved in a manner that minimizes muscle fatigue. For example, the best electrical control system would be inadequate if the muscle fatigued so rapidly that the person fell down.

To allow the limbs to move, it is further necessary to control muscle contractions when the muscle changes length while lifting a constant load. During these isotonic contractions (contractions exerted during phases of walking), it is necessary to control the velocity of movement of muscle very precisely and again to do this in a manner that minimizes muscle fatigue. With these criteria before us, we set out to design a computer control system capable of controlling movement in muscle.

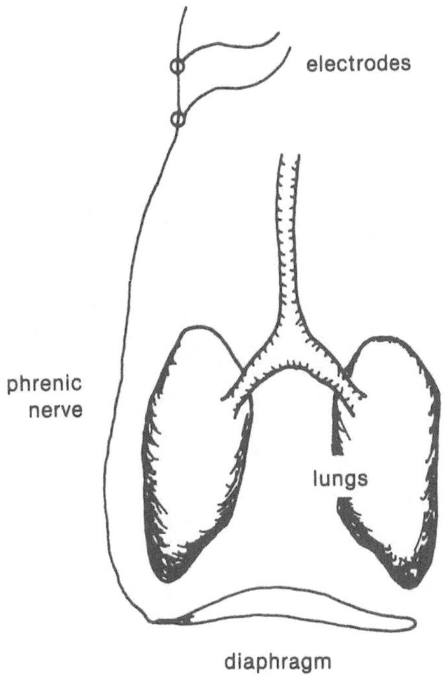

FIGURE 1.2. Example of electrophrenic respiration utilizing on–off control.

B. A Microprocessor-Controlled Stimulator for Isometric Contractions

The microprocessor-controlled stimulator system was developed using an Intel 8080A microprocessor. To stimulate muscle in a manner analogous to the technique normally used by the body during voluntary activity, an interface was designed that would allow the computer to control the frequency of firing of the motor nerves and would allow us to vary the number of active motor units (recruitment). Normally, when a muscle develops a tension, the tension is varied by altering both the frequency of motor unit discharge and the number of active motor units (temporal and spatial recruitment, respectively). These two variables were controlled by developing an electrode array capable of varying the frequency that the motor nerves were firing at over the range of 10–100 hertz (Hz) and capable of varying the number of motor units that were firing (in increments of 0.5%) up to 100% of all the motor units (Petrofsky, 1978).

In the initial experiments, stimulation was accomplished through the spinal cord. This involved using electrode arrays (Petrofsky, 1979). To vary the number of motor units that were firing (spatial recruitment), the stimulation voltage was increased. Since some nerves are more sensitive to electrical stimulation than others, increasing the stimulation voltage made possible the recruitment of more and more motor units. To vary the frequency of discharge, the frequency of firing of the electrodes was increased (temporal recruitment). Electrode stimulation involved a monophasic square wave stimulus of variable amplitude and a width of a tenth of a millisecond (0.1 ms). The program used by the computer was based on the recruitment patterns used in man to sustain isometric contractions (Milner-Brown and Stein, 1975). To achieve a given tension, motor units were recruited at a frequency of 10 Hz. Recruitment was varied to achieve the target tension. If recruitment alone did not result in the development of enough tension in a muscle, the frequency of stimulation was increased (Fig. 1.3).

In studies dating back to Galvani's late eighteenth century experiments, only a single electrode pair was generally used to stimulate nerves or muscle. Although tension could be developed with only a single electrode pair, very high frequencies (> 200 Hz) of stimulation were necessary to achieve smooth contractions. However, such high frequencies of stimulation result in rapid muscle fatigue. To reduce the frequency of stimulation and enhance the controllability of tension, the motor nerves going to the muscle were therefore divided into three populations. If each of these populations is stimulated sequentially inside the spinal cord, a muscle will develop a smooth maximum contraction at frequencies of only 50 or 60 Hz (Rack and Westbury, 1969; Lind and Petrofsky, 1978). This is because the increments of tension (which is

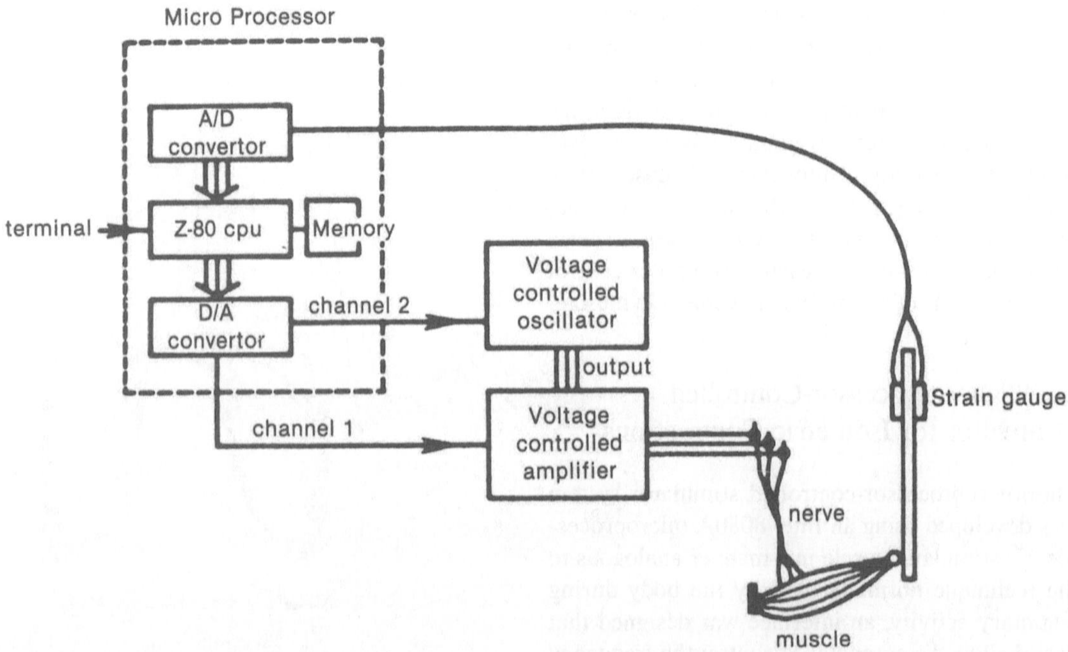

FIGURE 1.3. An Intel 8080A microprocessor-controlled stimulator.

developed by contraction of each of the three parts of the muscle) overlap and make the contraction smoother. This is well within the range that the brain normally uses to control muscle movement.

In our initial experiments, we found that with this stimulation system we could in fact control tension in muscle. However, even at a set stimulation voltage and frequency, there is a great deal of variation in muscle tension. Therefore, we used feedback tension from the muscle to aid the computer in controlling tension. This was accomplished by connecting the muscle to an isometric strain gage bar (Fig. 1.3). The electrical output (of a Wheatstone bridge from the strain gages) was fed into the computer through an analog-to-digital converter. This closed-loop feedback system maximized the controllability of tension.

In addition, the fatigability of the muscles was quite low. One way of measuring fatigability during isometric contractions is to measure endurance: that is, the ability of a muscle (e.g., handgrip muscle) to maintain an isometric tension smoothly at a set load. Loads can be set at various percentages of the muscle's maximum strength and plotted against the endurance times (in seconds), as described by Rohmert (1968). It is then found that the endurance (in minutes) is inversely proportional to the percentage of maximum muscle strength.

For low tension isometric contractions, the endurance is infinite. In contrast, when the tension is increased above 15% of the muscle's maximum strength, endurance is only a few seconds (i.e., contractions above 70% of the muscle's maximum strength cannot be sustained). When we measure the endurance of cat skeletal muscle utilizing our microprocessor (feedback) control system and compare the results to the endurance of human subjects, the similarity is striking. It appeared, therefore, that we had achieved our design objective of developing a computer-controlled stimulator system whereby tension could be developed smoothly in muscle and with low fatigability.

In these early experiments, stimulation was done in the spinal cord. Spinal cord stimulation, however, provides a pathway for bacteria to enter the spinal cord and cause fatal infections when electrodes are placed on the spinal cord. Furthermore, the spinal cord is very sensitive, and surgery often permanently traumatizes the neurons. Therefore, we developed a system whereby we could stimulate

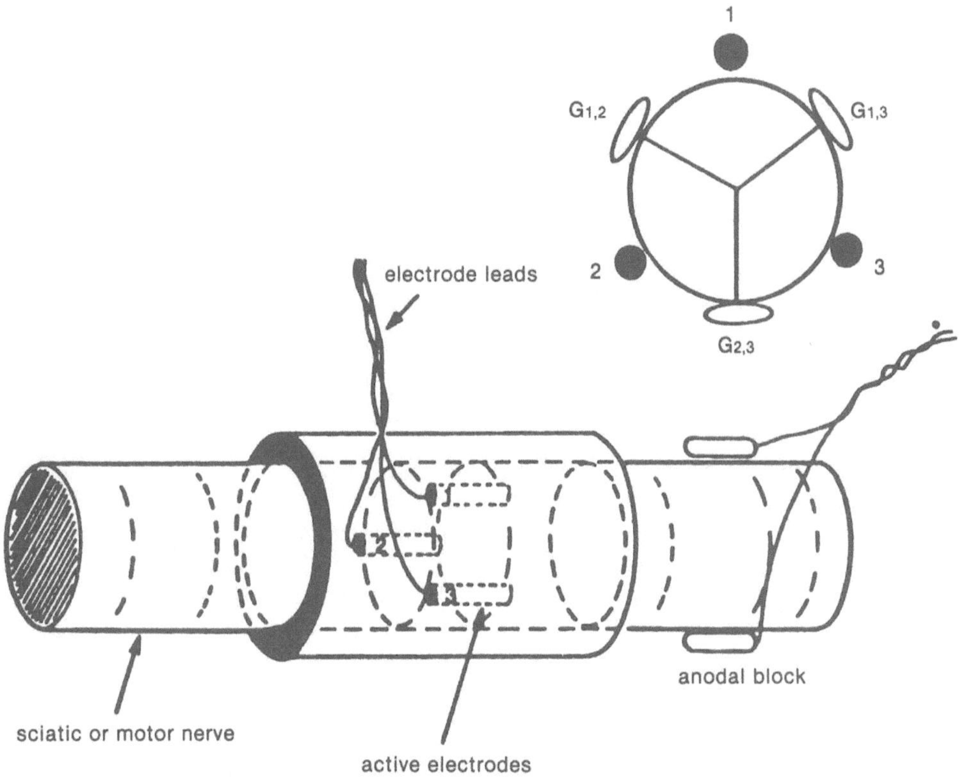

FIGURE 1.4. Representation of the sleeve electrode. (From Petrofsky and Phillips, 1979a, with permission of the Institute of Electrical and Electronic Engineers, © 1979 IEEE.)

muscle through peripheral motor neurons, and called it a focused field stimulation system (Petrofsky and Phillips, 1979a; 1981a).

An electrode array was developed that could be placed around the peripheral motor nerve (Fig. 1.4). As was the case with stimulation through the spinal cord, we wish to mimic the normal asynchronous patterns of stimulation that occur during voluntary activity. Placing three electrodes around a motor nerve allows many nerves to be stimulated twice. Therefore, another series of three electrodes was added. These three electrodes were anodal blocking electrodes. The application of a reversed-bias voltage to these electrodes prevented stimulation of the adjacent nerves. By combining voltages on the stimulation electrodes and the blocking electrodes, the stimulation field could be focused in a manner similar to that by which the beam of a cathode ray tube is focused.

By using this focused field electrode system, groups of motor neurons could be stimulated sequentially through peripheral motor nerves. This type of surgery is far less traumatic than spinal surgery. In fact, it can be accomplished fairly easily in most situations. With the focused field electrode array and the sequential method of motor unit stimulation, we had achieved our initial design objective for isometric contractions. Specifically, we had developed a method of stimulating muscle to control tension and to do it with low fatigability (Fig. 1.5).

C. Computer Control and Movement During Isotonic Contractions

The problem of controlling movement during isotonic contractions proved to be far more complex. The programs described above for proportional control of tension required feedback control of tension to adjust the computer program to the proper stimulation voltage and frequency. Although muscles

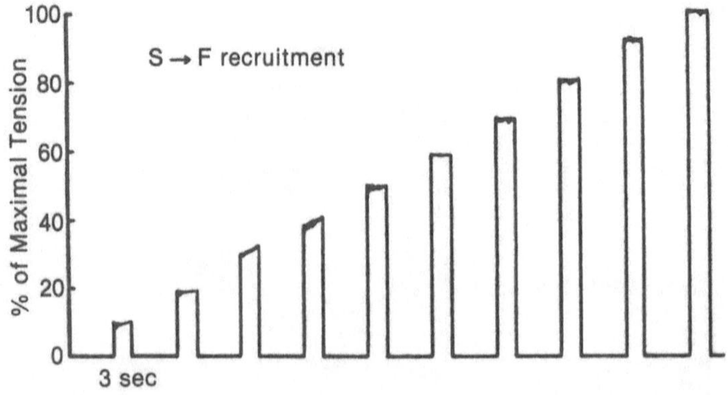

FIGURE 1.5. The tension developed during brief isometric contractions: recruitment proceeding from the slowest to fastest units. (From Petrofsky and Phillips, 1979a, with permission of the Institute of Electrical and Electronic Engineers, © 1979 IEEE.)

develop tension very rapidly when stimulated to contract isometrically, when muscles are stimulated and allowed to shorten, there are long time constants involved, as indicated in Figure 1.6 (Hill, 1938; Petrofsky and Phillips, 1979b). Following stimulation, the muscle does not reach its peak velocity of movement for about 30 ms. Thus when using a feedback control program such as the one described above, seconds of continual adjustment of the stimulation voltage might be necessary before the muscle reached its target velocity. However, during normal voluntary movement in man, muscles must reach their target velocity within about a tenth of a second (if a specific muscle action is to be coordinated with that of the other muscle groups). For this reason, simple feedback control is entirely inadequate for this type of muscle stimulation.

Consequently, we developed a different system (Fig. 1.7). The purpose of this computer program was to stimulate a known number of motor units (in this case, 10% of the motor unit population) and then measure the target velocity 30 ms later. Once this had been accomplished, the number of motor units to be recruited could be calculated mathematically and the proper increase in stimulation voltage applied to the electrode, thus allowing the muscle to move at the desired velocity. This program should require approximately 60 ms to operate and bring the muscle to its proper target velocity. The accuracy of the program depends on the accuracy of the equations modeling the relationship between the motor units in the muscle and their movement. The complex series of experiments that was necessary to mathematically model the variables involved the load on the muscle, the maximum strength of the muscle, and the velocity of the contraction of the muscle. The equation could then be empirically derived.

In a series of experiments, the number of motor units in the medial gastrocnemius muscle of the cat

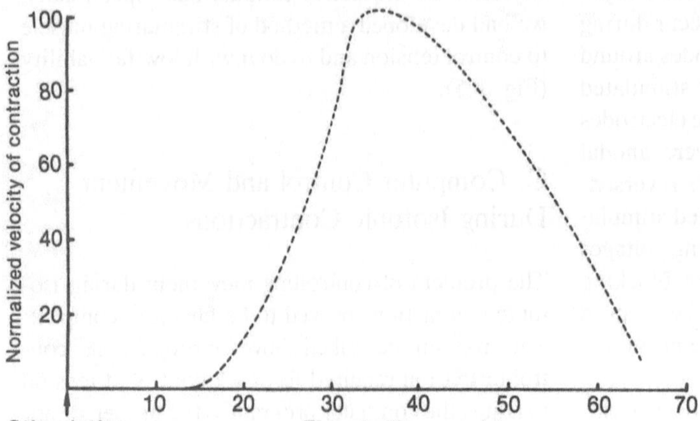

FIGURE 1.6. Velocity of muscle following stimulation.

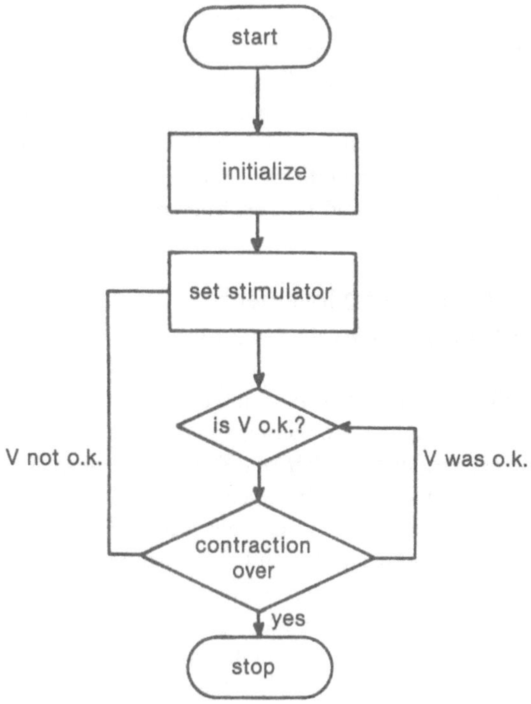

FIGURE 1.7. Computer algorithm for muscle stimulation. (From Petrofsky and Phillips, 1979b, with permission of the International Federation for Medical and Biological Engineering.)

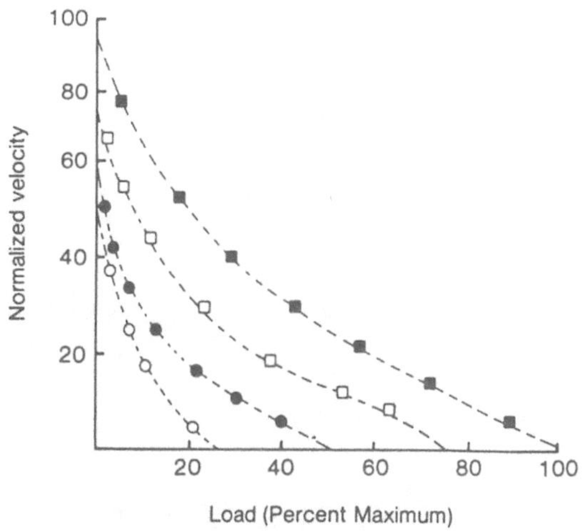

FIGURE 1.8. Force–velocity relationship in cat medial gastrocnemius muscle at four levels of activation corresponding to recruitment when 25, 50, 75, and 100% of the motor units are active.

was varied between a few motor units and the recruitment of all the motor units in the muscle. The velocity of contraction was measured during these episodes. By repeating these experiments at a wide variety of loads up to 100% of the muscle's maximum strength, the empirical data base was established as shown in Figure 1.8. It was then simply a matter of determining the equation relating activation, velocity of contraction, and the maximum strength of the muscle. This equation was originally based on the equation of A.V. Hill (1938). Hill measured the force–velocity relationship in frog skeletal muscle and discovered that in most muscles it could fit the following basic equation:

$$V = \frac{(P - p)b}{P + a} \qquad (1.1)$$

where a is the Hill A coefficient, b the Hill B coefficient, V the velocity of contractions of muscle, p the current load on the muscle, and P the maximum strength of the muscle. Our current work, however, indicates that this equation may need to be modified to account for the increased passive stiffness of the cat skeletal muscle (Phillips and Petrofsky, 1981).

Using Equation (1.1), we added the term A to represent the activation level of muscle. This term is defined as the fraction of the number of motor units that were firing in a given muscle, such as the medial gastrocnemius in the cat. The new equation for the force–velocity relationship including activation is (Petrofsky and Phillips, 1980):

$$V = \frac{(P_{mx} - P)\,(0.571 + 6.007A - 2.673A^2)}{P + 0.657 + 1.734A} \qquad (1.2)$$

where V is the velocity of contraction of muscle, P the current load on the muscle, P_{mx} the maximum strength of the muscle, and A the activation level.

Control equations for the velocity of muscle contraction have subsequently been expanded to include fiber composition (Phillips and Petrofsky, 1980) and muscle temperature and initial length (Petrofsky and Phillips, 1981b), in addition to muscle load and activation level. Such equations would make it possible to control a variety of muscles performing active movement in a dynamic environment.

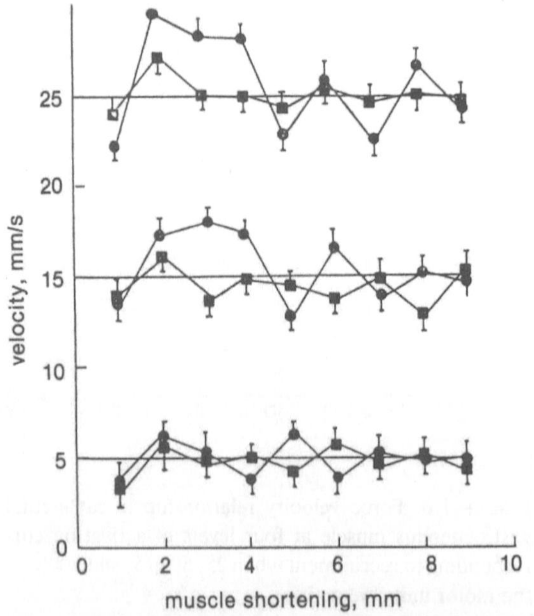

FIGURE 1.9. Velocity of shortening of the medial gastrocnemius muscle during isotonic contractions against no load. Three test velocities were set as targets (5, 15, and 25 mm/s). Each data point illustrates the mean of four measurements, plus or minus the respective standard deviations for no antagonist activity (●) at 5 and 10% antagonist activity (■). (From Petrofsky and Phillips, 1979b, with permission of the International Federation for Medical and Biological Engineering.)

Using Equation (1.2) with the algorithm (Fig. 1.7), we have been able to control the velocity of muscle movement. For example, in the cat medial gastrocnemius muscle, we set the computer to control target velocities of 5, 10, and 15 millimeters per second (mm/s) through a shortening distance of 10 mm.

The computer was able to track the target velocity fairly accurately by using the algorithm. As mentioned above, one important criterion in electrically stimulating muscle is minimizing the degree of fatigability in the muscle. As was the case for static exercise, the fatigability of the muscle to sustain workloads at various percentages of the muscle's maximum work capacity with electrical stimulation was minimal. In fact, during electrically induced stimulation, the muscle could work at up to half of its maximum work capacity for an indefinite period of time with no indication of muscle fatigue. This is similar to the pattern of fatigability that is seen in man during work on the bicycle or in running. Typically in man, work can be sustained at levels at up to 50–75% of the maximum working capacity for indefinite periods of time. Above this point, the muscle begins to produce lactic acid and fatigues very rapidly.

D. Control of Movement in Multiple Muscle Groups

Although stimulation of a single muscle was a first step, movements are accompanied by the synergistic activity of a number of different muscles. Two different muscles are involved for the movement of any one joint. The muscle that moves the joint in the direction desired is called the agonist muscle, and the muscle that moves the joint in the opposite direction is called the antagonist muscle. In general activity, movement is a combination of these muscle groups. Although most of the movement is controlled by the agonist muscle, some antagonist activity is necessary to dampen agonist activity. This allows us to precisely control that movement.

To study the relationship between agonist and antagonist muscle activity, we performed a series of experiments in which we stimulated the medial gastrocnemius muscle of the cat leg (an agonist that extends the cat's foot). We also (simultaneously) dampened the activity of this muscle with its antagonist (tibialis anterior). The experiments were similar in design to the ones described above in which we altered the velocity of contraction of muscle. Here, however, the medial gastrocnemius muscle was stimulated to shorten over a distance of 10 mm at target velocities of 5, 10, and 15 mm/s. During this shortening, the actual velocity of the muscle was measured with no activity of the antagonist muscle and with the computer program set to supply 5 and 10% of the maximum strength of the tibialis anterior muscle to dampen the overshoots in the velocity of the medial gastrocnemius muscle.

The program then was very straightforward. At the onset of the contraction, 10% of the motor units were recruited. The muscle was then stimulated, and 30 ms later the target velocity was measured. At that point, the actual number of motor units required to bring the muscle to the proper target tension was calculated and the stimulation voltage

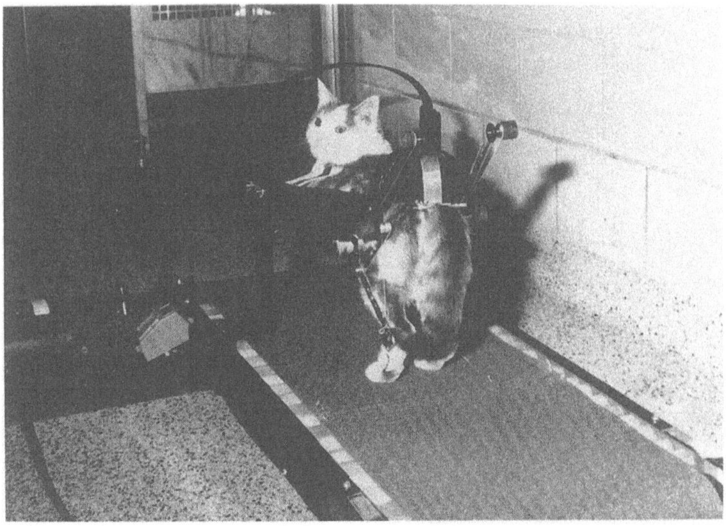

FIGURE 1.10. Cat walking with a sensor harness.

was increased appropriately. Once the stimulation voltage had been increased, the velocity was again measured. If the velocity was below that of the target, stimulation was increased. However, if the stimulation voltage was too great, there was both a reduction in stimulation voltage and a small burst of activity at either 5 or 10% of the muscle's maximum strength in the antagonist muscles to dampen the overshoot. The results of these experiments were significant. Using antagonist activity allowed the muscle to track the target velocity even better (at any three of the velocities examined), as shown in Figure 1.9.

These studies show the feasibility of using computer-aided stimulation of multiple muscle groups to return movement of paralyzed muscles. However, complex movements such as walking require many muscle groups to be relaxing and contracting out of phase with one another. In man, as many as 40 muscles are involved in simply walking on a level surface. We have conducted our initial experiments on multiple muscle unit movement in the cat. Basically, we then proceeded to model the movement of cat skeletal muscle during walking. To minimize the number of muscles involved in walking, we developed the harness array illustrated in Figure 1.10 (Petrofsky and Phillips, 1979a). The harness accomplishes two purposes. First, it contains a number of sensors, which sense the position of the joints in the cat's leg. These sensors incorporate ten-turn potentiometers. The force necessary to move the potentiometers is approximately 0.5 gram. The output of the potentiometers goes into a Wheatstone bridge, and the output of this bridge is fed into the computer for feedback positioning of the joints. This required some modification in the computer program because the position of a joint was used as feedback.

By stabilizing the movement of a potentiometer, the effectiveness of the isometric contraction can be assessed. To measure velocity of movement, it is necessary only to differentiate the output of the Wheatstone bridges. In addition to the potentiometers, a footswitch is located at the bottom of the cat's foot so that the computer can determine when the foot has made contact with the treadmill.

The second function the harness array performs is to minimize the number of muscles involved in walking. Forty muscles would have been far too many to stimulate in our initial studies. Therefore, the harness array serves the same purpose as the braces worn by polio patients. Normally the legs move in three axes: they move forward and backward, rotate sideways right and left, and finally move up and down. The harness array stabilizes the sideward rotation so that a single plane of movement is forward and back and up and down. We inserted wire electrodes into the skeletal muscle in the cat's legs in order to measure the electrical activity. We then trained the cat to walk on the treadmill. We found that instead of 40 muscles being active during walking, the harness array

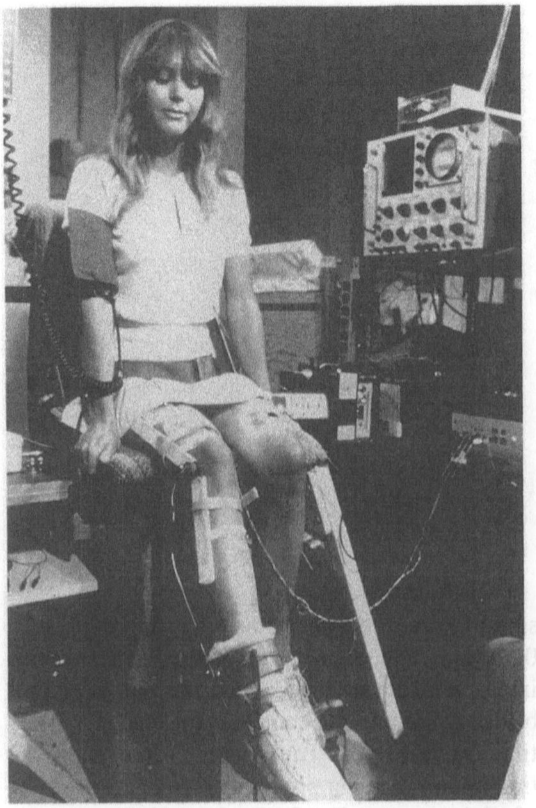

FIGURE 1.11. Computer-controlled leg exercise utilizing positional feedback.

stabilized the leg joints so that only 12 muscles were active during walking. This is a much more manageable number and has practical advantages when stimulating human skeletal muscle to induce walking in a paralytic.

II. The Early Human Studies

Electrical stimulation results in muscle movement due to the development of nerve impulses or action potentials on nerve and muscle. This follows the application of electrical current to the skin, motor nerves, or the muscle directly. To the active muscle, there is little difference between the normal nerve impulse and that generated by electrical stimulation. Although a number of investigators in Europe and the United States have attempted development of systems to stimulate paralyzed muscle, problems have arisen.

The first problem is that the degree of movement and the speed of movement of the muscle (associated with a given burst of electrical activity applied with a pair of electrodes) is a function of fatigue, position of the electrode, impedance of the electrode–electrolyte interface, muscle temperature, and a large number of other parameters (Petrofsky et al., 1979; Phillips and Petrofsky, 1980, 1981). Consequently, we developed a mathematical model to control movement during various types of activity in cat skeletal muscle (Petrofsky and Phillips, 1979b, 1980, 1981b; Phillips and Petrosfky, 1980, 1981, 1983). The development of such control algorithms allowed precise control of movement in cat skeletal muscle during electrical stimulation and has been reviewed in Section I.

A second problem is associated with evaluating the physical effects on the body of spinal cord injury and finding ways to reverse those effects. Following spinal cord injury, parts of the body cannot be used much, if at all. Therefore muscles decondition and atrophy, and muscular strength reverts to a fraction and its initial value. Atrophy of muscle and poor circulation also predispose the patient to pressure sores, which can be further incapacitating. Finally, the cardiovascular and respiratory systems decondition as a result of disuse, and blood volume is also reduced. Consequently, subjects who have high level spinal cord injuries may have orthostatic hypotension. These are the issues that needed to be addressed in man by both evaluating the stress to the body (associated with stimulation) and finding an approach for reconditioning the body.

In the past few years, several systems have been developed and are reviewed in detail elsewhere in this book. One such system is a mini-exercise gym involving resistance exercise (Petrofsky and Phillips, 1983a). Although such systems are commonly found in gymnasiums used by able-bodied people, the systems developed were different in that they could be utilized by paralyzed individuals. Special seat backs had to be used. Special feedback sensors (on the exercise trainers) were used to give the computer information regarding the position of the muscles. This would allow control of movement.

Resistance exercise has been shown to be a most effective means of training for muscular strength. An early example of a leg exercise trainer with computerized control of movement through elec-

FIGURE 1.12. Exercise bicycle ergometer utilizing functional electrical stimulation.

trical stimulation has been described (Petrofsky and Phillips, 1983a) and is shown in Figure 1.11. The leg exerciser was not only able to increase leg strength rapidly (leg strength increased after one month of training), but also resulted in an increase in resting blood pressure and a more stable blood pressure response to exercise after only a few weeks of physical training.

However, resistance exercise systems are a very poor means of conditioning muscle endurance and the cardiovascular system. The best way to condition these systems is by aerobic activities such as bicycling or running. Therefore, an indoor bicycle was developed to be used by paralyzed individuals (Phillips et al., 1984). An early commercial version is shown in Figure 1.12. The indoor bicycle also used special sensors and a special seat back to hold the user in place. The users (under computer control) could pedal the bicycle and experience muscular endurance training and cardiovascular fitness training. With these procedures, the mean resting blood pressure underwent an average increase after one month of workout in four quadriplegic subjects.

In addition to the significant increase in mean resting blood pressure (ameliorating the problem of syncope with changes in body position), mean exercising blood pressure underwent an average reduction in peak systolic pressure (Phillips et al., 1984). Consequently, the bicycle and the leg trainer were found to increase cardiovascular stability and also to result in the regaining of muscle size.

In summary, techniques have been developed to recondition the body following a spinal cord injury. These systems are significant because they can be used for the treatment of any type of spinal cord injury. When a cure for spinal cord injury (either chemical or surgical) is found, it can be applied successfully because we now know how to recondition the body.

The ultimate goal—to restore walking in the paralyzed individual—has been successfully accomplished in our laboratory for paraplegic and quadriplegic subjects (Petrofsky and Phillips, 1983b). The major problem in walking was control. To create ambulation, 10 muscle groups were chosen, with a view to controlling specific joint motions. The result represents our early attempts at ambulation in paraplegic and quadriplegic subjects. For more detail, see Chapter 2.

III. The Later Human Studies

A. Introduction

Computer-controlled walking by functional electrical stimulation (FES) at Wright State University in Dayton, Ohio, has historically used closed-loop control (Petrofsky and Phillips, 1979b, 1983b); for a review, see Petrofsky and Phillips (1985). Closed-loop control utilizes the data from sensors to modify the output of a controller. To help coordinate walking, sensors were placed on the hips, knees, and/or ankles in various experiments to provide positional data for a computer controller (Petrofsky and Phillips, 1983b; Petrofsky et al., 1984). This type of control was developed in animal experiments in our own lab (Petrofsky et al., 1976; Petrofsky and Phillips, 1979b) and subsequently was partially described by others (Solomonow et al., 1978; Crago et al., 1980). Such methods of control provide just enough stimulation to achieve a given movement.

Two major problems have limited the use of FES technology to achieve ambulation of the spinal-cord-injured individual. First, there is a high energy cost of movement induced by FES which results in rapid muscular fatigue. FES is optimally applied during the "swing phase" of gait and results in movement of the individual. However, FES is also applied during the "stance phase" of gait to maintain an upright posture. This does not result in movement, but the tonic (isometric) contraction of the paralyzed muscle contributes significantly to fatigue.

A second major problem is that the subject is not protected from falling. Postural instability is associated with stimulating only a few muscle groups (out of the total population of paralyzed muscles). In addition, electronic components can periodically malfunction, resulting in partial or total loss of the tonic electrical activity (which maintains the patient upright).

The reciprocating gait orthosis (RGO) was used to help resolve some of these problems and is described more fully later. The RGO is a long leg brace with a hip joint, pelvic band, and thoracic supports constructed of plastics and lightweight aluminum (Douglas et al., 1983). In our opinion, the pelvic band and thoracic support offer better postural stability for mid- and high-level paraplegics than other braces (e.g., Craig-Scott).

A significant feature of the RGO is its cabling system. By using two cables attached to the anterior and posterior aspects of the hip joints, a reciprocating action at the hips is produced. Thus upper body weight is shifted with posterior movement of the shoulder. This shift, which results in the forward thrust of the pelvis with hip extension, is followed by a transfer of weight onto the arms (and onto a walker or crutch handle) and a slight rotation of the main axis of the body. For full ambulation (when weight is on the contralateral hip), the FES can provide contralateral hip extension such that the posteriorly placed ipsilateral leg is moved forward. The operation and application of this "motor efferent" neural prosthesis is described in detail elsewhere (Petrofsky et al., 1985). This FES-RGO (computer-directed FES combined with an RGO) walking system was utilized to obtain the results described below and is shown in Figure 1.13.

We have noted that the paralyzed individual (who uses the FES-RGO walking system) must rely on some degree of sensory feedback, and this is generally visual. Basically the subject looks down at the floor, visually determines the sequence of walking, and adjusts his upper body accordingly. Although this is satisfactory in a controlled laboratory setting, it is not practical for routine walking in other environments. The person's visual senses should be directed toward more important activities (such as interacting with the environment), to facilitate appropriate reactions. Auditory feedback would be one approach, but we believe that this sensation should also remain unrestricted. The person can then better interact with other environmental stimuli. Consequently, our approach has been to employ sensory feedback by means of tactile stimuli.

Numerous investigators have demonstrated the usefulness of sensory feedback through tactile stimulation. This has frequently involved electrocutaneous stimulation, whereby the skin is stimulated with electrical impulses (via surface electrodes). For example, it has been used for sensory feedback with an arm prosthesis (Shannon, 1979; Scott et al., 1980). Sensory information to the skin has been very beneficial to individuals with sensory disabilities (Pfeiffer, 1968). A few studies have also been conducted to systematically study different tactile codes that might be utilized for electrocutaneous stimulation (Rollman, 1969; Szeto, 1982).

Our approach differs from this previous work in that we have employed vibrocutaneous stimulation. A four-channel stimulator system and complete carrier suppression are also features of our

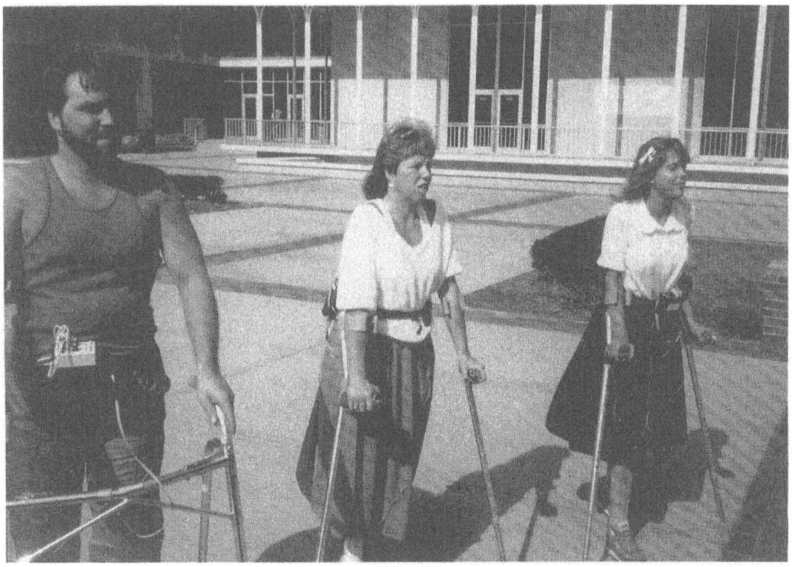

FIGURE 1.13. The FES-RGO walking system.

apparatus. A preliminary description of our system has been reported (Phillips and Petrofsky, 1985a). This "total neural prosthesis" could be utilized for the acquisition and maintenance of an upright posture and balance in a paraplegic subject (Phillips and Petrofsky, 1985b).

A human being is a rather tall structure, balanced on a relatively small base. The "center of gravity" (CG) of the body is quite high and is located below the small of the back (Vander et al., 1975). To maintain posture and balance, a set of complex, counteracting reflexes is necessary. There are no "center of gravity receptors." Instead, information concerning the location of a person's CG is obtained by integration of (a) interno- and externoreceptors (i.e., afferent signals from the muscles, joints, and skin), (b) the vestibular system, and (c) the eyes. Integration of this information is done in the subcortical coordinating centers, which map the position of the body in space. The neurophysical control of posture and balance is quite complex, and the interested reader is referred for details to standard textbooks on neurophysiology (e.g., Eyzaguirre and Fidone, 1975).

In this posture and balance control system, it is not easy to assign a percentage of importance to any one of the three afferent systems mentioned above. However, we do know that vision is probably the most important system, and that an individual can maintain upright posture and balance as long as two of the three systems are functional. When walking in the dark, for example, a person's gait is uncertain and halting. Balance and upright posture are significantly improved by the simple act of letting a fingertip touch and move along the wall. The fingertip provides little or no physical support, but it does add considerably to sensory afferent information provided to the subcortical integrating centers and cerebellum. On the other hand, a blindfolded person who also lacks a vestibular system cannot maintain an upright posture and balance. When tilted from an upright position, the trunk does not make the appropriate motion to counteract the tilt. However, the limbs will do so as a result of the stimulation of joint position receptors (Vander et al., 1975).

The purpose of this section is to describe a "total neural prosthesis" with respect to the acquisition and maintenance of an upright posture and balance in a T4 level (complete) paraplegic individual.

B. Methods and Results

Closed-loop control of paralyzed skeletal muscle by means of a combined functional electrical stimulation (FES) and orthosis system was integrated

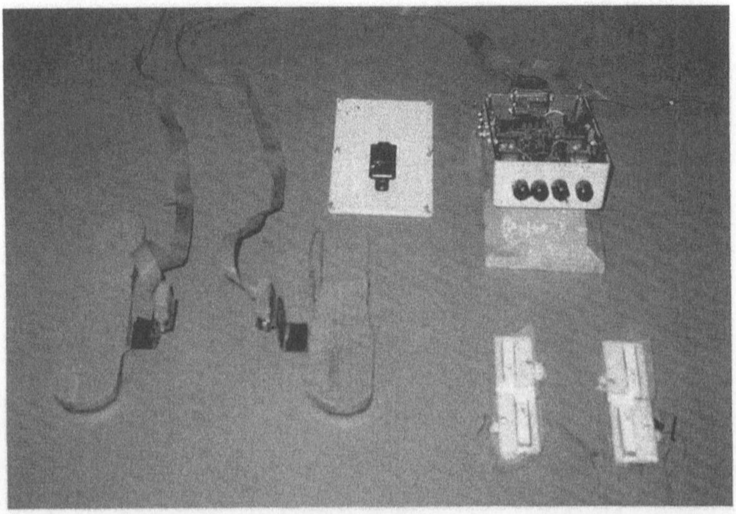

FIGURE 1.14. The sensory feedback system (SFS).

with a sensory feedback system (SFS) that allowed stand-up, upright posture, maintenance of balance, and sit-down. The SFS consists of foot-load transducers, electronic signal conditioning, and vibrocutaneous interface (Fig. 1.14). The results were obtained for a T4 level paraplegic subject. The subject first stood up using the FES-RGO. While upright and blindfolded, there was an initial period of 30 ±15 seconds during which the subject continuously adjusted his posture (and inclination at the hip varied up to ±5° peaks). This was followed by a 5-minute (average) period of prolonged standing (during which hip inclination varied up to ±1° peaks). In half of the experimental trials, the subject then sat down using the FES-RGO. In the other half of the experimental trials, the SFS was deactivated while the subject was still blindfolded. Within a period averaging 5 ±2 seconds, the subjects lost their balance (at which time hip inclination exceeded ±10°).

C. Discussion

One of the interesting aspects of these results (as reported in detail by Phillips and Petrofsky, 1986) is that a high level paraplegic was able to maintain an upright posture and balance despite the paralysis of the major trunk muscles as well as leg muscles. When a normal person sways forward, the extensors of the trunk and the flexors of the leg contract sufficiently to restore balance. When a normal person sways backward, the recti abdominis and the leg extensors contract. When the sway is sideways, the contralateral external oblique muscle responds (Keele et al., 1982).

That a high level paraplegic was able to adjust upright posture and balance is certainly due to the interaction of the subject, the FES-RGO, and the sensory feedback system. We do not feel we would have obtained similar results with the Craig-Scott braces. Some unique design features of the RGO (pelvic band, thoracic support, and the reciprocating cables at the hip) appear to be essential to the successful interaction between the subject's upper body position changes and the appropriate body weight changes on the foot plate sensors.

Perhaps the most interesting aspect was that upright body posture and balance can be maintained in a high level paraplegic without the use of visual feedback (Fig. 1.15). As stated earlier, upright posture and balance can be maintained with vestibular and tactile feedback alone. However, we certainly do not anticipate that future paraplegic subjects will rely exclusively on these two sensory modalities. Visual feedback will remain the most important sense for maintaining upright posture and balance. However, it does appear that the current system provides enough tactile informa-

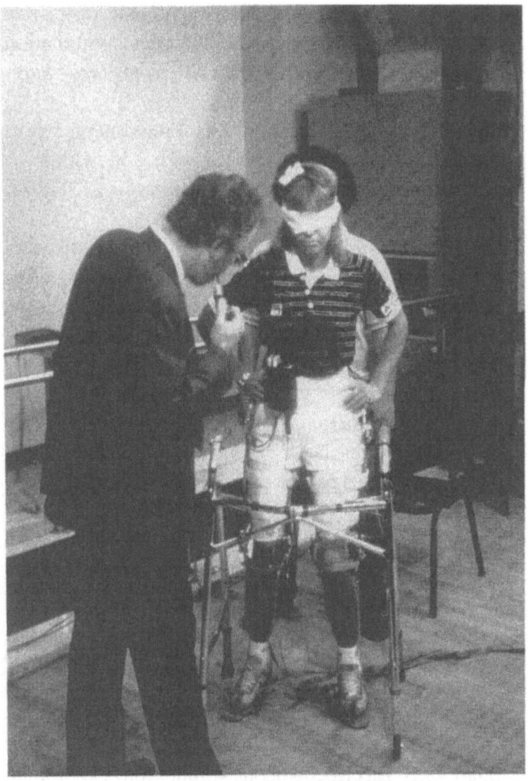

FIGURE 1.15. SFS utilized by a paraplegic subject.

tion to permit upright posture and balance to be maintained when, for example, a sudden power failure leaves a paraplegic standing with an FES-RGO system in a darkened area.

It is more likely, however, that a sensory feedback system (as an integral part of a "total neural prosthesis") will alleviate the continual necessity of having ambulating paraplegics look down at their legs and feet. Certainly, a greater amount of their visual sense could be redirected toward other more interesting and useful ways of interacting with their environment. A sensory feedback system could take up a large part of the sensory workload associated with upright posture and balance.

This section has described a "total neural prosthesis" and has reported its capabilities with respect to four activities: standing up, obtaining an upright posture, maintaining balance, and sitting down. However, more work must be performed before a final ambulation system can be realized. Walking is a repetitively phasic activity (alternating one leg and the other between a "stance" phase and a "swing" phase). The FES-RGO has successfully accomplished this in excess of one mile continuously for various subjects (Petrofsky et al., 1985). Although walking is a dynamic activity, the sensory feedback system (as described so far) has addressed only the issues of posture and balance. These are basically static activities. To extend SFS from static to dynamic activities, additional studies have been conducted on the sensory feedback system. For more detail, see Chapter 10.

References

Crago, P.E., Mortimer, J.T., and Peckham, P.H. Closed-loop control of force during electrical stimulation of muscle. *IEEE Trans. Biomed. Eng.* BME-27:306–312, 1980.

Douglas, R., Larson, P.F., D'Ambrosia, R., and McCall, R.F. The LSU Reciprocating Gait Orthosis. *Orthopedics*, 6:834–839, 1983.

Eyzaguirre, C., and Fidone, S.J. *Physiology of the Nervous System*, 2nd ed. Year Book Medical Publishers, Chicago, 1975.

Guttmann, L. *Spinal Cord Injuries, Comprehensive Management and Research.* Blackwell Scientific Publications, Oxford, 1976.

Hill, A.V. The heat of shortening and dynamic constants of muscle. *Proc. Roy. Soc. Ser. B.*, 126:136–195, 1938.

Keele, C.A., Neil, E., and Joels, N. *Samson Wright's Applied Physiology*, 13th ed., Oxford University Press, New York, 1982.

Lind, A.R., and Petrofsky, J.S. Isometric tension from rotary stimulation of fast and slow cat muscles. *Muscle Nerve*, 1:213–218, 1978.

Milner-Brown, H.S., and Stein, R.B. The relation between the surface electro-myogram and muscular force. *J. Physiol.* 246:549, 1975.

Petrofsky, J.S. Control of the recruitment and firing frequencies of motor units in electrically stimulated muscles in the cat. *Med. Biol. Eng. Comput.* 16:302–308, 1978.

Petrofsky, J.S. Sequential motor unit stimulation through peripheral motor nerves. *Med. Biol. Eng. Comput.* 17:87–93, 1979.

Petrofsky, J.S., and Phillips, C.A. Microprocessor stimulation of paralyzed muscle. *IEEE NAECON Record*, 79:198–210, 1979a.

Petrofsky, J.S., and Phillips, C.A. Constant velocity contractions in skeletal muscle of the cat. *Med. Biol. Eng. Comput.* 17:593–592, 1979b.

Petrofsky, J.S., and Phillips, C.A. Influence of recruitment order and fiber composition on the force–velocity relationship and fatigability of skeletal muscles in the cat. *Med. Biol. Eng. Comput.* 18:381–390, 1980.

Petrofsky, J.S., and Phillips, C.A. Impact of recruitment order on electrode design for neural prosthesis of skeletal muscles. *Am. J. Phys. Med.* 60:243–253, 1981a.

Petrofsky, J.S., and Phillips, C.A. The influence of temperature, initial length, and electrical activity on the force–velocity relationship of the medial gastrocnemius muscle of the cat. *J. Biomech.* 14:297–306, 1981b.

Petrofsky, J.S., and Phillips, C.A. Active physical therapy, a modern approach for rehabilitation of the disabled. *J. Neurol. Ortho. Surg.* 9:165–173, 1983a.

Petrofsky, J.S., and Phillips, C.A. Computer-controlled walking in the neurological paralyzed individual. *J. Neurol. Ortho. Surg.* 4:153–164, 1983b.

Petrofsky, J.S., and Phillips, C.A. Closed-loop control of movement of skeletal muscle. *CRC Crit. Rev. Biomed. Eng.* 6:35–94, 1985.

Petrofsky, J.S., Rinehart, J.S., and Lind, A.R. Isometric strength and endurance in slow and fast muscles in the cat. *Fed. Proc.* 35:291, 1976.

Petrofsky, J.S., Guard, A., and Phillips, C.A. The influence of fatiguing isometric contractions on the twitch characteristics of skeletal muscle. *Life Sci.* 24:2285–2292, 1979.

Petrofsky, J.S., Phillips, C.A., and Heaton, H.H. Feedback control system for walking in man. *Comput. Biol. Med.* 14:135–149, 1984.

Petrofsky, J.S., Phillips, C.A., Larson, P., and Douglas, R. Computer synthesized walking–an application of orthosis and functional electrical stimulation (FES). *J. Neurol. Ortho. Med. Surg.* 6:219–230, 1985.

Pfeiffer, E.A. Electrical stimulation of sensory nerves with skin electrodes for research, diagnosis, communication and behavioral conditioning: A survey. *Med. Biol. Eng.* 6:637–651, 1968.

Phillips, C.A., and Petrofsky, J.S. Velocity of contraction of skeletal muscle as a function of activation and fiber composition: A mathematical model. *J. Biomech.* 13: 549–558, 1980.

Phillips, C.A., and Petrofsky, J.S. The passive elastic force–velocity relationship of cat skeletal muscle: Influence upon the maximal contractile element velocity. *J. Biomech.* 14:399–403, 1981.

Phillips, C.A., and Petrosfky, J.S. System equations for functional electrical stimulation in the cat. *IEEE NAECON Record*, 83:1193–1199, 1983.

Phillips, C.A., and Petrofsky, J.S. Cognitive feedback as a sensory adjunct to functional electrical stimulation (FES) neural prosthesis. *J. Neurol. Ortho. Med. Surg.* 63:231–238, 1985a.

Phillips, C.A., and Petrofsky, J.S. Preliminary report on balance/stance in paraplegia with and without a cognitive feedback system. *Physiologist*, 28:302, 1985b.

Phillips, C.A., and Petrofsky, J.S. A total neural prosthesis for spinal cord injury rehabilitation: The cognitive feedback system with a functional electrical stimulation (FES) orthosis. *J. Neurol. Ortho. Med. Surg.* 7:225–234, 1986.

Phillips, C.A., Petrofsky, J.S., Hendershot, D.M., and Stafford, D. Functional electrical exercise: A comprehensive approach for physical conditioning of the spinal cord injured patient. *Orthopedics*, 7:1112–1123, 1984.

Rack, P.M.H., and Westbury, D.R. The effects of length and stimulus rate on tension in the isometric contraction of the cat soleus muscle. *J. Physiol.* 204:443–460, 1969.

Rohmert, W. Rechts-links-vergleich bei isometrischem armuskel Training mit verschiedenem Trainingreiz bei archtjahrigen Kindern. *Arbeitsphysiologie*, 26:363–374, 1968.

Rollman, G.B. Electrocutaneous stimulation: Psychometric functions and temporal integration. *Percept. Psychophys.* 5:289–293, 1969.

Scott, R.N., Brittain, R.H., Caldwell, R.R., Cameron, A.B., and Dunfield, V.A. Sensory-feedback system compatible with myoelectric control. *Med. Biol. Eng. Comput.* 18:65–69, 1980.

Shannon, G.F. A myoelectrically controlled prosthesis with sensory feedback. *Med. Biol. Eng. Comput.* 17:73–80, 1979.

Solomonow, M., Foster, J., Eldred, E., and Lyman, J. Proportional control of paralyzed muscle. *Fed. Proc.* 37:215, 1978.

Szeto, A.Y.J. Electrocutaneous code pairs for artificial sensory communications systems. *Ann. Biomed. Eng.* 10:175–192, 1982.

Vander, A.J., Sherman, J.H., and Luciano, D.S. *Human Physiology–Mechanisms of Body Function*, 2nd ed. McGraw-Hill, New York, 1975.

2
Control of Extremity Motor Prostheses: The Motor Feedback Applications*

I. Introduction and Overview

Control systems are classically divided into two categories depending on the configuration of the system elements. The first category is a feed-forward or "open-loop" control system where the signal path (or paths) all proceed in one direction (Fig. 2.1a). The system, G (e.g., a motor or other actuator), is driven by a controller, K. With an open-loop system, a change in the output signal (θ_o) is proportional to a change in the reference (driving) signal (θ_r) multiplied by the gain of the controller (K) and the system (G) (Fig. 2.1a).

The second category is a feedback or "closed-loop" control system where one (or more) signal paths proceed in an opposite direction from the other signal path (or paths) (Figure 2.1b). Diagrammatically this involves a third element, called a sensor (H), which transduces the output signal (θ_o) into a signal that can be compared to the reference signal (θ_r). When the feedback signal and the reference signal add, we have a system called "positive" feedback. However, this chapter will focus on control systems where the feedback signal is subtracted from the reference signal, a system called "negative" feedback.

With a negative-feedback system, a change in the output signal (θ_o) is proportional to a change in the reference (driving) signal (θ_r) *multiplied* by GK (as in the open-loop system) and *divided* by one plus the product of GKH (Figure 2.1b). In the case where H is approximately unity, and GK is significantly greater than unity, then a change in the output signal (θ_o) is proportional (equivalent) to a change in the reference signal (θ_r). Proportional control could be done with an open-loop system if the GK product were near unity.

Unlike an open-loop system, however, a negative-feedback system is capable of regulation and can be used to regulate the performance of a motor. The motor can assume a variety of different shapes and forms. Nonbiological systems include brush-driven motors, servomotors, and hydraulic motors. In biological terms, the body's own motor is skeletal muscle. Whatever type of mechanism is serving as the actuator, the appropriate control signal must be applied to vary its movement. A sensor then measures the extent of that movement. In different types of closed-loop control systems, this sensor can be responsive to position, velocity, acceleration of movement, or any combination of these, as shown in Figure 2.2.

The common denominator is that the position or degree of movement of the actuator is measured to provide information for the controller that is used to regulate function. The controller then provides a control voltage to either vary the movement of the actuator or modify its movement from a preset level. Controllers of this type can be relatively simple or extremely complex. For example, such a system can be as simple as a position sensor and a controller that modifies the voltage to a motor such that it turns to different preset positions. The control system could also be very complex, involving

*From J.S. Petrofsky and C.A. Phillips, Closed-loop control of movement of skeletal muscle. Reprinted with permission from *CRC Critical Reviews in Biomedical Engineering,* 16:35–96, 1985. Copyright CRC Press, Inc., Boca Raton, FL.

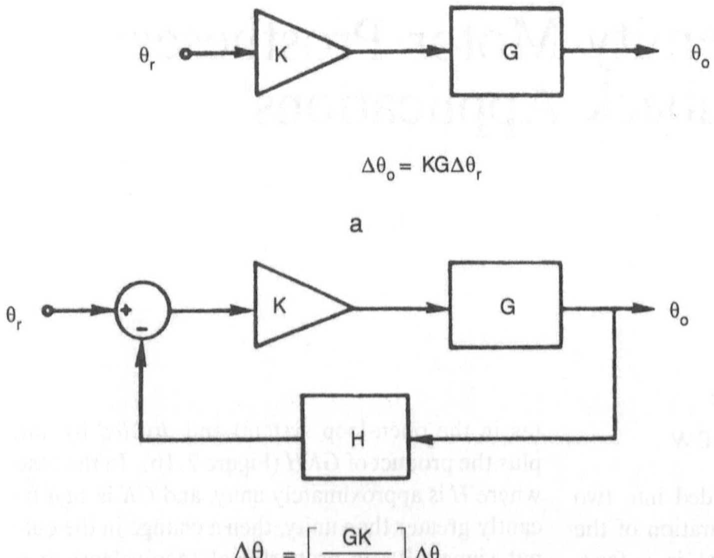

FIGURE 2.1. Categories of control systems: (a) feedforward (open-loop) system and (b) feedback (closed-loop) system.

inputs from many different sensors (such as position detectors or accelerometers), and it could follow multiple programmed inputs (such as might be encountered in an industrial robotics system).

Ideally, actuator control systems would involve only linear functions. However, for many systems such as robotic controllers, servomotor controllers, and radar systems, the controllers can become highly complex due to nonlinear control functions. One type of control mechanism used to control robotic systems (or complex decision-making circuits) is one that determines the current position of a robot arm, its velocity, and acceleration. The controller then looks ahead to see where the system should be over the next time period and makes appropriate calculations for control movement to reach this desired point. Sometimes the past history of the control variable can also be used to modify the controller's output.

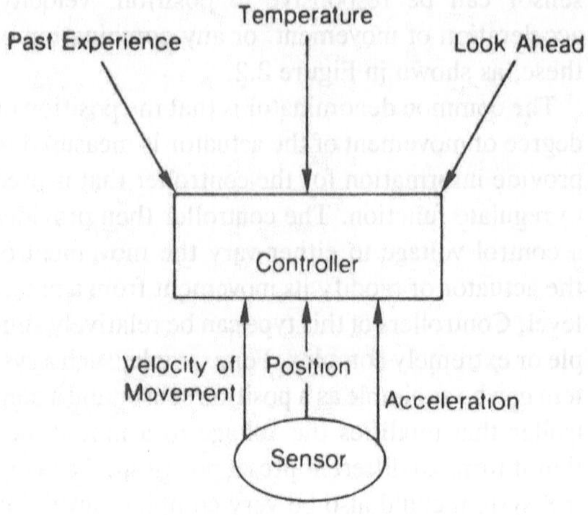

FIGURE 2.2. Various potential inputs that might influence a controller in a closed-loop system.

In man, an actuator (muscle) moves following the arrival of one or more nerve impulses. Specifically, the actuator is the skeletal muscle of the body and forms a living motor transducing chemical energy into mechanical work. The human body uses its skeletal muscle as part of a closed-loop control system. In the body, the sensors are biological sensors that measure the position of muscle and tendons, the length of muscle, and such inputs as pain and pressure on the skin. The sensors in muscle and tendons are Golgi tendon organs and muscle spindles. This information (as well as information from the brain and other active muscles) is integrated by the spinal cord to make decisions concerning the extent of activity in skeletal muscle.

Unlike electrical controllers where electrical wires connect the various elements of the system, the human body uses nerves that conduct an electrochemical wave (termed a nerve impulse) to provide communication for the closed-loop control system. For a number of years the body's own closed-loop control system has been analyzed. It has been only in the past few years, however, that closed-loop control theory has been used to electrically control movement in human muscle from outside the body.

Obviously, the simplest form of electrically controlling movement would be an open-loop control system. This would be analogous to placing two electrodes on the skin and delivering an electric impulse to cause the muscle to move. However, this type of stimulation has inherently poor control and causes the muscle to fatigue rapidly (Brown and Burns, 1949; Petrofsky, 1978; Lind and Petrofsky, 1979). Therefore, there has been a great deal of interest in closed-loop technology to control skeletal muscle with electrical stimulation. Such technology can provide coordinated movement, precise control, and minimal stress on the body. This chapter reviews the engineering aspects of closed-loop control relevant to the body and reviews the literature concerning closed-loop control and its application to rehabilitation engineering. Because the topic of closed-loop control and rehabilitation engineering is so broad, only those aspects dealing with electrical stimulation of paralyzed muscle will be covered.

This chapter continues with three additional sections. In Section II, both the structure of skeletal muscle and the neuromuscular system will be reviewed. Next (Section III), the basic technique for modeling of movement and control will be discussed and applied to the closed-loop control of movement. Finally (Section IV), closed-loop control of paralyzed skeletal muscle in paraplegic and quadriplegic subjects will be addressed.

II. The Biological System

Before any discussion of the operation of either a hypothetical or actual control system involving skeletal muscle can proceed, it is necessary to understand the structure of that system and its characteristics. This section will deal with the structure and operation of skeletal muscle. Obviously, such a discussion would also be necessary before reviewing any other biological control system. A pacemaker system that uses feedback of the amount of blood the heart is pumping would be another example of a closed-loop control system. Since the system must take into consideration the special operating characteristics and time delays associated with contraction of the muscle that pumps blood (cardiac muscle), it would be similarly necessary to examine this tissue. Since the examples used in this chapter deal exclusively with skeletal muscle, our review will consider only the anatomy, physiology, and engineering aspects of skeletal muscle. The anatomy, physiology, and engineering aspects of cardiac muscle are treated in a separate book (Phillips and Petrofsky, 1983b).

To move the various joints in the body, a tissue called skeletal muscle is utilized. Skeletal muscles are organized into muscle groups attached to bone to provide a common movement. A muscle (or group of muscles) that provides a given movement is called an agonist muscle (Fig. 2.3). A muscle

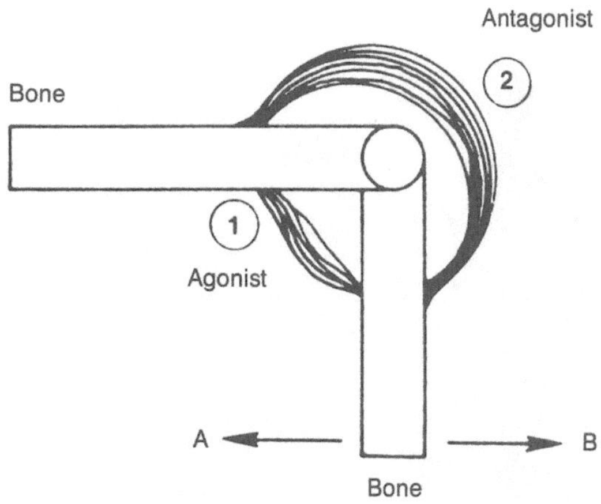

FIGURE 2.3. Typical example of an agonist–antagonist muscle. When the bone is moved toward position A, muscle 1 contracts while contraction of muscle 2 resists that movement. Muscle 1 therefore is called the agonist, while muscle 2 is called the antagonist. If the situation were reversed and the bone were to move toward B, the contraction of muscle 2 would promote that movement while muscle 1 would be against that movement. In this case then muscle 2 would be the agonist and muscle 1 would be the antagonist.

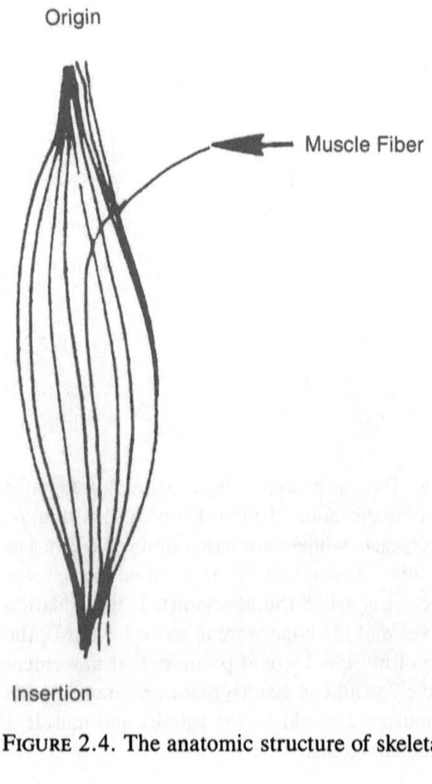

FIGURE 2.4. The anatomic structure of skeletal muscle.

group that opposes that movement is called an antagonist muscle group. The exact extent of antagonist activity during movement of any joint varies with the degree of physical training. For the untrained individual, a great deal of antagonist activity is present in normal movement during exercise (Astrand and Rodahl, 1977). In the trained individual, the antagonist muscle activity is reduced, thereby increasing the efficiency of exercise. When a joint moves in the opposite direction, the role of agonist and antagonist is obviously reversed (as shown in Fig. 2.3).

Muscle itself is a discrete unit covered by a tough connective membrane. Muscle is formed of cells called muscle fibers, which run from the beginning to the end of the muscle. The two ends of the muscle are called the origin and insertion of the muscle (Fig. 2.4). Each muscle cell can be quite long. These multinucleated cells can be as much as several feet long in some of the muscles in the body. Muscle fibers themselves are connected together in small bundles by another membrane. Although all muscle fibers run from the origin to the insertion of the muscle, the actual positions of the origin and insertion with respect to the direction of movement of the muscle are not always in a parallel direction.

Muscle movement may be unaligned with the direction of the arrangement of the fibers, as shown in the top sketch of Figure 2.5. In a muscle like the biceps, muscle fibers run from the upper arm to just below the elbow and the fibers contract in the direction of movement of the muscle. Since muscle fibers can shorten by as much as 57%, the muscle length can be shortened by about half. This means that muscles with a long fiber arrangement (such as the biceps) will move bones significant distances. However, since the actual power developed by the muscle is proportional to the cross-sectional area of the power-generating units of the muscle (discussed below), power is limited by this fiber arrangement.

Muscles can also be arranged in what is termed a pennate arrangement (Fig. 2.5, bottom). In this

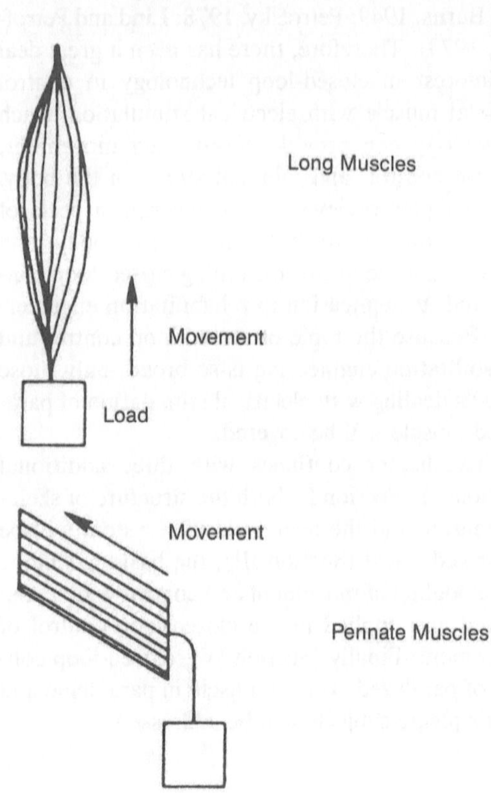

FIGURE 2.5. The long and pennate arrangements of a skeletal muscle. In both cases, the muscle fibers run from the origin to the insertion of the muscle. The major difference is the relative direction of movement of the origin and insertion relative to the workload.

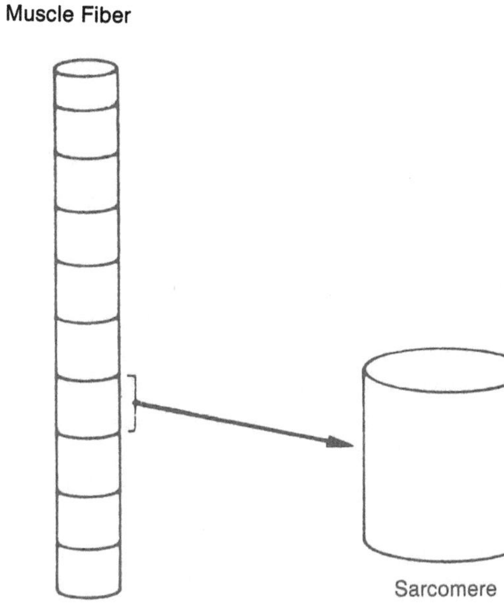

FIGURE 2.6. The structure of a muscle fiber.

arrangement, the direction of movement of muscle is partially or almost fully perpendicular to the arrangement of the muscle fibers. The actual displacement of the origin and insertion relative to the direction of movement may be very small, while the strength developed by the muscle may be very large (because of the relatively high cross-section area in relation to the direction of movement). One such example would be the deltoid muscles, where the shoulder moves very little but develops high power. There are a number of permutations of these arrangements (such as bipennate and other arrangements) that alter the strength development of muscle versus its ability to shorten. Therefore, muscle is far from a simple actuator, and its structural design has a great deal to do with the mechanics of power development and the degree of movement and contraction.

Within any one muscle fiber, the smallest repeating subunit is called the sarcomere. The sarcomere (Fig. 2.6) is generally only a couple of micrometers (μm) in length, but contains the actual contractile machinery of the muscle. The sarcomere is composed predominantly of two types of protein known as actin and myosin. In much smaller quantities, there are a number of other proteins present (such as troponin, tropomyosin, and other structural proteins) that are involved in the regulation of activity of skeletal muscle.

A series of helical filaments extending from the ends of the sarcomere (Z-line) toward the center of the sarcomere (Fig. 2.7) is composed predominantly of the muscle protein myosin. Myosin strings themselves are arranged in a helical structure (Fig. 2.7) with protruding heads that extend toward thin filaments. On these heads are proteins (light chains), each with an approximate molecular weight of 20,000 daltons. The head with its light chains has the ability to split the energy-producing compound adenosine triphosphate (ATP) in muscle. The thin filaments are composed predominantly of actin. Actin is found in a globular form and is arranged in pairs of two globular molecules in parallel forming a slowly winding chain of adjacent molecules that extend toward the end of the sarcomere.

Interposed between the actin molecules is a long, tapering protein tropomyosin, and at every half-rotation of the actin chain are two troponin molecules. Troponin contains proteins that regulate the activity of actin. Without the presence of calcium, interaction is prevented between myosin, actin, and ATP, and the muscle is relaxed. However, in the presence of very minute quantities of calcium, the regulator proteins on troponin no longer prevent this interaction. In this case, there is an interaction between the myosin and actin molecules so that the ATP molecule is split into

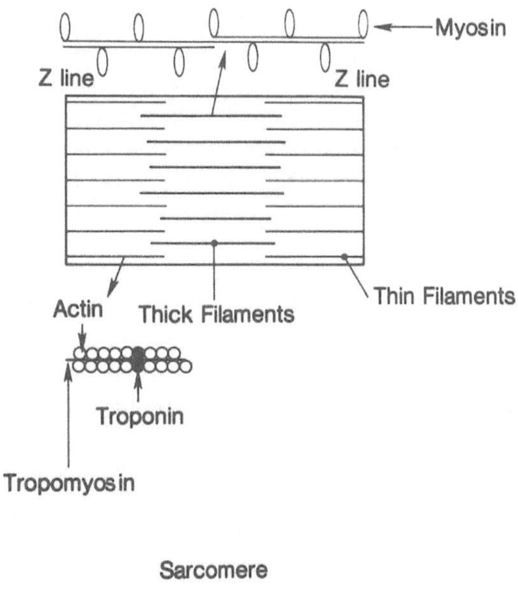

FIGURE 2.7. The structure of the sarcomere.

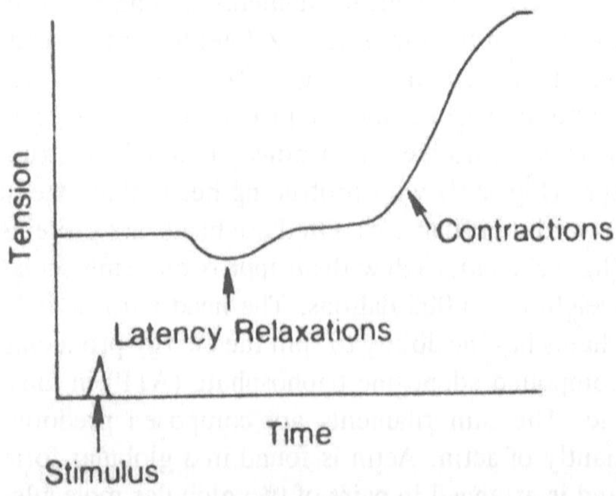

FIGURE 2.8. The time course of activation of skeletal muscle showing the latency period, and the latency relaxation that occurs before muscle contraction.

adenosine diphosphate and inorganic phosphate with the production of energy and heat. This reaction is surprisingly efficient. Depending on the type of exercise that is being done (as described below), as much as 35% of the energy derived from the breakdown in glucose can be turned into mechanical energy. Considering that the reduction of glucose into ATP is only about 50% efficient, the actual muscle (as a motor) is extremely efficient in the final breakdown of ATP.

Activation of actin by the presence of calcium is accomplished by the release of calcium from the sarcoplasmic reticulum (SR) surrounding the sarcomere. The T-system of transverse tubules allows the nerve impulses (action potentials) to be conducted deep within the muscle. These action potentials are transferred by electrotonic conduction to the SR. At the terminal cisternae of the SR, the depolarization causes calcium to be released into the sarcoplasm.

The electromechanical coupling at the muscle (from the nerve impulse to the production of mechanical energy) is an electrochemical and not an electronic process. There is an inherent time delay during which a nerve impulse travels across the sarcotubular system of skeletal muscle and calcium is released and diffuses onto the thin filaments inside the sarcomeres. This differs from an electrical motor or hydraulic actuator, since a time delay is always found. The mechanical time delay from the point of excitation to the time of muscle contraction is called the latency of muscle contraction and must be considered in any evaluation of control. This and other time delays (described below) enter into the equations for accurately controlling movement in skeletal muscle. Although the process is poorly understood, once a muscle has been electrically stimulated and the nerve impulse has begun to travel through the muscle, there is a brief period of time (following its arrival) when the muscle relaxes before it contracts (Fig. 2.8). This is called latency relaxation.

The nerve impulse that travels to a skeletal muscle fiber comes from a specialized type of nerve called a motor nerve composed of alpha motor neurons. Each muscle fiber is electrically insulated from all the other muscle fibers found within the muscle. A motor nerve leaves the spinal cord and then innervates anywhere from a few (3–6) to more than 100 muscle fibers. This is accomplished by the motor nerves splitting into a number of different branches as they enter skeletal muscle. One motor neuron and all the muscle fibers that it innervates is called a motor unit. There are several time delays in this particular scheme. The first of these delays occurs with the conduction of a nerve impulse from the spinal cord. Mammalian motor nerves range in diameter from about 12 to 20 μm. These motor nerves vary in their speed of impulse conduction from 70 to 120 m/s. Even in the large myelinated motor nerves, conduction velocity is low. For example, a motor nerve leaving the spinal cord going into the hand would have an inherent delay of 10 ms (assuming the best-case maximum conduction velocity of 120 m/s). With nerves innervating muscle closer to the spinal cord, these conduction delays would be minimized. For nerves innervating muscle farther away from the central nervous system (hence longer), there would be a greater time delay.

Motor conduction is only one of a number of delays. A delay also occurs at the junction between the nerve and muscle (the neuromuscular junction). The electrochemical nerve impulse is transduced into the release of a chemical substance called a neurotransmitter, which then must diffuse across to the sarcolemma of skeletal muscle. On the muscle side (postsynaptic side) of the neuromuscular junction, the neurotransmitter reacts with receptors, where a new nerve impulse is generated. This action potential then travels through the skeletal muscle.

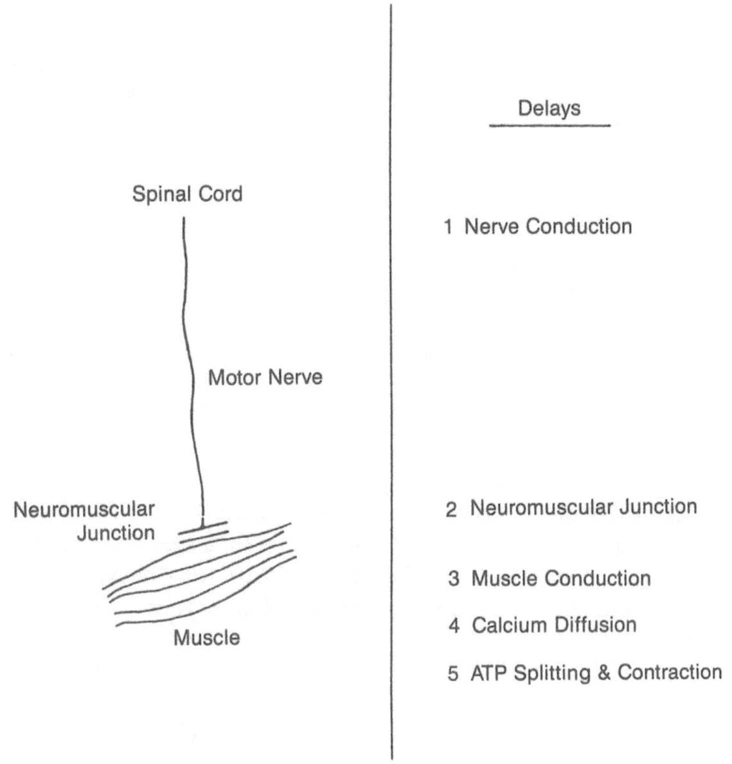

FIGURE 2.9. The time delays associated with contraction of skeletal muscles.

Activation of skeletal muscle thus involves a number of different time delays before the muscle itself begins contracting.

Since contraction of skeletal muscle is a chemical process, contraction itself does not develop instantaneously but develops over tens of milliseconds until contractile power peaks. After a muscle begins contracting, it takes a finite time before the muscle stops contracting due to the removal (by metabolically coupled pumps) of calcium from the inside of the sarcomere. These delays are summarized in Figure 2.9. When electrodes are applied to a motor nerve near the spinal cord, there is a considerable latency period before the muscle actually contracts, followed by a long period of tension development (Fig. 2.10). This simple model, however, vastly

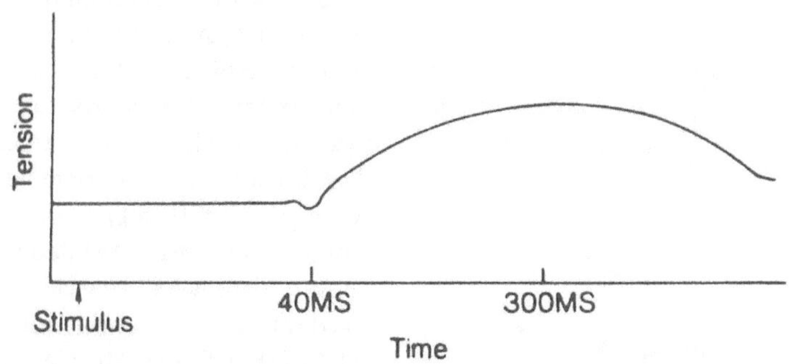

FIGURE 2.10. Typical time delay that might be encountered in the cat soleus muscle following electrical stimulation in the spinal cord. After 30–40 ms delay during which the muscle shows latency relaxation, the muscle slowly begins to contract, with peak tension being developed some 300 ms after the onset of the stimulus.

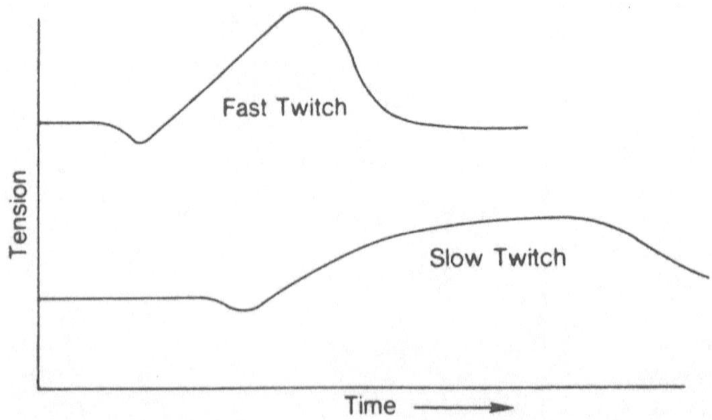

FIGURE 2.11. Comparison of the latency and contraction speeds of fast-twitch and slow-twitch motor units.

oversimplifies the complexities of the neuromuscular system in terms of actuation.

Genetically, two types of identifiable muscle fiber are often described. One of these is called a fast-twitch muscle fiber and the other one a slow-twitch muscle fiber. Fast-twitch muscle fibers are muscles with a specific type of myosin that allows the rapid splitting of ATP and contraction of the muscle. These types of muscle fiber are innervated with rapidly conducting nerve fibers and therefore have a short latency as shown in Figure 2.11. In contrast, slow-twitch muscle is often used for endurance activities and consists of slowly conducting nerves and very slowly contracting muscle. Slow-twitch muscle takes hundreds of milliseconds to respond and therefore represents a very low velocity conducting and contracting system. Fast-twitch muscle fibers are used for rapid activity and developing strength, while slow-twitch muscle fibers are used by the body for endurance activities such as prolonged standing.

Most muscles of the body are of various mixtures of these two types of cell. However, physical training can alter the composition of the muscles and make muscles look histochemically, biochemically, and physiologically like either fast-twitch muscles (with weight training) or slow-twitch muscle (with endurance training).

Since the process of muscle contraction is metabolic, the speed of contraction and tension developed by muscle is also altered by a number of physiological and environmental parameters. For example, a significant parameter of skeletal muscle contraction is muscle temperature. As muscle temperature is reduced, muscle strength falls and the speed of muscle contraction and endurance is reduced (Fig. 2.12). Muscle contraction speed and the conduction velocity of action potentials on the motor nerves are reduced by 50% for every 3–5°C reduction in tissue temperature (Close, 1972;

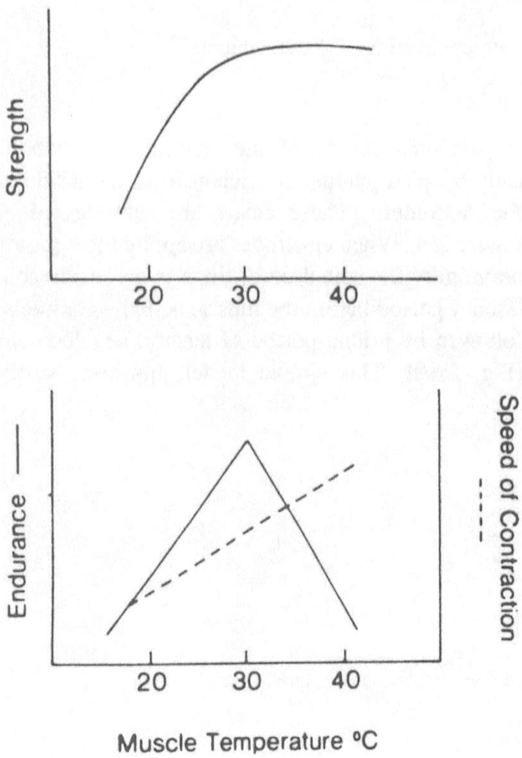

FIGURE 2.12. Speed of muscle contraction, endurance for exercise, and strength of muscle as a function of muscle temperature.

Petrofsky and Lind, 1981). To complicate matters even further, muscle temperature differentially affects fast-twitch and slow-twitch muscle fibers in terms of both strength and speed of contraction (Close, 1972; Petrofsky et al., 1978). Finally, a number of other factors such as muscle fatigue alter the speed of contraction of skeletal muscle (Edwards et al., 1972; Petrofsky et al., 1979). Therefore, skeletal muscle must be carefully modeled in terms of its mechanical and physiological characteristics before any equations can be developed to control movement of skeletal muscle in a closed-loop setting. This concept will be discussed further in Section III.

Sensors are required for a closed-loop control system, and the body has its own natural sensors. These sensors are typified by the Pacinian corpuscle, the muscle spindle, and the Golgi tendon organ. Golgi tendon organs transmit information about force, and the Pacinian corpuscles transmit information about the pressure to the central nervous system. For example, the Pacinian corpuscle (Fig. 2.13) is an onion-shaped device with a free nerve ending in the center. Pressure on this device causes ionic leakage in the free nerve endings. This results in trains of nerve impulses being developed.

However, most biological sensors are not ideal sensors. In fact biological sensors suffer from a phenomenon called accommodation. If pressure is applied to the Pacinian corpuscle over a long period of time, the receptor will eventually change shape and adapt to the pressure so that no output will appear on the nerve (although the pressure is still being exerted). Different sensors in the body adapt with different time constants. When the pressure is finally released, nerves will generally fire again until they accommodate to the new baseline pressure. For this reason, accommodation is an important variable and certainly an important part of normal closed-loop control of the human body. For example, pressure receptors from the skin show accommodation. When we put our clothes on in the morning, we can feel their weight on our skin. Over a period of time, the pressure receptors accommodate to the weight of the clothes and we no longer feel them. This is a typical example of accommodation.

The length of muscle is determined by a complex receptor called a muscle spindle. The muscle spindle itself has a contractile component of intrafusal fibers at its center. At its periphery is a gamma efferent motor nerve. This particular nerve fiber is employed by the central nervous system to adjust the baseline length of the muscle spindle to various levels. If a muscle is stretched, the central portion begins to accommodate. The gamma efferent nerve fiber can be actively stimulated, causing the peripheral intrafusal fibers to contract. This results in a further stretching of the central portion of the muscle spindle (and so the spindle readjusts its length). The system itself is a complex integrated closed-loop control system that operates within certain reflex arcs of the central nervous system (called the gamma efferent system) to adjust to the demands of the control systems in the body. A complete discussion of the muscle spindle would be impossible in this section because of its complexity. The reader should refer to a standard neurophysiology text for a complete description of this and other sensors (e.g., see Stratton, 1981).

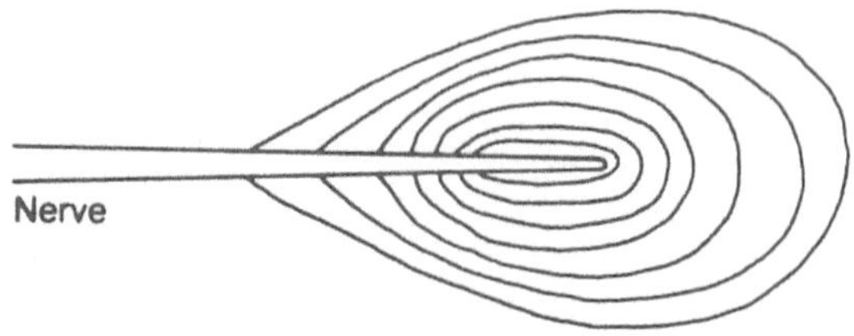

FIGURE 2.13. Structure of a typical biological sensor: the Pacinian corpuscle.

To understand the operation of the central nervous system, it is necessary to look at the integrating unit in the spinal cord where decisions are made. This integrating system is the cell body of the alpha motor neuron. The alpha motor neuron (Fig. 2.14) itself receives connections from many other parts of the body. The average alpha motor neuron has been estimated to have as many as 10,000 connections to other nerves. These other nerves form synapses very similar to that of the neuromuscular junction described above. When activated with a nerve impulse, some of these synapses excite the alpha motor neuron to develop its own nerve impulse. Synapses of this type are called excitatory.

Generally speaking, a single nerve impulse arriving at an excitatory synapse is insufficient to cause the nerve to develop its own action potential. Action potentials are developed at a region of the

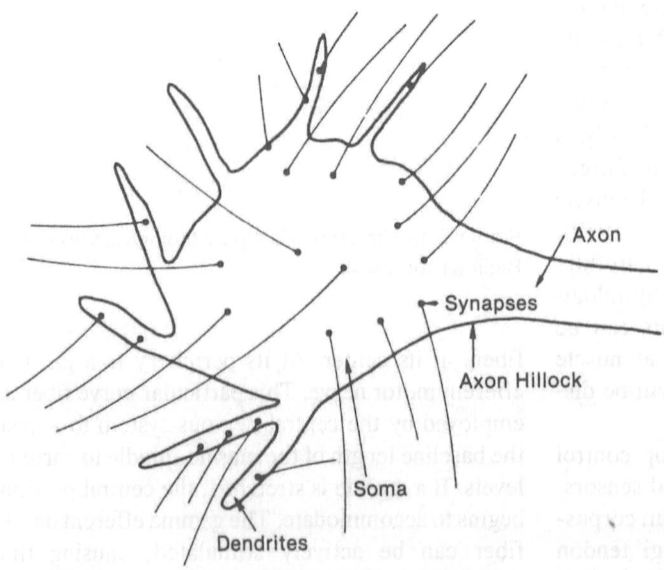

FIGURE 2.14. Typical alpha motor neuron of the spinal cord.

nerve (Figure 2.14) called the axon hillock. The axon hillock is the integrating area where the final decision is made as to whether or not the axon of the neuron conducts. To excite the axon hillock to develop an action potential, either a nerve must fire continuously (temporal summation) or a number of nerves must fire in roughly the same time domain (spatial summation). Therefore, if a number of different inputs are being applied to the alpha motor neuron (all of which are tending to make it excitatory), then action potentials will develop and the muscles will begin contracting. However, some of the synapses are called inhibitory synapses because, when nerve impulses fire across these synapses, they reduce the excitability of the nerve at the axon hillock. Synapses of this type therefore ultimately inhibit movement.

It is the summation of all excitatory and inhibitory activity that determines whether a muscle contracts. This type of controller is, in fact, a nonlinear controller with multiple inputs from various parts of the body. The effect of each of the synapses is also not constant. A synapse applied close to the axon hillock has a greater effect than a synapse farther distant. This is because reduction in current is proportional to the distance on the nerve cell body, and also due to the high electrical resistance of the membrane. The alpha motor neuron cell body therefore receives inputs for pain, temperature, voluntary control, and level of activity from muscle spindles, Pacinian corpuscles, and Golgi tendon organs. The final decision is made by integration of all the data at the axon hillock as to whether a muscle contracts or does not contract and the extent of any such contraction.

Obviously, this is only a brief glimpse of the complexity of the controller involved in the central nervous system. In addition to these spinal cord integrating centers located at the various segments of the spinal cord (such as the alpha motor neurons), there are many higher centers involved throughout the brain to integrate information and provide control of skeletal muscle. For example, there are inputs from the inner ear for balance. A complete description of the neurophysiology of movement and control of movement would take an entire textbook and may be reviewed by the reader using standard neurophysiology texts (see, e.g., Brobeck, 1979; Stratton, 1981; Ruch and Patton, 1982). This brief description is meant only to acquaint the reader with the complexity of the system and some of the control problems associated in modeling such a complex biological system.

III. Mathematical Modeling and the Closed-Loop Control of Skeletal Muscle

This section reviews mathematical modeling approaches for the feedback control of skeletal muscle and summarizes some of the salient issues.

First, a discussion of mathematical models of muscle in general is presented. Second, mathematical models applied to the closed-loop control problem of skeletal muscle are surveyed. The reader must appreciate that we are dealing with a necessarily limited view of muscle mechanisms (in general) and the motor control problem (in particular). Partridge and Benton (1981) estimated that the literature on the mechanical aspects of muscle function contained approximately 100,000 items. That review article treats the more general issue of muscle as a transformer that converts nerve signals (input) into mechanical action (output).

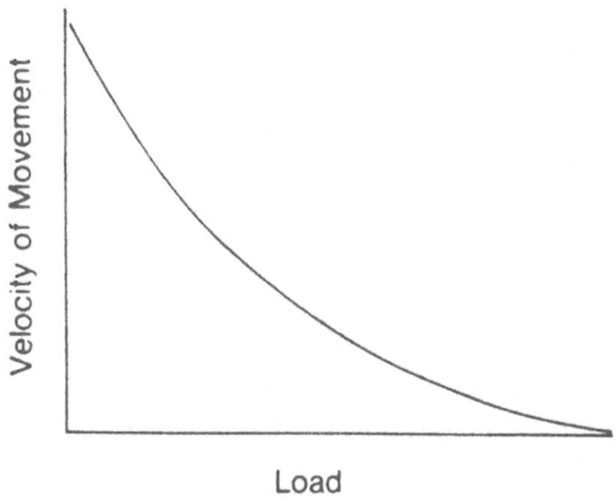

FIGURE 2.15. Typical Hill force–velocity relationship.

A. Mathematical Models of Muscle

Partridge and Benton (1981) also noted that at least 15% of the papers they reviewed used some mathematical representation of muscle properties that might be called a formal model. If this is extrapolated to their prior estimate (see above), then there are about 15,000 items currently in this category alone. Therefore, the following comments should be viewed only as a simplistic overview of the area.

What is a model? Why make a model? Any time a set of biological data is described by an equation (either in numerical or graphical form), we have what may be termed a mathematical model of the system. The significance of such models is that they allow us to quantify the relationship between variables. Such quantification is very useful in permitting us to define and explain the system, but equally important, to analyze the system for predicting and even controlling its performance. Consequently, when dealing with any control problem, our results will be no better (and often worse) than the basic model (equations) that we use to define the system being controlled.

Generally speaking, equations involved in biological work can be classified into two broad categories: empirical and theoretical. An *empirical* equation can be defined as one that has been fitted into experimental data in order to describe a relationship that has actually been observed between two or more variables. As Riggs (1970) pointed out, a particular set of data can be explained by any number of empirical equations. When no guidance from theory is available, the law of parsimony requires that the simplest equation that adequately describes the data be utilized.

A *theoretical* equation is derived not from a particular group of experimental observations, but from some theory or hypothesis about the fundamental nature of the biological system. An example of an empirical equation is the Hill (1938) relationship. When A.V. Hill's data for the rate of extra energy liberation are plotted in the form of a force–velocity curve, an inverse hyperbolic relationship is found (Fig. 2.15). Classically, the ordinate (velocity) is treated as the dependent variable, and the abcissa (load) is treated as the independent variable. The relationship is for a single muscle (type), shortening under a constant load, constant level of activation, and constant temperature. Here the velocity is the peak initial velocity (assumed to very close to the muscle's initial length). Despite the fact that the ordinate is velocity, the "Hill" model is *static*, since any point on the curve is independent of time with respect to any other point on the curve.

This "Hill" curve is based on measurements of heat and work and not on direct measurement of muscle-shortening velocity. However, it was found by Hill (1938) to be identical to that obtained a few years earlier by Fenn and Marsh (1935), who measured directly the dependence of muscle-shortening velocity on load. It is easy to see how this was subsequently interpreted as a theoretical model, especially since Hill could relate the curve-fitting parameter "*a*" as the coefficient for heat of muscle shortening. However, the curve-fitting parameter "*b*" had no theoretical physiological explanation. In fact, "*b*" was and still is an empirical parameter utilized to fit the inverse hyperbolic relationship to a set of force–velocity data points. It is, therefore, no wonder that the sequela of the Hill force–velocity relationship have been to challenge it regarding a number of important physiological observations it does not explain (Fung, 1981). This includes, for example, tension–time–heat (Mommaerts, 1969).

In summary, a well-founded theoretical equation should explain all sets of data and, more importantly, explain why the observed relationship exists. Subsequently, Hill (1970) redefined his model (a is replaced by α). Nonetheless, the original Hill model (1938) has been very helpful to a generation of muscle physiologists. The basic model has been refined and extended to help formulate our understanding of a host of muscle physiology problems that would be too complex to solve nonmathematically (e.g., Phillips and Petrofsky, 1981, 1982).

Accepting the primarily empirical nature of the Hill (1938) equation, we have extended the force–velocity relationship to develop a series of subsidiary empirical equations (Petrofsky and Phillips, 1980, 1981; Phillips and Petrofsky, 1980). Basically, the original Hill equation relates shortening velocity (V) and isotonic load (P) as discussed above:

$$V = f(P) \tag{2.1}$$

For control of skeletal muscle movement, this force–velocity relationship must also be defined in terms of recruitment magnitude (activation, A), fraction of slow-twitch (vs. fast-twitch) fiber composition (X), initial length (L), and muscle temperature (T), as well as the load (P) on the muscle (Petrofsky and Phillips, 1983c):

$$V = f(A, X, L, T, P) \tag{2.2}$$

Even such a complex relationship can provide only a "first approach" to the skeletal muscle control problem.

With respect to a controlled system model, a mathematical representation must still be made of the neuromuscular interface (excitation–contraction coupling), as well as the joint dynamics themselves (Alexander, 1981).

Concentrating on the muscle "compartment" itself as the actuator, a number of issues need to be addressed. In addition to the shortening velocity defined above, a similar (although more simplified) relationship needs to be defined for lengthening. Some first attempts have been made in this area (Partridge and Glaser, 1960; Rack, 1966), including the interesting observation that ATP is resynthesized as a consequence of muscle lengthening (Petrofsky and Fitch, 1980).

Lengthening equations (in parallel with muscle-shortening equations) recognize that muscle operates as agonist–antagonist pairs in moving a joint (Alexander, 1981). As mentioned above, lengthening equations can be simpler than shortening equations. While shortening represents an active contractile process, lengthening can be viewed as a series passive-dampening process. Our original experiments (Petrofsky and Phillips, 1979a) on isovelocity contractions of the cat gastrocnemius muscle included selective activation of its antagonist (tibialis anterior) (Fig. 2.16). Inspection of this figure shows a significant dampening of velocity overshoots and undershoots during the isovelocity contractions. Complicating simple muscle lengthening equations (for the slower, isovelocity movements) would be (a) pulsed damping of the agonist by the antagonist as the joint position comes to rest after a rapid change in rotational angle and (b) reduction of the degree of passive series dampening by the antagonist in the athletically conditioned individual.

Other aspects of the muscle compartment that need to be modeled include the contribution of activation of the muscle's passive properties in the maintenance of tone. Muscular stiffness is now recognized to be composed of an "active" series elastic component (Huxley and Simmons, 1971) and an "active" parallel elastic component (D.K. Hill, 1968).

Finally, the complete muscle compartment model must recognize the influence of the physical arrangement of muscle fibers, differentiating long fiber arrangements from pennate fiber arrangements. The necessary biological data have not yet been acquired to permit the formulation of any such comprehensive empirical model of the muscle compartment as we have described above.

Theoretical equations of muscle mechanics must also be pursued, however. This is especially true if we are ever to reconcile the molecular process in muscle with its mechanical properties. In contrast to the 1938 elastic matrix model of A.V. Hill (a contractile element combined with a series elastic element), other theoretical models have been proposed. Bergel and Hunter (1979) have described

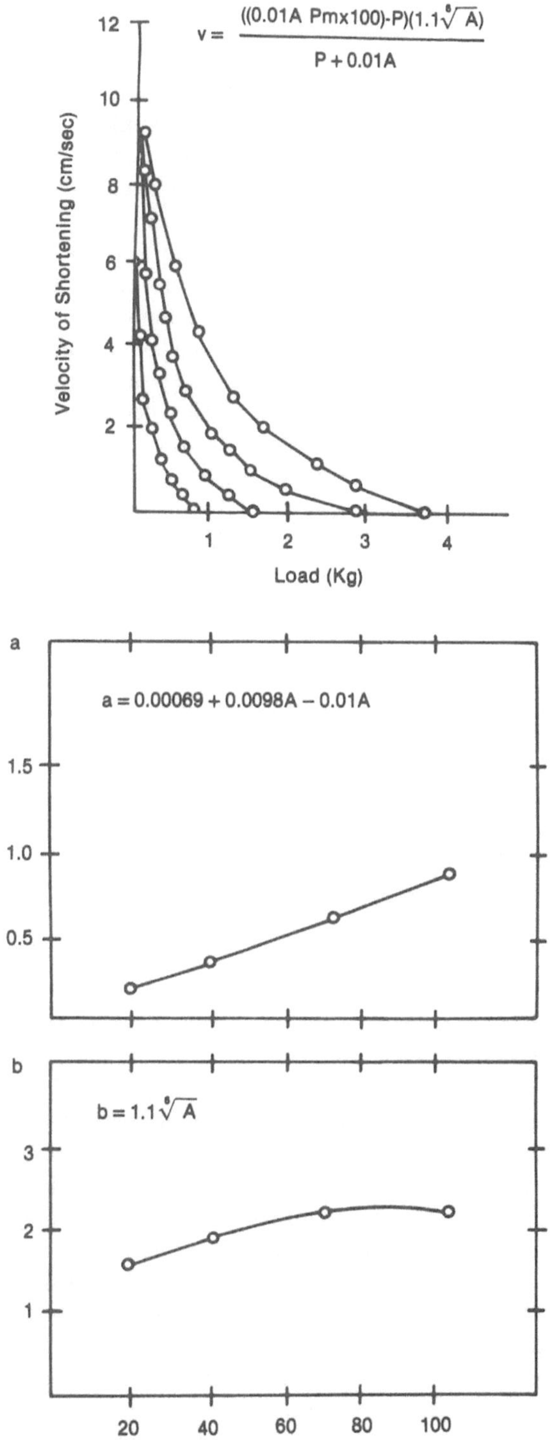

FIGURE 2.16. The influence of activation on contractile properties of the cat gastrocnemius muscle; symbols as given in Chapter 1 in connection with Equations (1.1) and (1.2).

phenomenological approach. Huxley (1957, 1974) has postulated a mixed mechanical–thermodynamic basis for muscular contraction. Iwazumie (1970) theorizes an electromagnetic field as the basis for muscle movement. Finally, Caplan (1966) first applied the rigor of irreversible thermodynamics to understand the mechanism of muscle contraction.

An essential quality of any of these models should be to reconcile molecular details (either theoretical or factual) with mechanical reality. An extension of the Caplan (1966) model proposed by Bornhorst and Minardi (1970a,b) appears to fulfill this requirement. First, the model can be generalized from skeletal muscle to also include cardiac muscle (Phillips et al., 1979a,b).

Second, the molecular details of the Bornhorst and Minardi (1970a,b) model are contained in operationally defined quantities called transport coefficients, which appear in the resulting thermodynamic equations. The real value of the thermodynamic equations is that they relate (through the transport coefficients) the different types of experimental measurement (mechanical, chemical, and heat) and are therefore complementary to a molecular approach (Phillips and Petrofsky, 1983b). For example, molecular knowledge might be used to provide a theoretical basis for an estimation of the variation of transport coefficients for whole muscle. Variations of the transport coefficients are not dependent on this view, however, since they could be determined directly from experimental observation. Clearly, if the molecular view does agree with experimental variation of these transport coefficients, then a greater understanding of muscle is obtained than if the variations of these coefficients are based on macroscopic observations alone.

Molecular processes in muscle, which in various cases appear to be closely related to motor properties, constitute a major topic and cannot be reviewed sufficiently here. The interested reader is referred to Sandow (1970) for a list of review papers up to 1970.

B. The Closed-Loop Control Problem and Skeletal Muscle

An open-loop (or feedforward) control system is shown in Figure 2.17a. In such a system the output variable, θ_o, is related to the reference value, θ_r, by the gain (G) of the system:

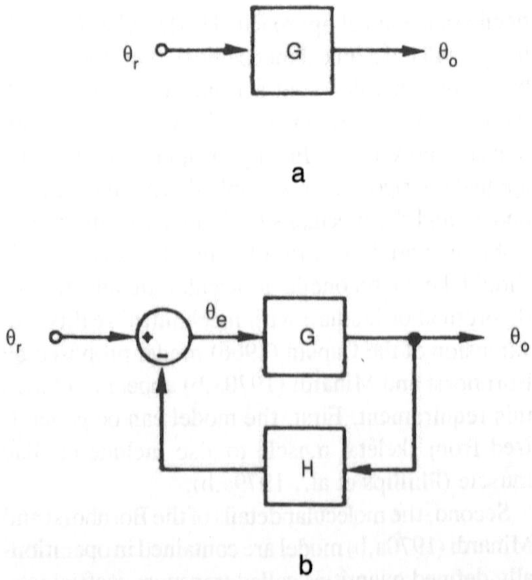

FIGURE 2.17. Examples of (a) an open-loop (feedforward) control system; and (b) a closed-loop (feedback) control system.

$$G = \frac{\theta_o}{\theta_r} \quad (2.3)$$

In simple terms the output simply "follows" the input as some function of G.

A closed-loop (negative feedback) control system is shown in Figure 2.17b. In such a system the output variable, θ_o, is related to an error variable, θ_e, by the gain (G) of the system:

$$G = \frac{\theta_o}{\theta_e} \quad (2.4)$$

In turn, the error variable, θ_e, is related to the reference variable, θ_r, as follows:

$$\theta_e = \theta_r - H\theta_o \quad (2.5)$$

Combining Equations (2.4) and (2.5) shows that for such a system the output variable, θ_o, is related to the control value, θ_r, by the following expression:

$$\frac{\theta_o}{\theta_r} = \frac{G}{1 + GH} \quad (2.6)$$

Under the correct conditions (see above), such a system is capable of regulation. Automatic control by open-loop (feedforward) control will not be addressed in this review. The open-loop control of muscle using functional electrical stimulation has been reviewed (Vodovnik et al., 1981).

A general discussion of the automatic control problem reduces itself to two fundamental elements: accuracy and stability. The basic principle of closed-loop (feedback) control is that such operation produces accurate performance, since the control system continually endeavors to correct any error that exists. However, such corrective action can result in unstable operation when used with control elements having a large amplification and significant delays in time response (Nyquist, 1932). A control system is unstable when any temporary disturbance causes the control variable to deviate from the reference value, and to continue to deviate after the removal of that disturbance. An unstable system is no longer able to maintain the control value very near the reference value. Rather, large sustained oscillations or erratic control will develop.

Accuracy of the control system increases with system gain. However, if adequate measures are not taken to assure stable operation, the advantages of closed-loop control are useless. Stability alone is not sufficient; one must have a system that demonstrates an adequate margin of stability (Nyquist, 1932). Especially in neuromuscular control, the system must recover rapidly and smoothly from the stress of irregular inputs and severe disturbances.

The requirements of stability and accuracy are mutually incompatible. To provide greater accuracy, the tolerable error for corrective action must be smaller, and full corrective action must be initiated sooner. This higher accuracy is acquired through higher system gain. Time delays such as those occurring in neuromuscular excitation–contraction coupling which were not significant at low gain may become significant in a system having high levels of gain. Since there are significant time delays in the neuromuscular control problem (see Section II), current closed-loop muscle control systems exhibit low gains (Partridge, 1972; Crago et al., 1980; Solomonow et al., 1983b).

Two categories of feedback control system need to be defined when surveying this broad field as applied to muscle control. The nature of the variable controlled seems to determine the choice of systems.

A *regulator* or *regulating system* is a feedback control system in which the reference input or

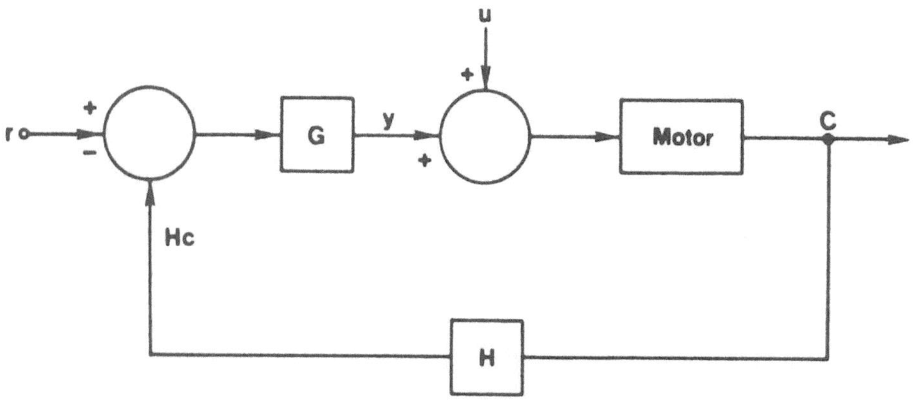

FIGURE 2.18. A regulator feedback control system.

command is constant, often for long periods of time (Fig. 2.18). This system is common when the controlled variable is isometric muscle force (i.e., under conditions of a constant muscle length). Crago et al. (1980) have demonstrated a family of isometric force plateaus using linear analog feedback control. Petrofsky (1978, 1979) had earlier demonstrated a family of isometric force steps using nonlinear digital feedback control. It is generally recognized that the muscle control problem is nonlinear (Thomson, 1961; Demieville and Partridge, 1980). Except for an occasional homogeneous muscle, most muscles significant in gait are of mixed fast-twitch and slow-twitch populations (Ariano et al., 1973) and therefore nonlinear regarding their strength–recruitment profile (Petrofsky and Phillips, 1979b; 1980).

Only Solomonow et al. (1983a) and Solomonow and Scopp (1983) have found a linear strength–recruitment relationship over the entire recruitment range. However, their studies examined only isometric force and were not extended to muscle undergoing velocity (length) changes. Second, their system involved a second ("high frequency") blocking electrode distal to the stimulating electrode. The authors can only speculate as to what this blocking system is doing at a biophysical level (Solomonow et al., 1983a). Third, their system has very large time delays in adjusting to a target tension (Solomonow and Morse, 1983). Finally, their system requires implantation around the motor nerve, and the long-term effects on the motor nerve itself have not been evaluated.

Nonlinearities introduce significant complications to the automatic control problem. First, the effective range over which the variable can be controlled is significantly reduced. Second, and more important, standard stability criteria (e.g., Routh, 1930; Evans, 1954) cannot be applied to nonlinear systems (Cosgriff, 1958). For these reasons, we have utilized piecewise linear analysis to mathematically describe system variables (Petrofsky and Phillips, 1981). This is rather straightforward when employing digital control, as it requires only the addition of appropriate software commands.

A *servomechanism* or *servomotor system* is a power-amplifying feedback control system in which the controlled variable is mechanical position (Fig. 2.19) or a time derivative of position such as velocity or acceleration. This system is common when the controlled variable is isotonic muscle length (i.e., under conditions of constant muscle force). Servomechanisms are divided into types 0, 1, and 2. The type number depends only on the number of poles at the origin of the open-loop transfer function, not on the time constants of the function (Davis, 1966). When the controlled variable is isotonic muscle length, then there is a type 0 servomechanism. Petrofsky and Phillips (1979a) have developed and analyzed such a system using nonlinear digital feedback control. When the controlled variable is isotonic muscle acceleration, we have a type 2 servomechanism. No such system has been reported in the extant literature. Moreover, there is evidence that such a system is not necessary in closed-loop muscle control. Partridge (1966) has demonstrated that inertial (acceleration) compensation occurs in skeletal muscle. He demonstrated a constant amplitude

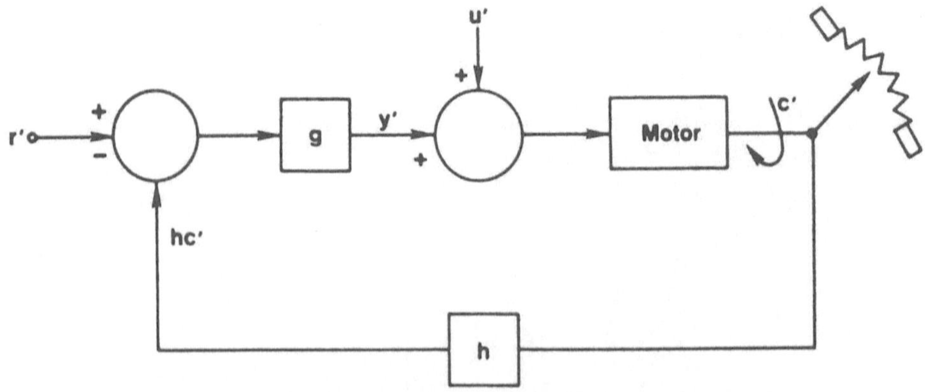

FIGURE 2.19. A servomotor feedback control system.

movement of loads (even over a 28-fold change in system inertia).

The preceding classification system is important to our understanding of the closed-loop muscle control problem. By placing such feedback control systems into the appropriate category, we can then interpret the physiological experimental results in terms of control system science. It is now accepted that control systems engineering spans not only the entire breadth of all engineering sciences, but the biological and social sciences as well (DiStefano et al., 1967). Adequate classification of various biological control systems *in the terminology* of control system science will prevent us from comparing "apples to oranges."

As seen above, muscle is usually tested in the experimental laboratory under either isometric or isotonic conditions. In contrast, physiological muscle action has long been recognized to be auxotonic, with both force and length changing simultaneously (Santesson, 1892, 1893; Rosemann, 1921). This higher order problem must still be appropriately classified in terms of conventional control system engineering if we are to adequately utilize the mathematical armamentarium of control system science.

For walking, such a system could be a type 0 servomechanism; that is, we select joint position (in part a function of muscle length) to be a controlled variable. We now require a higher level of regulating system design involving multiple-loop and multiple-input feedback system theory (Chestnut and Mayer, 1959). There would be separate inputs for muscle force (F) and muscle length (L). There would be separate feedback loops for muscle length (L) and its first derivative, muscle velocity (V). Knowing velocity allows predictive control, whereby future positions (after an interval of time) can be anticipated. A qualitative discussion of the overall problem has appeared (Hemami and Stokes, 1983). No such comprehensive system yet exists, but a "first approach" was initiated for a cat agonist–antagonist muscle pair (*in situ*) (Petrofsky and Phillips, 1979a), the hind limb of a cat (in vivo) (Petrofsky and Phillips, 1979c; Phillips and Petrofsky, 1983a), and standing and walking in a human paraplegic subject (Petrofsky and Phillips, 1983b). The latter utilized computer control algorithms described by Petrofsky et al. (1984b) and is an example of nonlinear digital control of human gait in the paralyzed individual (see Section IV).

The mathematical approach to closed-loop control of movement in paralyzed muscle is certainly complex, but complexity per se should not discourage us from taking the first steps.

IV. Human Studies on Muscle Control

A. Introduction

The initial human experiments involving closed-loop stimulation began with control of a single muscle group, the quadriceps muscle group. This muscle group extends the lower leg and is one of the prime muscles involved in walking and in

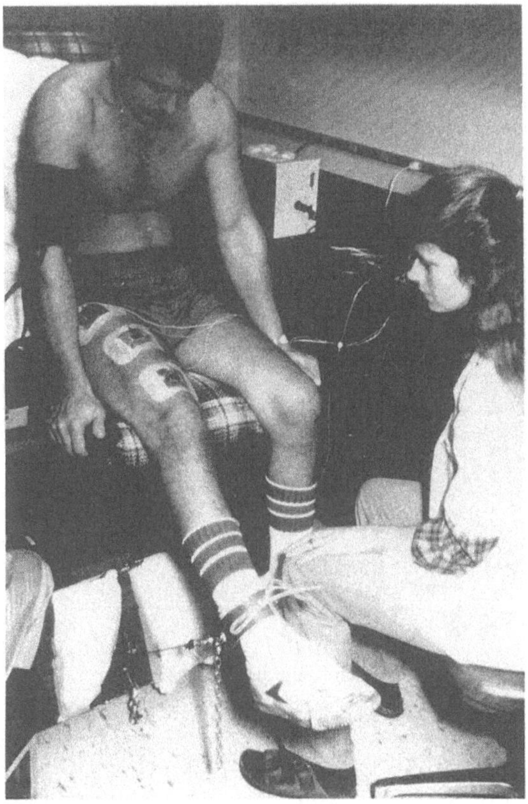

FIGURE 2.20. Stimulation of the quadriceps muscle with sequential stimulation for control during weight lifting.

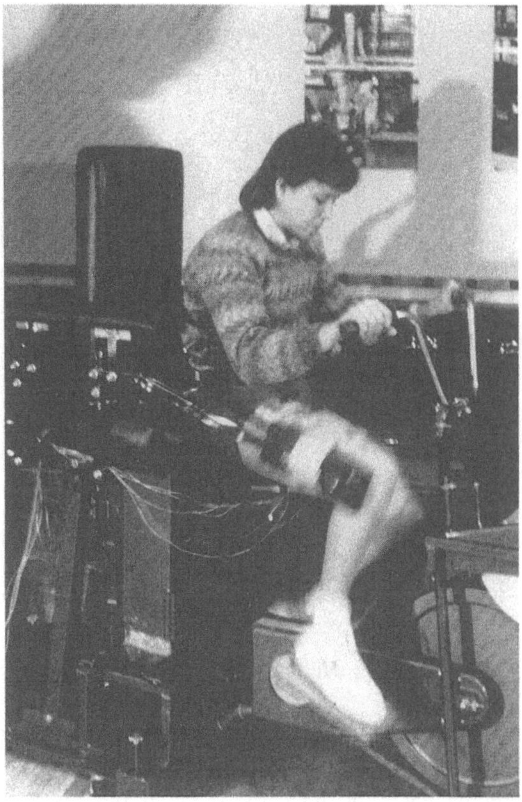

FIGURE 2.21. A paralyzed patient on a computer-controlled bicycle.

activities such as standing and bicycling. The quadriceps muscle is stimulated through sequential stimulation as shown in Figure 2.20. Three surface electrodes were applied to the quadriceps muscle and activated in alternate pairs at a base frequency of 50 Hz. Strength of contraction of the muscles was adjusted by changing pulse amplitude. The amplitude with surface electrodes varied between 5 and 45 V. Pulse width was kept constant at 300 μs.

The initial object was to apply the control equations of dynamic activity to a single muscle. To avoid rapidly solving complex control equations, control equations were developed so that the program followed a profile of predetermined muscle contractions. The program was set so that the lower leg moved through an arc of 60° with 3 seconds for the lower leg to slowly lift up, and 3 seconds for the lower leg to slowly go down (at a constant velocity). This profile was generated by an analog circuit, which then was compared to the actual position of the knee joint sensed by a potentiometer on the chair (Figure 2.20). By simply comparing the sensor output to the profile output, stimulation could be appropriately increased or decreased (Petrofsky and Phillips, 1983b). Only a small voltage correction is necessary if the muscle is close to the profile. A proportionately large increase or decrease in stimulation voltage was used for large errors.

The results of these early experiments showed that muscle tension could in fact be controlled successfully with a fairly small error using profile generation with electrical stimulation and sequential activation of the muscle in man (Petrofsky et al., 1984a; Phillips et al., 1984).

As with the cat experiments, the investigators then proceeded to develop multijoint control. An exercise bicycle was developed for this purpose (Petrofsky et al., 1984c; Phillips et al., 1984) (Fig. 2.21) and offered the advantage of supporting the

FIGURE 2.22. Patient pedaling exercise tricycle ergometer outside the laboratory.

body weight. The muscles to pedal the bicycle would have to function in the same type of coordinated pattern that we would later employ with walking. To close the loop for stimulation, a sensor (360° single-turn potentiometer) was placed in series with the bicycle pedals to indicate the position of the pedals to the computer.

The four muscles for bicycling were the quadriceps muscle and iliacus muscle of each leg. These four muscles were stimulated (as described above for the leg trainer) by sequentially activating a given muscle through three electrodes placed superficially over the belly of the muscle. Stimulation frequency was set at 50 Hz and stimulation pulse width was 300 μs, while amplitude was used to adjust recruitment. To control movement during bicycling, the computer algorithms were set to vary the stimulus amplitude to the various muscles based on a model of the normal amplitude profiles of the relevant muscles during voluntary bicycling (in nonparalyzed subjects). Since spinal-cord-injured (SCI) individuals, especially quadriplegics, have difficulty with control of posture, this system was also modified with a special seat on the back of the bicycle and contained a harness and seat strap, allowing the subject to sit comfortably.

Paraplegic and quadriplegic subjects were able to pedal the bicycle successfully at speeds of 15 km/h. From this initial development, the system was further modified to be used outdoors by adding a small Z80-based microprocessor system and carried in the basket of a tricycle (Petrofsky et al., 1983a). The tricycle (Fig. 2.22) offered a stable platform so that balance was not a problem while cycling. The tricycle was provided with a throttle control (somewhat like that used on a motorcycle) so that the subjects had voluntary control of the degree of stimulation to their legs. The programming for this system was similar to the stationary exercise bicycle described above. The system was a battery-operated and rechargeable system. The entire program was stored on an erasable–programmable read-only memory (EPROM) with a memory of 2 kilobytes.

It appeared from these initial experiments that the technology was available to have paralyzed individuals stand and walk. Ambulation in paralyzed individuals has been investigated extensively since Liberson's first demonstration that standing could be achieved in a paralyzed man (Liberson et al., 1962). Subsequent studies for gait restoration dealt with the use of FES in partially paralyzed

FIGURE 2.23. Sensory harness used for human patients.

FIGURE 2.24. Paraplegic standing with the computer control system.

man for conditions such as drop foot (Gracanin et al., 1967; Waters et al., 1975). Standing has been recently reexamined in paraplegic patients (Bajd et al., 1982) as well as walking (Kralj, 1983). The common denominator in these studies is that movement is induced by "open-loop" electrical stimulation. Movement of muscle is not monitored by any type of sensor.

To properly coordinate movement and to achieve free balance, sensors must be incorporated and appropriate "closed-loop" control algorithms must be designed. In our own work, we have expanded some of the cat studies by building a sensory harness (Fig. 2.23) that could determine the position of the hip, knee, and ankle joints of paraplegics and quadriplegics and provide feedback to the computer for control of standing and walking in man (Petrofsky and Phillips, 1983b). Special shoes were also developed with pressure transducers on the sole to indicate contact of the foot with the floor and thus properly coordinate the standing phase of the gait.

The initial system was applied to two paraplegics (Petrofsky and Phillips, 1983b) and was very successful in allowing the paraplegics to stand under computer control (Fig. 2.24). A safety harness was initially worn (Fig. 2.25) for medical safety reasons. Since these initial experiments, subjects have been able to walk without the safety harness (Fig. 2.26). They currently can perform four-point walking with canes (Fig. 2.27). Quadriplegic subjects are now able to walk and stand with a safety harness (Fig. 2.28).

The system involves stimulation through 24 separate channels of different leg and postural muscles (Petrofsky et al., 1983b), with balance input provided through solid-state level detectors. A portable version of the computer system (Fig. 2.29) was used to allow a paraplegic subject to walk to graduation (Fig. 2.30) and receive her degree. This system used many of the feedback control principles described for the cat work cited above but involved additional complexity. The computer control system is described in detail below.

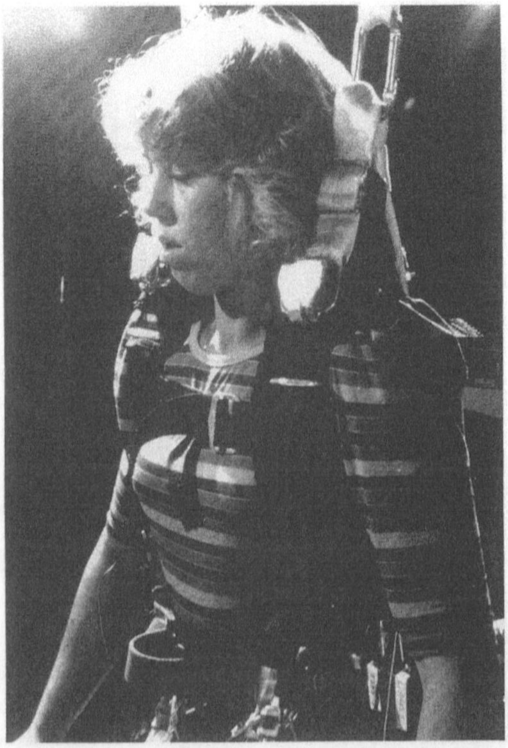

FIGURE 2.25. Paraplegic in a safety harness.

FIGURE 2.27. Paraplegic walking with canes.

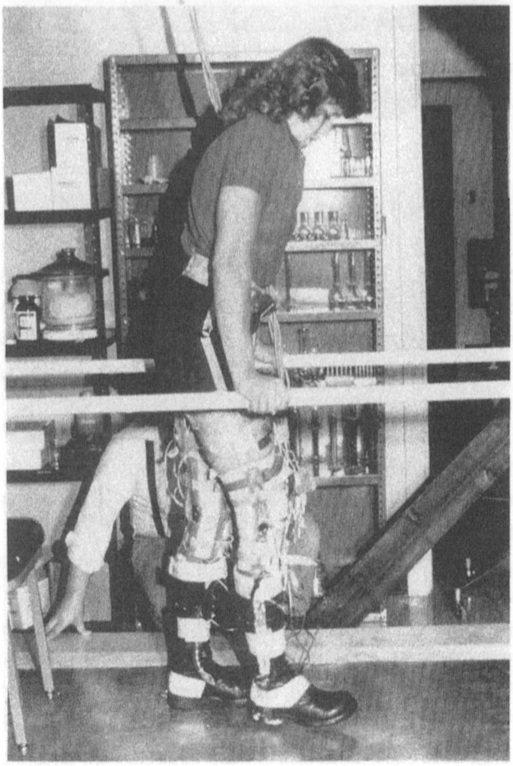

FIGURE 2.26. Paraplegic walking without a safety harness.

B. Computer Software for Controlling Movement

The software for controlling movement was divided into three sections. The first of these is "stand-up," the next one is "stance," and the final section is "walking." Walking is a modification of the software for the stance routines. The stand-up routine is the simplest. The muscles most extensively involved in standing up are the quadriceps muscles. To stand up, the upper body is shifted forward and the quadriceps muscles are activated; then the lower legs are extended until the knees are in a locked position. Contraction strength during this activity is successfully reduced. The computer algorithm for standing up appears in Figure 2.31.

This algorithm initially involves starting the quadriceps voltage in both muscles at zero stimulation voltage. It is slowly increased in a ramp while checking to see if the knees are locked. Stimulation voltages range from 0 to 255 units. This represents the range from zero to complete saturation of an eight-bit register. If the knees are within 5° of full extension (as indicated by the sensor), the quadriceps voltage is reduced to 120 units. The output voltage (associated with a unit output of 120)

FIGURE 2.28. Quadriplegic walking.

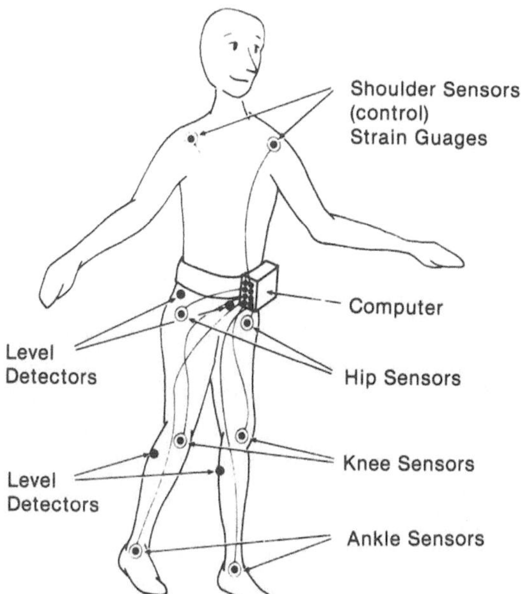

FIGURE 2.29. Portable stimulator system to allow a paraplegic to walk to graduation.

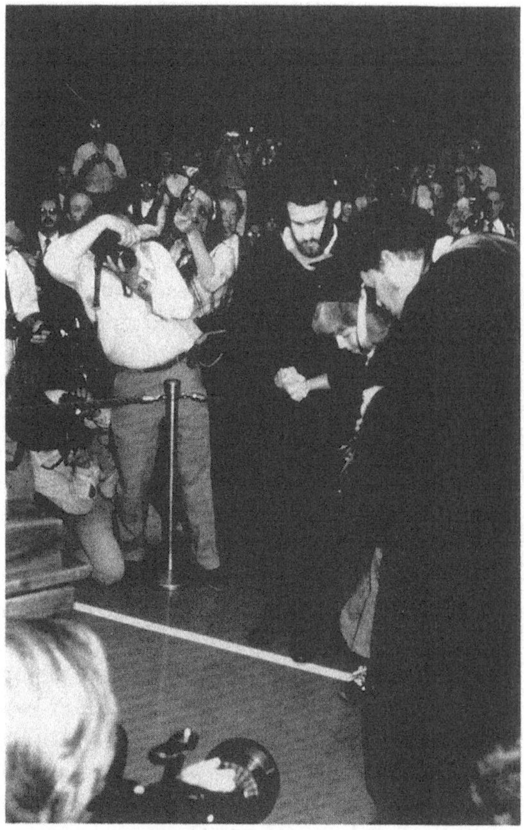

FIGURE 2.30. Paraplegic receiving her degree with computer-controlled movement.

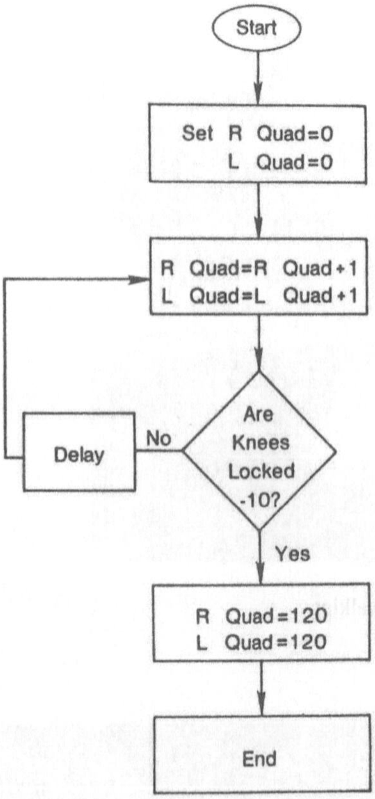

FIGURE 2.31. The control algorithms required to have a paraplegic or quadriplegic stand up.

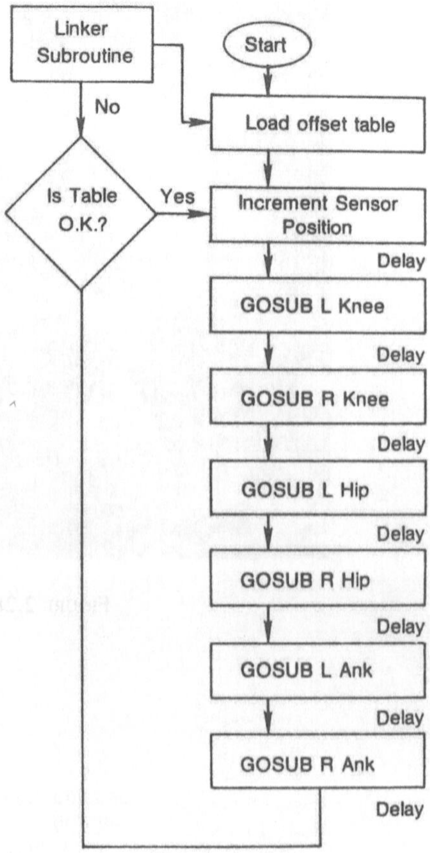

FIGURE 2.32. The "stance" subroutines.

actually represents 30–35 V measured at the electrodes. The result of the program is a rapid increase in stimulation voltage causing the quadriceps muscles to contract at the knee (extending the lower leg), followed by a reduction in muscle stimulation once the knee is locked. An output of 120 units was sufficient to develop about 10% of the quadriceps muscle strength. For most subjects, this resulted in knee lock.

The program at this point was automatically switched to the stance program, which then checked for knee lock and adjusted the voltages appropriately to maintain that knee lock. The stance subroutines are shown in Figure 2.32. Once "stand-up" was accomplished, the "stance" program automatically began to scan all six sensors (i.e., hip, knee, and ankle of each leg) to check for proper standing. The analog-to-digital converter (a Cromenso A/D converter board) is a two's complement A/D converter ranging from 0 to ± 2.544 V full range. The sensor outputs were buffered through a series of operational amplifiers with balance controls. Each sensor (potentiometer) read 0 V when the hips, knees, and ankles were neutral in-line (with the right leg backward). This was the basic stance subroutine.

Assuming all sensors read 0 V, entry into the "walking" subroutines was initiated by loading an offset table into the computer. The offset table was a table of values (used in the walking program) that allowed the sensors to vary under software control by an offset. The zero point (i.e., the point where the joint was in the proper position) could be varied from the neutral joint position (to either side of the neutral position) by adding or subtracting an offset to the sensor position. The offset table for standing was simply a series of zeros, since no offset was required. Therefore, when the offset table was added to the sensor positions, the output would

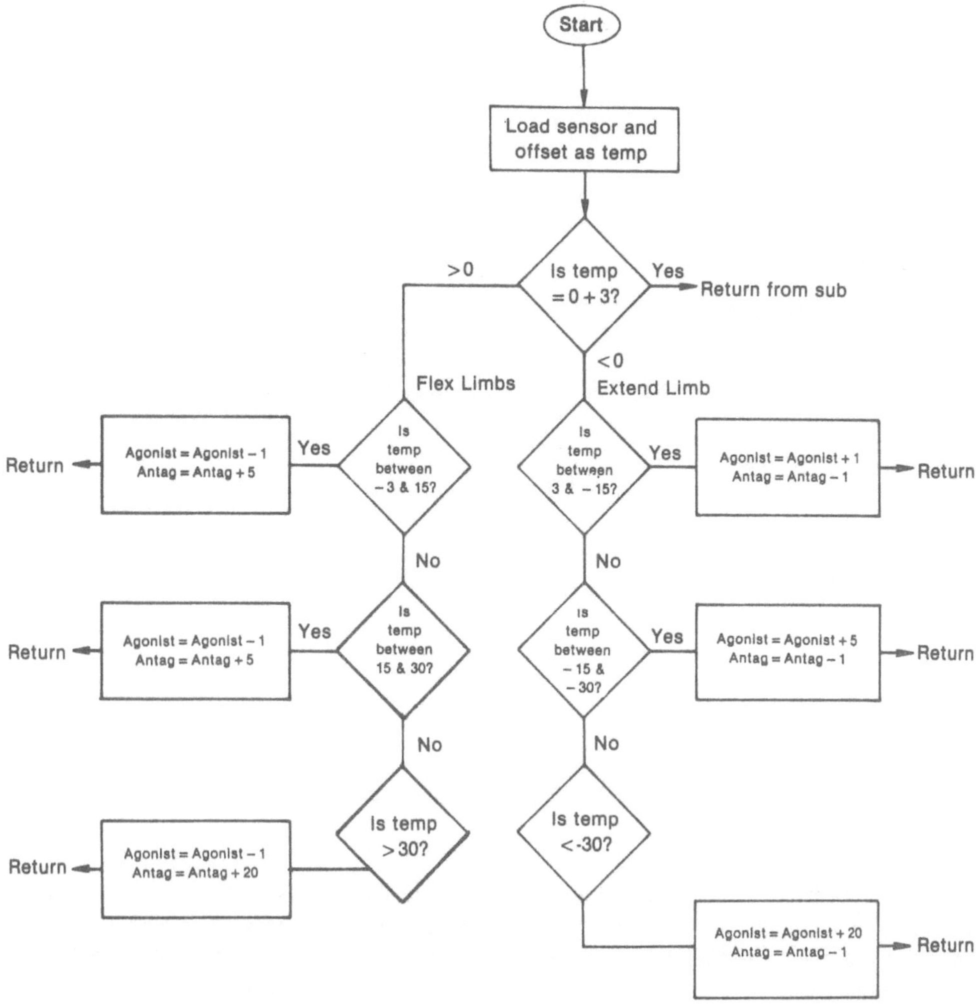

FIGURE 2.33. The sequence of subroutines used at an individual joint to control movement at that joint. This algorithm shows the flow diagram of a typical subroutine for one joint such as the hip or ankle joint. The linker subroutines are used to determine what the person is thinking about doing from shoulder activity.

have been a series of zeros if both knees were in the neutral position. After this particular subroutine (and in between all succeeding subroutines), a small delay was initiated. This delay was under software control (i.e., a timing loop) to slow down or speed up the subroutines appropriately.

Six subroutines were then run through: left knee, right knee, left hip, right hip, left ankle, and right ankle. All six subroutines essentially have the same format. As shown in Figure 2.33, a typical joint subroutine program involved looking at the current sensor position and the sensor offset and determining if the sensor difference was zero (desired joint position). If the sensor with offset was not zero, further action of the subroutine was required. Otherwise, the subroutine would return from the joint subroutine (Figure 2.33).

Using the knee as an example, a number less than zero (at this point in the program) would show that the lower leg was flexed too far and needed to be extended. This was accomplished by stimulating the agonist muscles extending the knee (quadriceps group) and reducing activity of the antagonist (hamstring group).

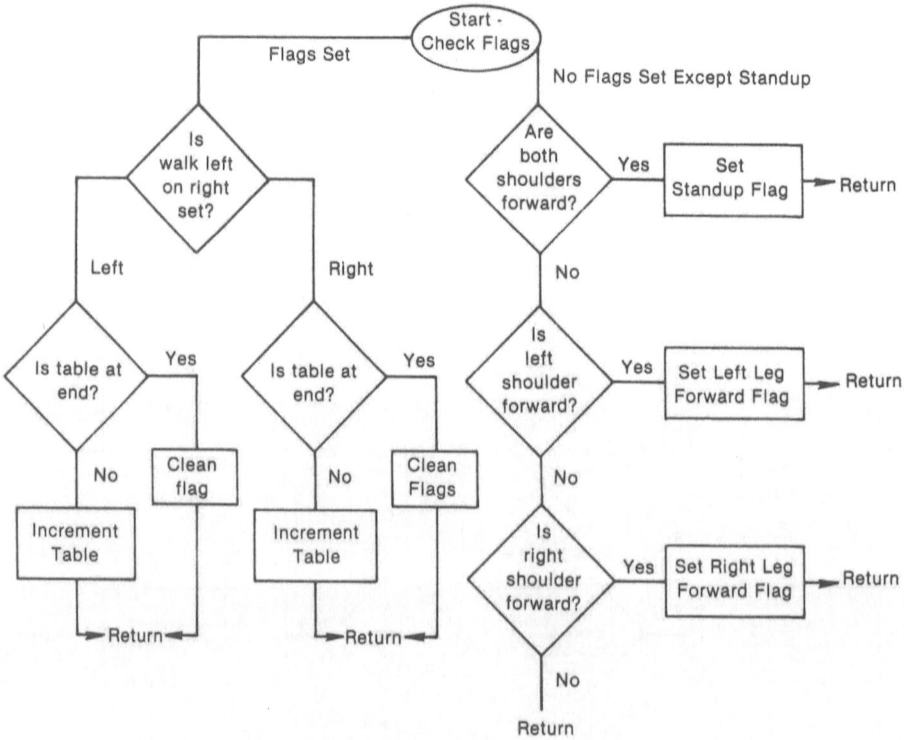

FIGURE 2.34. The linker subroutines.

Action of the antagonist and agonist muscles was accomplished by a technique called "successive windowing." For example, if the joint sensor position was close to that required (i.e., null position), voltages were increased and decreased by only a single volt. If the joint sensor position was moderately off, voltages were increased or decreased to a moderate degree. If the joint sensor position was substantially different from that programmed, then there was a large change in voltage. In this manner, stimulation voltage was applied in a logarithmic function to adjust the muscle activation to the proper level (appropriate for the target). If the limb was extended too far (i.e., the joint sensor position was greater than zero), the opposite would occur as shown in Figure 2.33.

These subroutines were accomplished in machine code until the termination of all six subroutines. At this point, the program entered the "linker" subroutine (Fig. 2.34).

The purpose of the linker subroutine was to determine whether the subject desired to continue to stand or to walk. The linker subroutine can be set in one of two ways. The linker subroutine (in the current application) was set under software control to look for movement of the right or left shoulder forward. The system was also programmed (for future use) to adjust the size of the step as a function of the degree of extension of the shoulder.

This was accomplished by allowing the subroutines to use analog inputs (from elastic resistors) to determine the length of the stride as well as the digital input described below. In their present form, the linker subroutines were used to set flags in the computer if one shoulder or the other was set forward. When flags were set (for the shoulder that moved forward), the subroutines jumped into the "walking" subroutines. Once a flag was set, the "stance" subroutines could still be used. However, walking was initiated by triggering a sequence of rapid changes in the offset table added serially and temporally to the initial sensor positions. As discussed previously, stance was accomplished by setting all sensors to zero. The computer would then do a seek and search at the joints to keep all the joint sensors nulled in this position. This was used to initiate walking.

Walking was divided into four phases. Each phase was obtained by rapidly changing the offset table for the joint sensor positions and scanning through the modified stance table. The first activity was to raise the right heel. This was accomplished by offsetting the hip, knee, and ankle sensor positions (through rapidly changing this offset table) until the right heel lifted off the floor. This terminated the first phase of walking. The next phase of walking involved lifting the hip. Again, by rapidly changing the joint sensor offset table, the leg was forced to move through a controlled right hip lift sequence with rate controlled by adjusting the delays in the stance subroutine. This continued until the right toe left the floor. When this was accomplished, the next phase of walking involved straightening the right leg by extending the lower leg at the knee.

Once the knee was extended and locked, the final subroutine involved moving the extended leg forward. This was accomplished by rotations at the opposite ankle through contracting the gastrocnemius muscle. This sequence of offset tables moved the body to its final position with the right leg forward and the left leg back. These programs of muscle and joint positions were patterned after Inman et al. (1952) for the movement of the joints during a standard gait cycle. When the left shoulder was moved forward, the opposite program was run so the left leg could move forward. Walking was basically divided into a piecewise linear analysis of the gait cycle as recorded by Inman and colleagues (1952). Each individual component of that piecewise linear analysis was under feedback control from the knee, ankle, and hip sensors. If any joint moved outside the required profile, appropriate action would be taken by the agonist and antagonist muscles.

C. Concluding Discussion

The algorithms just discussed have been tested only in a laboratory environment, where changes in muscle temperature and muscle fatigue were very limited. However, the control algorithms allowed limited movement for locomotion in man. Although speed of movement was reasonably close to that expected, the system can be improved in numerous ways. Rather than controlling the movement of the legs by position, velocity or acceleration may provide a better way to control movement. Since the shoulder can move back as well as forward, it should be possible to reprogram the computer to take a step in the backward direction also. By adding additional muscles (such as the sartorius and semimembranosis), it should be possible to take steps sideways as well, with inward and outward movement of the shoulder. With appropriate transducers, the system should be able to reproduce the three-dimensional leg movement.

Other adjustments are needed in the control algorithms. In practice, two paraplegic and two quadriplegic subjects were tested—all of whom could stand in excess of 15 minutes with the current control program. However, this was after extensive leg training (Petrofsky and Phillips, 1983a). It should be possible (with appropriate modification of the computer control algorithms) to get a more efficient assessment of joint activity and allow longer standing before the muscle fatigues. Certainly, one problem lies in sensing the position of the knee joint. Once the knee joint is in the lock phase, only a minimum amount of muscle tension should be required to keep the knee in that position. Since the sensors were position sensors and not tension sensors, once the knee is in the lock position, it is hard to deduce whether the muscle has 10 or 100% of the fibers contracting. This has a substantial effect on muscle fatigue. With many muscle fibers firing, substantial muscle fatigue could result.

Another problem lies in the ankle joint sensors. The knee and hip joint sensors were quite reproducible. Movement of these sensors through an arc of $\pm 45°$ and back to the original position resulted in a reproducibility error of either joint sensor of less than 2%. However, this was not the case with the ankle sensor. With repeated use of the ankle sensor, as much as 10% error would develop in the readout of the ankle sensor for a given ankle position. Therefore, ankle joint sensor technology needs to be improved.

Once the above problems are solved, the system must be miniaturized to a large-scale integrated (LSI) circuit. The eventual LSI circuit could be positioned on the surface of the body and radiofrequency-coupled to internal electrodes implanted inside the body. Surface electrodes were good for testing the initial system but are probably impractical for final use. They cause subject inconvenience, and high voltages and currents are necessary to excite deep muscles in the leg. Some muscles are also so deep within the leg that they cannot be excited sufficiently from the skin surface.

To make the system more practical, internal electrodes must be developed and either placed above the muscle, in the muscle, or preferably near the motor nerve to stimulate the muscle. Electrodes placed on the motor nerve require substantially lower currents and voltages than for surface electrodes. The average stimulation parameters for surface electrodes were a voltage of 60 V and a current as high as 50 mA.

For electrodes placed around motor nerves, we have demonstrated that voltages as low as 200 mV and currents of a few microamperes are sufficient to tetanize a muscle (Petrofsky, 1978; Petrofsky and Phillips, 1983d). Future development of the system would thus entail development of implantable electrodes. This would make the system smaller and more efficient, and it would have a lower power consumption.

References

Alexander, R.M. Mechanics of skeleton and tendons. In *Handbook of Physiology–The Nervous System, Vol. II*, V.B. Brooks, Ed. American Physiological Society, Bethesda, MD, 1981.

Ariano, M.A., Armstrong, R.B., and Edgerton, V.R. Hindlimb muscle fiber populations of five mammals. *J. Histochem. Cytochem.* 21:51, 1973.

Astrand, P.O., and Rodahl, K. *Textbook of Work Physiology*, 2nd ed. McGraw-Hill, New York, 1977.

Bajd, T., Kralj, A., and Turk, R. Standing-up of a healthy subject and a paraplegic patient. *J. Biomech.* 15:1, 1982.

Bergel, D.A., and Hunter, P.J. The mechanics of the heart. In *Quantitative Cardiovascular Studies*, N.H.C. Hwang, D.R. Gross, and D.J. Patel, Eds. University Park Press, Baltimore, 1979.

Bornhorst, W.J., and Minardi, J.E. A phenomenological theory of muscular contraction. I. Rate equations at a given length based on irreversible thermodynamics. *Biophys. J.* 10:137, 1970a.

Bornhorst, W.J., and Minardi, J.E. A phenomenological theory of muscular contraction. II. Generalized length variations. *Biophys. J.* 10:155, 1970.

Brobeck, J.R. *Best and Taylor's Physiological Basis of Medical Practice,* 10th ed. Williams & Wilkins, Baltimore, 1979.

Brown, G.L., and Burns, B.D. Fatigue and neuromuscular block in mammalian skeletal muscle. *Proc. R. Soc. London, Ser. B*, 136:182–195, 1949.

Caplan, S.R. A characteristic of self-regulated linear energy converters. The Hill force–velocity relation for muscle. *J. Theor. Biol.* 11:63, 1966.

Chestnut, H., and Mayer, R.W. *Servo Mechanisms and Regulating System Design*. Wiley, New York, 1959.

Close, R.I. Dynamic properties of mammalian skeletal muscle. *Physiol. Rev.* 52:129, 1972.

Cosgriff, R.L. *Nonlinear Control Systems*. McGraw-Hill, New York, 1958.

Crago, P.E., Mortimer, J.T., and Peckham, P.H. Closed-loop control of force during electrical stimulation of muscle. *IEEE Trans. Biomed. Eng.*, BME-27: 306, 1980.

Davis, S.A. *Outline of Servomechanisms*. Regents, New York, 1966.

Demieville, H.N., and Partridge, L.D. Probability of peripheral interaction between motor units and implications for motor control. *Am. J. Physiol.* 238:R119, 1980.

DiStefano, J.J., III, Stubberud, A.R., and Williams, I.J. *Theory and Problems of Feedback and Control Systems*. McGraw-Hill, New York, 1967.

Edwards, R.H.T., Harris, R.C., Hultman, E., and Nordesjo, L.O. Phosphagen utilization and resynthesis in successive isometric contractions, sustained to fatigue, of the quadriceps muscle in man. *J. Physiol.* 224:400, 1972.

Evans, W.R. *Control-System Dynamics*. McGraw-Hill, New York, 1954.

Fenn, W.O., and Marsh, B.S. Muscular force at different speeds of shortening. *J. Physiol.* 85:277, 1935.

Fung, Y.C. *Biomechanics: Mechanical Properties of Living Tissues*. Springer-Verlag, New York, 1981.

Gracanin, F., Prevec, T., and Tronteli, J. Evaluation of use of functional electronic personal brace in hemiparetic patients. In *Proceedings of the International Symposium on External Control of Human Extremities*, 1967.

Hemami, H., and Stokes, B.T. A qualitative discussion of mechanisms of feedback and feedforward in the control of locomotion. *IEEE Trans. Biomed. Eng.* BME-30:681, 1983.

Hill, A.V. The heat of shortening and the dynamic constants of muscle. *Proc. R. Soc. London, Ser. B*, 126:136, 1938.

Hill, A.V. *First and Last Experiments in Muscle Mechanics*. Cambridge University Press, Cambridge, 1970.

Hill, D.K. Tension due to interaction between the sliding filaments in resting striated muscle. The effect of stimulation, *J. Physiol. (London)* 199:637, 1968.

Huxley, A.F. Muscle structure and theories of contraction. *Prog. Biophys. Biophys. Chem.* 7:255, 1957.

Huxley, A.F. Muscle contraction. *J. Physiol. (London)* 243:1, 1974.

Huxley, A.F., and Simmons, R.M. Mechanic properties of the cross-bridges of frog striated muscle. *J. Physiol. (London)* 218:59P, 1971.

Inman, V.T., Ralston, H.J., Saunder, C.M., Feinstein, B., and Wright, E.W. Relation of human electromyogram to muscular tension. *Electroencephalogr. Clin. Neurol.* 4:187, 1952.

Iwazumie, T. *A New Field Theory of Muscle Contraction.* Ph.D. thesis, University of Pennsylvania, Philadelphia, 1970.

Kralj, A. *Rehabilitation Engineering Center for Progress Reprint.* National Institute of Handicapped Research, Washington, DC, 1983.

Liberson, W.T., Dondey, M., and Asa, M. Brief repeated isometric maximal exercises. *Am. J. Phys. Med.* 41:3, 1962.

Lind, A.R., and Petrofsky, J.S. Amplitude of the surface electromyograms during fatiguing isometric contractions. *Muscle Nerve*, 2:257, 1979.

Mommaerts, W.F.H.M. Energetics of muscular contractions. *Physiol. Rev.* 49:427, 1969.

Nyquist, H. Regeneration theory. *Bell Syst. Tech. J.* 2:126, 1932.

Partridge, L.D. Signal-handling characteristics of load moving skeletal muscle. *Am. J. Physiol.* 210:1178, 1966.

Partridge, L.D. Interrelationship studies in a semibiological "reflex." *Am. J. Physiol.* 223:144, 1972.

Partridge, L.D., and Benton, L.A. Muscle, the motor. In *Handbook of Physiology–The Nervous System*, Vol. II, V.B. Brooks, Ed. American Physiological Society, Bethesda, MD, 1981.

Partridge, L.D., and Glaser, G.H. Adaptation in regulation of movement and posture. A study of stretch response in spastic animals, *J. Neurophysiol.* 23:257, 1960.

Petrofsky, J.S. Control of the recruitment and firing frequencies of motor units in electrically stimulated muscles in the cat. *Med. Biol. Eng. Comput.* 16:302, 1978.

Petrofsky, J.S. Sequential motor unit stimulation through peripheral motor nerves in the cat. *Med. Biol. Eng. Comput.* 17:87, 1979.

Petrofsky, J.S., and Fitch, C. Contractile characteristics of skeletal muscle depleted of phosphocreatinine. *Eur. J. Physiol.* 384:123, 1980.

Petrofsky, J.S., and Lind, A.R. The influence of temperature on the isometric characteristics of fast and slow muscle in the cat. *Eur. J. Physiol.* 389:149, 1981.

Petrofsky, J.S., and Phillips, C.A. Constant velocity contractions in skeletal muscle of the cat. *Med. Biol. Eng. Comput.* 17:583, 1979a.

Petrofsky, J.S., and Phillips, C.A. Determination of the contractile characteristics of the motor units in skeletal muscle through twitch characteristics. *Med. Biol. Eng. Comput.* 17:525, 1979b.

Petrofsky, J.S., and Phillips, C.A. Microprocessor controlled stimulation in paralyzed muscles. *IEEE NAECON Record*, 79:198, 1979c.

Petrofsky, J.S., and Phillips, C.A. The influence of recruitment order and fiber composition on the force–velocity relationship of skeletal muscle in cat. *Med. Biol. Eng. Comput.* 18:381, 1980.

Petrofsky, J.S., and Phillips, C.A. Influence of temperature, initial length and recruitment order on the velocity of contraction of skeletal muscle in the cat. *J. Biomech.* 14:297, 1981.

Petrofsky, J.S., and Phillips, C.A. Active physical therapy: A modern approach to rehabilitation therapy. *J. Neurol. Orthop. Surg.* 4:165, 1983a.

Petrofsky, J.S., and Phillips, C.A. Computer controlled walking in the neurological paralyzed individual. *J. Neurol. Orthop. Surg.* 4:153, 1983b.

Petrofsky, J.S., and Phillips, C.A. The force–velocity relationship in skeletal muscle. In *Mechanics of Skeletal and Cardiac Muscle*, C.A. Phillips and J.S. Petrofsky, Eds. Charles C. Thomas, Springfield, IL, 1983c.

Petrofsky, J.S., and Phillips, C.A. The impact of recruitment order on electrode design for neural prosthesis. *Am. J. Phys. Med.* 60:243, 1983d.

Petrofsky, J.S., Guard, A., and Lind, A.R. The influence of muscle temperature on the contractile characteristics of fast and slow muscle in the cat. *Physiologist*, 21:91, 1978.

Petrofsky, J.S., Guard, A., and Phillips, C.A. The effect of muscle fatigue on the isometric contractile characteristics of skeletal muscle in the cat. *Life Sci.* 24:2285, 1979.

Petrofsky, J.S., Heaton, H.H., and Phillips, C.A. Outdoor bicycle for exercise in paraplegics and quadriplegics. *J. Bioeng.* 5:292, 1983a.

Petrofsky, J.S., Phillips, C.A., and Heaton, H.H. Computer control of walking in paraplegic subjects. *IEEE NAECON Record*, 83:1162, 1983b.

Petrofsky, J.S., Heaton, H.H., and Phillips, C.A. Leg exerciser for training of paralyzed muscle by closed-loop control. *Med. Biol. Eng. Comput.* 22:298, 1984a.

Petrofsky, J.S., Phillips, C.A., and Heaton, H.H. Feedback control system for walking in man. *Comput. Biol. Med.* 14:135, 1984b.

Petrofsky, J.S., Phillips, C.A., Heaton, H.H., and Glaser, R.M. Bicycle ergometer for paralyzed muscle. *J. Clin. Eng.* 9:13, 1984c.

Phillips, C.A., and Petrofsky, J.S. Contraction of skeletal muscle as a function of electrical activation and fiber composition: A mathematical model. *J. Biomech.* 13:549, 1980.

Phillips, C.A., and Petrofsky, J.S. The passive element force–velocity relationship: Influence on maximal contractile element velocity. *J. Biomech.* 14:399, 1981.

Phillips, C.A., and Petrofsky, J.S. Reply to the comment on "The passive elastic force–velocity relationship of cat skeletal muscle: Influence upon the maximal contractile element velocity." *J. Biomech.* 15:903, 1982.

Phillips, C.A., and Petrofsky, J.S. A mathematic approach for functional electrical stimulation in the cat. *IEEE NAECON Record*, 83:1193, 1983a.

Phillips, C.A., and Petrofsky, J.S. Muscle: Structural and functional mechanics. In *Mechanics and Skeletal and Cardiac Muscle*, C.A. Phillips and J.S. Petrofsky, Eds. Charles C. Thomas, Springfield, IL, 1983b

Phillips, C.A., Grood, E.S., Scott, W.J., and Petrofsky, J.S. Cardiac chemical power. I. Derivation of the chemical power equation and determination of equation constants. *Med. Biol. Eng. Comput.* 17:503, 1979a.

Phillips, C.A., Grood, E.S., Scott, W.J., and Petrofsky, J.S. Cardiac chemical power. II. The application of chemical power, work and efficiency equations to characterize left ventricular energetics in man. *Med. Biol. Eng. Comput.* 17:510, 1979b.

Phillips, C.A., Petrofsky, J.S., Hendershot, D.M., and Stafford, D. Functional electrical exercise: A comprehensive approach for physical conditioning of the spinal cord paralyzed individual. *Orthopedics*, 7:1112, 1984.

Rack, P.M.H. The behavior of mammalian muscle during sinusoidal stretching. *J. Physiol. (London)* 183:1, 1966.

Riggs, D.S. *The Mathematical Approach to Physiological Problems*. MIT Press, Cambridge, MA, 1970.

Rosemann, R. *Landois' Lehrbuch der Physiologie des Menschen*, 17th ed. Urban-Schwarzenberg, Berlin, 1921.

Routh, E.J. *Advanced Part of the Dynamics of a System of Rigid Bodies*, Vol. 2, 6th ed. Macmillan, London, 1930.

Ruch, T., and Patton, H.D., Eds. *Physiology and Biophysics, Vol. VI, Excitable Cells and Reflex Control of Muscle*. Saunders, Philadelphia, 1982.

Sandow, A. Skeletal muscle. *Annu. Rev. Physiol.* 32:87, 1970.

Santesson, C.G. Studien über die allgemeine Mechanik des Muskels. *Scand. Arch. Physiol.* 3:381, 1891–1892; 46:98, 1892–1893.

Solomonow, M., and Morse, M. Time response of isometric muscle force during recruitment. *IEEE Frontiers Eng. Comput. Health Care*, Paper 9.8, page 184, 1983.

Solomonow, M., and Scopp, R. Frequency response of isometric muscle force during recruitment. *IEEE Frontiers Eng. Comput. Health Care*, Paper 9.7, page 179, 1983.

Solomonow, M., Eldred, E., Lyman, J., and Foster, J. Control of muscle contractile force through indirect high frequency stimulation. *Am. J. Phys. Med.* 62:71, 1983a.

Solomonow, M., King, A., D'Ambrosia, R., and Shoji, J. Dynamic control of muscle force by high frequency electrostimulation. *Proceedings of the First Vienna International Workshop on Functional Electrostimulation*, Bioengineering Laboratory, University of Vienna, 1983b.

Stratton, D.B. *Neurophysiology*. McGraw-Hill, New York, 1981.

Thomson, J.D. Dimensional and dynamic features of mammalian gastrocnemius muscle. *Am. J. Physiol.* 200:951, 1961.

Vodovnik, L., Bajd, T., Kralj, A., Gracanin, F., and Strojnik, P. Functional electrical stimulation for control of locomotor systems. *CRC Crit. Rev. Bioeng.* 6:63, 1981.

Waters, R.L., McNeal, D., and Perry, J. Experimental correction of footdrop by electrical stimulation of the peroneal nerve. *J. Bone Jt. Surg.* 57-A:1047, 1975.

3 Control of Extremity Motor Prostheses: The Sensory Feedback Applications*

I. Introduction and Overview

Control systems are categorized depending upon the arrangement of the system elements. One category is a feedforward or "open-loop" control system, where the signal path or paths all proceed in the same direction, as shown in Figure 3.1a. Another category is a feedback or "closed-loop" control system where one (or more) signal paths proceed in an opposite direction from the other signal path or paths, as shown in Figure 3.1b. Operationally, this requires an additional element, called a sensor (*H*), which transforms the output variable (X_o) into a signal that can be compared to the reference variable (X_i). In most control systems, the feedback signal is subtracted from the reference signal, and the system has "negative" feedback.

Unlike an open-loop system, a negative-feedback system is capable of regulation and can be used to regulate the performance of an actuator. The actuator can assume a variety of different shapes and forms. Nonbiological systems (such as electric-powered prostheses for amputees) can include ac or dc servomotors. In biological systems (such as functional electrical stimulation prostheses for a paralyzed extremity), the body's own actuator is skeletal muscle. Whatever type of mechanism is serving as the actuator, the appropriate control signal must be applied to vary its movements. A sensor then measures the extent of that movement. In different types of closed-loop control system, this sensor can be responsive to either position, velocity, or acceleration of movement. Childress (1973) extended this basic description of control systems to the more general case of a "man–machine" interface where the system (machine) interacts with the human operator (man). He specifically examined both body-powered and electric-powered prosthetic arms for the upper extremity amputee. These will hereafter be collectively referred to as *powered* prostheses.

In his development, Childress (1973) defined three categories of signal flow in the man–prosthesis system. Referring to Figure 3.2, he defined types A, B, and C feedback. Type A feedback was defined as visual and auditory, and it potentially exists in most prosthetic systems (except those of the blind and/or deaf). Type B feedback was defined as sensory or proprioceptive and is often directed at the surface of the body. Two commonly used modalities for stimulating skin receptors were mechanical vibration and electrical stimulation. Type C feedback involved the technical section of the system. Childress (1973) noted that most of the powered prostheses utilized in clinical facilities do not have technological feedback (type B or C). Therefore, the systems are considered "open-loop" and may exhibit and erratic control.

The body-powered prosthesis for the upper extremity amputee is an example of natural type A and type B feedback control (Fig. 3.3). It has the

*From C.A. Phillips, Sensory feedback control of upper- and lower-extremity motor prostheses. Reprinted with permission from *CRC Critical Reviews in Biomedical Engineering*, 16:105–140, 1988. Copyright CRC Press, Inc., Boca Raton, FL.

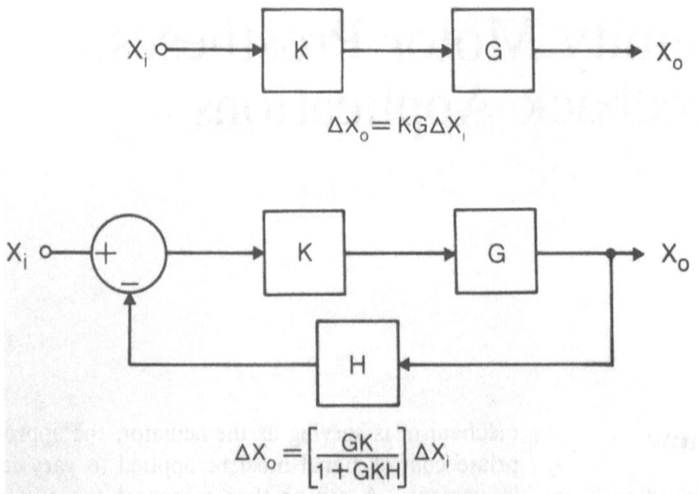

FIGURE 3.1. Categories of control sys tems: (a) feed forward (open-loop) and (b) feedback (closed-loop) systems.

advantage of being simple, and (if response characteristics are slow) it is visually controllable. It also provides a form of proprioceptive feedback, since cable position (which determines the joint angle of the prosthesis) is easily detected by the patient. Furthermore, forces that operate on the prosthesis are transmitted through the cable to the body. Consequently, body-powered prosthetic systems provide additional feedback (other than visual and auditory) from the prosthesis to the user's body.

Myoelectric control of electric-powered prostheses overcomes some disadvantages of body-powered prosthesis but with a loss of some natural feedback (particularly type B feedback). Therefore, Mann and Reimers (1970) have introduced the electric-powered prosthesis with technological feedback (Fig. 3.4). Proprioceptive force sensing in the Boston arm (Mann, 1968) occurs by electronically sensing force (*H* of Fig. 3.4) and feeding this parameter back to the limb servomechanism in the opposite direction of the body-powered electromyogram (EMG) signal. Kinesthetic (position) sensing was acquired by a unique cutaneous display resulting from electronic feedback of elbow angle (*I* of Fig. 3.4). In these cases, *H* and *I* represent the particular transfer functions characteristic of the specific type of feedback sensory system.

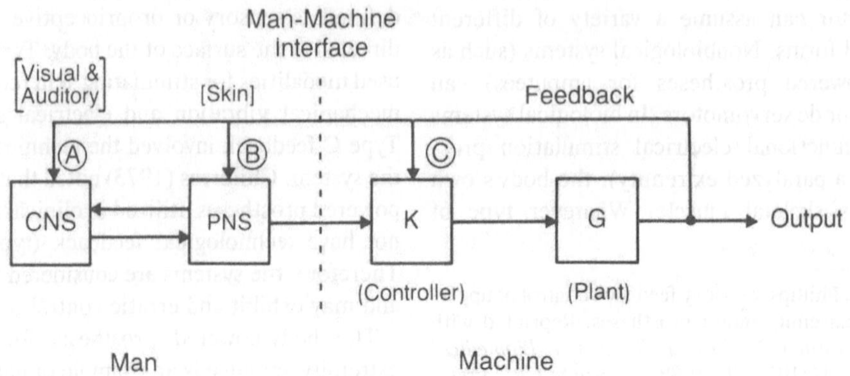

FIGURE 3.2. Diagram of potential information pathways for the human operator and an amputative prosthesis. Note that route *B* (via skin) can feed forward directly (reflex) or further feed back to the CNS (cognitive) before feedforward begins. CNS = central nervous system; PNS = peripheral nervous system. (Redrawn from Childress, 1973.)

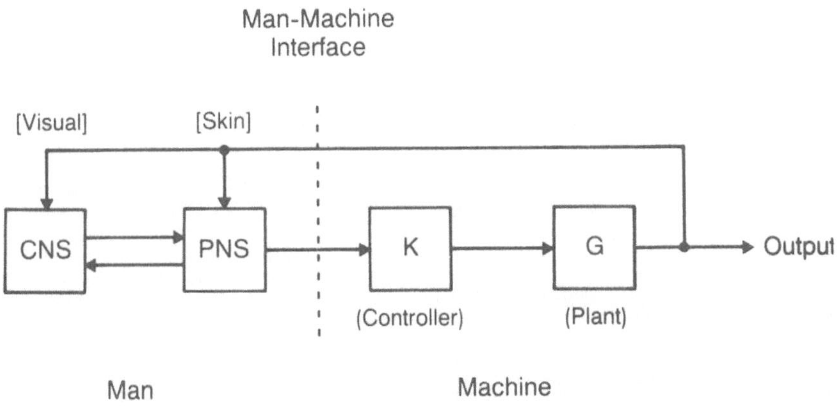

FIGURE 3.3. Diagram of information pathways for an upper extremity body-powered prosthesis with natural feedback control (see text).

Another category of prosthesis involves an intact extremity (or extremities), that is, one that has not been amputated. However, the extremity is functionally useless in that it is paretic and anesthetic. Such is the case in many neuromuscular disorders including spinal cord injury. One system to restore movement to the neuromuscularly injured extremity involves functional electrical stimulation and is defined as an FES prosthesis.

A diagram of the potential information pathways for the human operator and an FES prosthesis is shown in Figure 3.5. Although there is a man–machine interface, it is much less distinct than with a powered prosthesis. The FES prosthesis can also have type A, B, and/or C feedback, as with the powered prosthesis. The major distinction here is that for spinal cord injury, there is a cord lesion that interrupts communication between the upper motor neurons and the lower motor neurons (PNS 2) distally. However, when the alpha motor neurons (PNS 1) proximally below the lesion remain intact, they are capable of FES (S of Fig. 3.5) via a controller (K).

The plant (G) of a powered prosthesis is a mechanical arm or leg. The plant in an FES prosthesis is the human extremity with its peripheral nervous system and skeletal muscle. Distal to the spinal cord lesion, the neuromuscular plant is

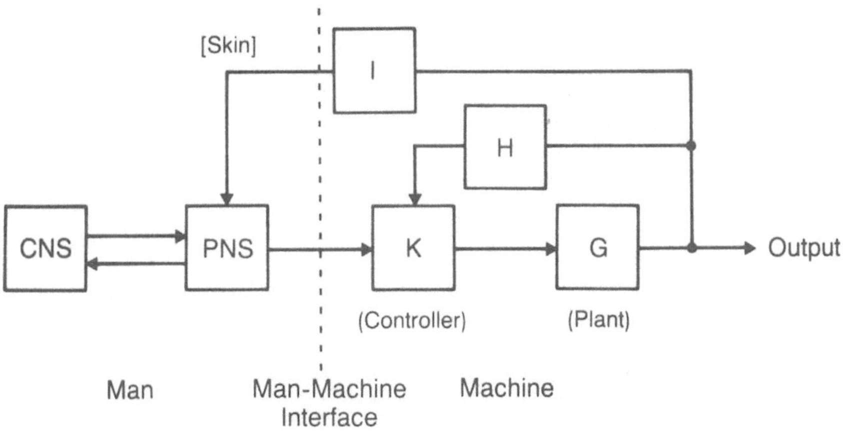

FIGURE 3.4. Diagram of information pathways for an upper extremity electric-powered prosthesis with technological motor feedback and sensory feedback (see text).

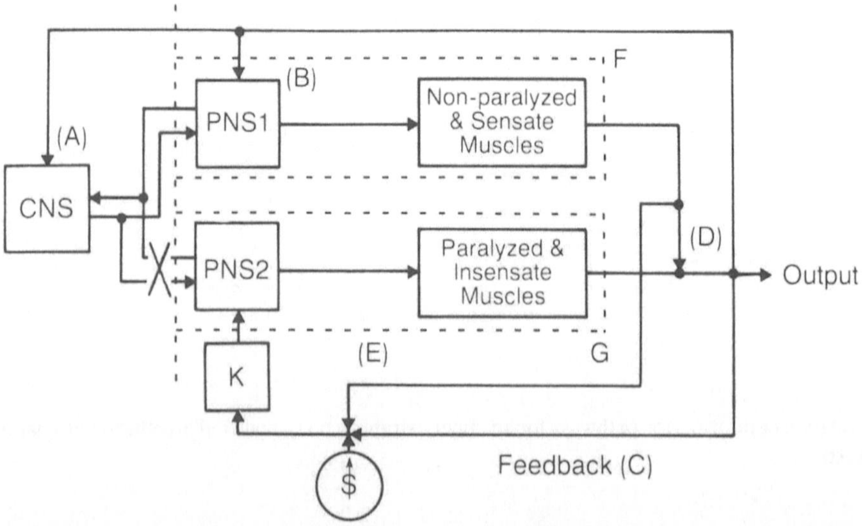

FIGURE 3.5. Diagram of potential information pathways for the operator and an FES prosthesis. Note the absence of a distinct nonmachine interface. *A*, *B*, and *C* are as in Figure 3.2; *D* and *E* are additional feedforward and feedback paths, respectively; *F* and *G* are the neuromuscular plant gains; *S* = stimulator (see text).

composed of paralyzed and insensate muscle (*G* of Fig. 3.5). Proximal to the spinal cord lesion, the neuromuscular plant is nonparalyzed and sensate skeletal muscle (*F*). This neuromuscular plant (*F*) receives type B feedback and its output can be divided into two information pathways. A feedforward pathway (*D* of Fig. 3.5) directly alters the output, while an additional feedback pathway (*E*) can alter the delivery of FES (*S*) to the controller. Examples of the feedforward pathway (*D*) are presented in Figures 3.6 and 3.7, below.

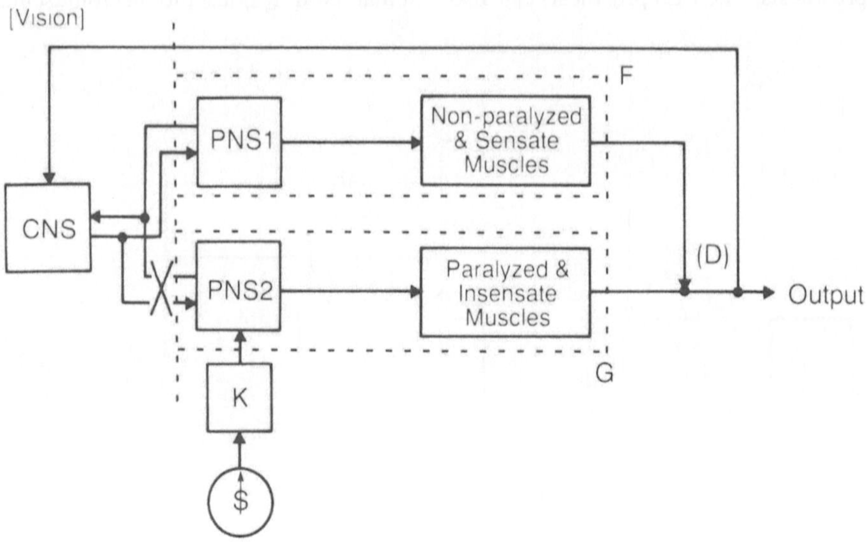

FIGURE 3.6. Diagram of information pathways for a lower extremity FES prosthesis with natural feedback control (see text).

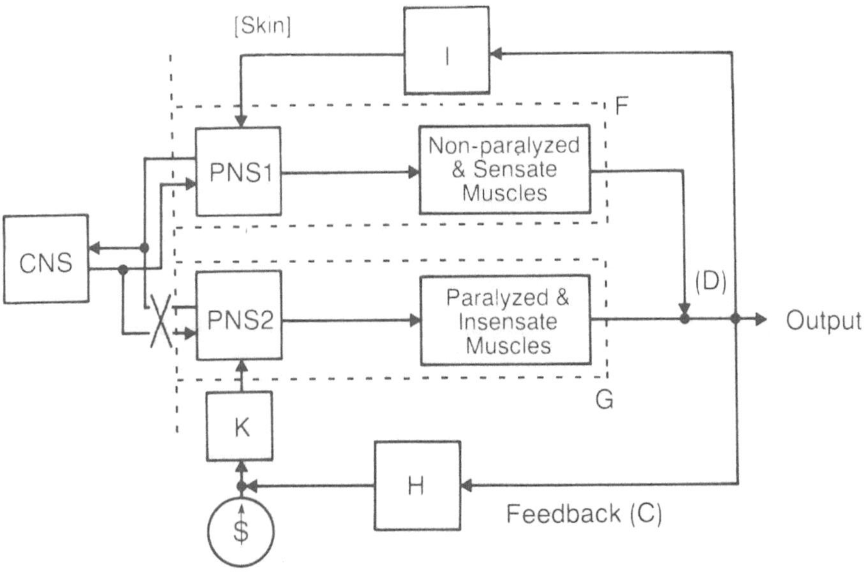

FIGURE 3.7. Diagram of information pathways for a lower extremity FES prosthesis with technological motor feedback and sensory feedback (see text).

An example of the feedback pathway (*E*) is the upper extremity FES prosthesis of Peckham et al. (1980). In this system, a shoulder-mounted switch alters FES delivery to the paralyzed upper extremity muscles. The switch itself, however, is under the voluntary control of the neuromuscular plant (*F*) so that switch position is adjusted via pathway *E* of Figure 3.5.

Kralj (1986) has developed an analytical model of a lower extremity FES prosthesis that uses natural feedback (type A). The FES prosthesis itself is open-loop (lower half of Fig. 3.6). The natural feedback is provided via visual cues, which result in the paraplegic subjects adjusting their upper body (under voluntary control) in order to adjust the output (balance during standing, for example).

Phillips and Petrofsky (1986) have described and analyzed an FES prosthesis in combination with an orthosis that utilizes technological feedback (Fig. 3.7). Lower extremity positional sensing occurs by electronically sensing knee angle (*H* of Fig. 3.7) and feeding back to the FES in order to control stand-up and sit-down in the FES–orthosis (Petrofsky et al., 1985).

Kinesthetic (foot-pressure) sensing was acquired by a cutaneous display resulting from electronic feedback of foot pressure (*I* of Fig. 3.7); *H* and *I* are the characteristic transfer functions of these particular sensory feedback systems. The system has allowed a paraplegic subject to stand up, sit down, and balance in the upright posture, even though blindfolded (i.e., no visual, type A feedback). The latter is done through voluntary upper body adjustments, which affect the output (balance) via pathway *D* (Fig. 3.7).

The remainder of this chapter is divided into three sections. In the next part (Section II), the upper extremity applications are reviewed. This section initially considers powered prostheses for both the arm and the hand, and then considers neural (FES) prostheses. In Section III, the lower extremity applications are reviewed. Both powered prostheses and FES–orthosis with a sensory feedback system are discussed. Section IV concludes with a systematic approach for effective sensory feedback control of extremity motor prostheses.

II. Upper Extremity Applications

Sensory feedback systems were initially utilized in conjunction with upper extremity powered prostheses. The primary objective was to improve the

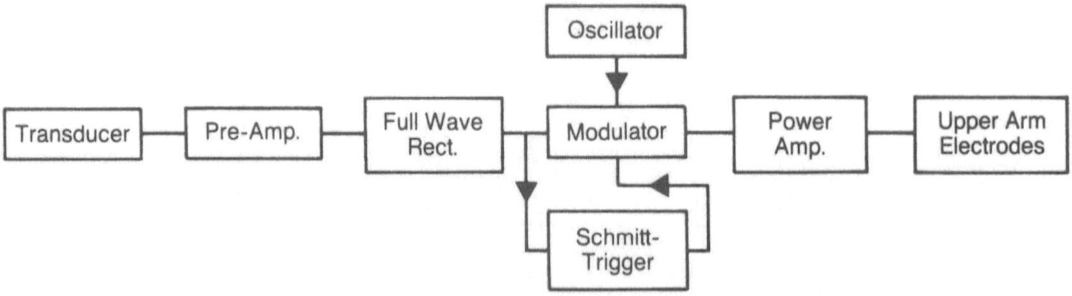

FIGURE 3.8. Block diagram of the electronic circuit for artificial touch developed by Beeker et al. (Redrawn from Beeker et al., 1967.)

function of an already developed and functional prosthesis. Therefore, these sensory feedback systems were a secondary development designed to supplement the upper extremity powered prosthesis.

More recently, upper extremity neural (FES) prostheses have appeared on the scene. The role of sensory feedback systems for such systems has also been evaluated. This section reviews in order the application of sensory feedback systems to upper extremity powered prostheses and to upper extremity neural (FES) prostheses.

A. Powered Prostheses

The first report of a sensory feedback system utilized with a powered prosthesis was made by Beeker et al. (1967), who described an artificial touch-sense for a hand prosthesis that was developed using a contact–pressure sensitive thumb (Fig. 3.8). Electrocutaneous stimulation was used to transmit signals back to the skin. The Beeker et al. report (1967) was essentially an engineering description, and no objective evaluation was provided regarding improved functional outcome. Pfeiffer et al. (1969) described the use of flexible pressure sensors (rather like the secretarial "rubber thumb") worn on the tip of the thumb and index finger of an anesthetic hand. This engineering-focused paper described the authors' mercury strain gage technology and a resistance-capacitance (*RC*) beat-frequency oscillator that emitted an audible sound whose frequency was a function of the pressure. Pfeiffer et al. (1969) mentioned the importance of objective evaluation of functional outcome (with and without sensory feedback), but their report did not provide any such information.

Mann and Reimers (1970) noted that the EMG-controlled arm as compared to the conventional cable arm provided no sensory (auditory and/or visual) feedback. They reported an innovative sensory feedback system for the Boston arm and provided objective psychophysical measurements (reaching) to evaluate functional outcome (with and without sensory feedback as compared to the cable controlled arm). This paper was in distinct contrast to previous reports in which functional outcome (with and without sensory feedback systems) had been reported as individual anecdotes and personal opinions. In the Mann and Reimers (1970) system, only the elbow angle was encoded and transmitted to the skin surface by vibrotactile stimulation. This was done at a constant amplitude and frequency (100 Hz), and the stimulus location was changed to reflect varying elbow angle. It was concluded that the EMG-controlled arm with vibrocutaneous sensory feedback performed as well as the conventional cable-controlled arm.

The work of Clippinger et al. (1974) involved direct stimulation of the median nerve. These investigators used a body-powered hand prosthesis activated in the conventional manner using a figure-8 harness and cable control. The locking cam was removed for manual operation and a sensory feedback system added. Like the majority of its predecessors, no functional testing was performed, to permit objective evaluation of any improvement in performance with the sensory feedback system.

In an initial report, Prior and Lyman (1975) reported on the assembling of hardware and

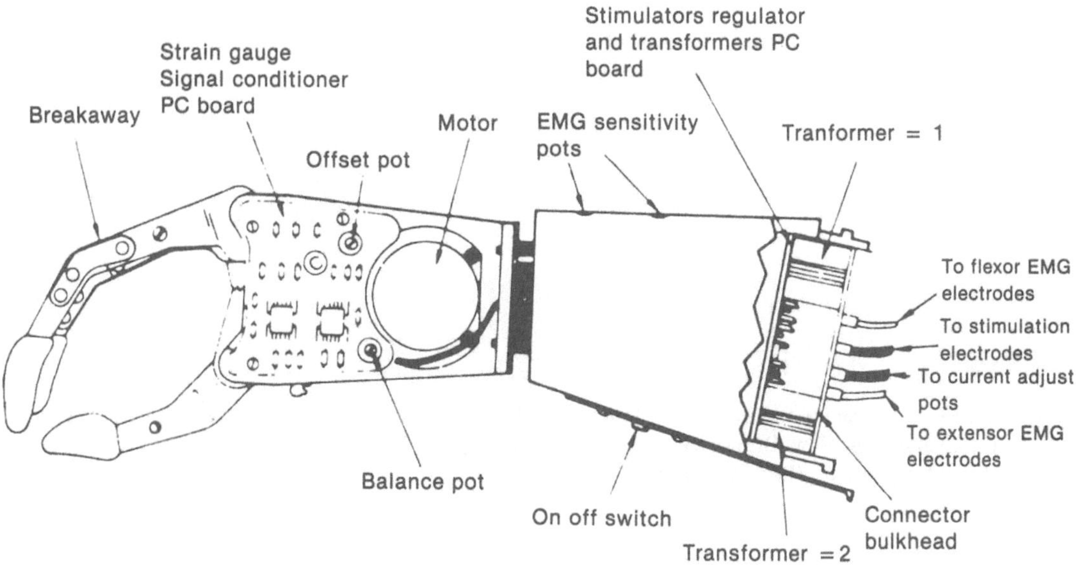

FIGURE 3.9. Sensory feedback system for the Veterans Administration/Northwestern University myoelectric hand using single electrodes, single channel, to monitor hand opening. (Reproduced from Prior et al., 1976b, with permission of Academic Press.)

finalizing of experimental design for sensory feedback systems (with powered upper extremity prostheses). They proposed block manipulation and grasp force as objective psychophysiological measurements of functional improvement using various generations of sensory feedback systems. In a later report, Prior et al. (1976b) defined supplemental sensory feedback for the amputee (Fig. 3.9). Electrocutaneous transmission of tactile information was defended as a physical parameter, but problems with psychological acceptance were noted. These investigators proposed that the myoelectrically controlled hand–arm could be used with electrocutaneous sensory feedback with appropriate isolation of the floating ground. They acknowledged that with second- and third-generation electrocutaneous sensory feedback systems (having two or more channels of information transfer), there were significant spatial discrimination problems.

In a subsequent paper, Prior et al. (1977) described a specific sensory feedback system using a potentiometer mounted over a hook for the below-elbow amputee. These investigators utilized a constant current stimulator, which produced an increasing pulse rate as a function of hook opening. Electrocutaneous stimulation was delivered by surface electrodes to the good arm. It was noted that the range of stimulation intensity was uncomfortable at the upper limit of hook opening. Furthermore, there was no discussion of the adaptation problem notoriously associated with electrocutaneous stimulation. A major strength of the Prior et al. (1977) paper was the excellent psychophysical measurements of the cable system activated hook–hand with and without supplement sensory feedback. Subjects judged the thickness of eight different blocks and demonstrated a 45% error rate with technological sensory feedback compared to a 60% error rate without technological sensory feedback.

The clinical application of powered upper limb prostheses was extensively reviewed by Lewis et al. (1975). Results of their two-year survey indicated that (a) external power does offer advantages to the arm amputee, (b) the components of the systems evaluated in their study were not without limitations and malfunctions, (c) many of the malfunctions were attributed to errors by prosthetists, for whom additional specialized education in the field of external power is necessary, and (d) many of the malfunctions were due to patient misuse.

A multifunctional sensory feedback system was described by Schmidl (1977) which incorporated an micropotentiometer in the thumb. In addition to the measurement of grip strength, contact switches

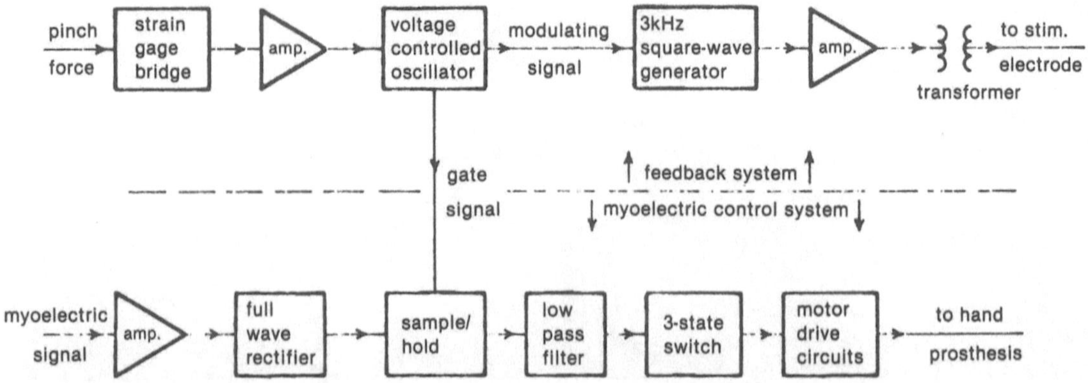

FIGURE 3.10. Block diagram of the feedback system and associated University of New Brunswick three-state myoelectric control system. (Reproduced from Scott et al., 1980, with permission of the International Federation for Medical & Biological Engineering.)

were mounted on the fingertips of the prosthetic hand. Upon making contact, the contact switches provided stimulation for only 300 ms. At any other time, stimulation indicated the level of grip strength. This allowed the two different functions to be conveyed via one channel of electrocutaneous stimulation. Schmidl (1977) gave a basic engineering description of the system, but only one subjective, qualitative functional report.

The gripping force between the thumb and index finger of a myoelectrically controlled hand was reported by Shannon (1979) and has been discussed previously as an application of strain gage technology. Scott et al. (1980) compared the University of New Brunswick three-state myoelectrical control system (when used with electrocutaneous sensory feedback of pinch force, Fig. 3.10) with the system described by Shannon (1979). Both papers were primarily engineering reports and neither provided objective, quantitative criteria to permit evaluation of functional improvement with the sensory feedback systems.

B. Neural (FES) Prostheses

Sensory feedback systems are by no means limited to upper extremity powered prostheses. They have also been applied to the functional electrical stimulation of paralyzed arms, although on a more limited scale. Peckham et al. (1980) obtained functional palmar prehension and release as well as lateral prehension and release in C5 quadriplegic patients. This was done via electrical stimulation applied chronically through percutaneous electrodes to the forearm muscles. The position of either the head or the shoulder was later used as the proportional controller. On–off and automatic control functions were provided by a myoelectrical signal. Peckham et al. (1980) reported that experience with their system demonstrated the need to provide the subject with information about the state of the control logic. Without sensory feedback, they found movements under FES control to be sometimes imprecise and cumbersome. Visual (type A) feedback was the subjects' primary means of assessing hand function. This group stated that they utilized additional type A (audio) feedback to signal the state of the control logic. Specifically, these audio tones were used to signal changes in the operating mode of the system. Unfortunately, no engineering details were given regarding such a system. The preliminary studies suggested that a patient's performance and confidence were improved with the addition of sensory feedback.

Petrofsky et al. (1984b) reported hand control in a C5 quadriplegic subject with FES applied to the forearm muscles (for finger flexion) using surface electrodes. This group utilized closed-loop (type C) technological feedback to coordinate the movement. In the Petrofsky et al. (1984b) system, elastomere sensors were woven into an undergarment. Movement of the shoulder in the up-and-down or forward-and-backward direction provided positional information to the computer regarding the desired hand position. Specifically, finger flexion and thumb movement were elicited. Additional elastomere pressure pads were placed on the fingertips of the subject (as a glove or simply taped on). These provided technological feedback to the computer which was utilized to coordinate movement.

Petrofsky et al. (1948b) reported the computer control algorithm for this system. However, no objective, quantifiable data were provided by either study (Peckham et al., 1980; Petrofsky et al., 1984b) to indicate any functional improvement in FES hand control with and without the use of either type A or type C sensory feedback.

Riso and Ignani (1986) described coding algorithms for use with an electrocutaneous sensory feedback system. Their type B feedback system provided information regarding machine state and the output command signals generated by a shoulder position controlled FES–orthosis as described by Peckham et al. (1980). A linear array of five electrodes was placed on the skin of the upper back (Fig. 3.11) and served as the output display of feedback information through the C5 or C6 quadriplegic individual using FES hand control. Riso and Ignani (1986) noted that subjects rapidly identified the absolute locus of stimulation when electrodes were spaced 4 cm apart. This finding was consistent with the observations of Phillips and Petrofsky (1985a) for two-point spatial discrimination of electrocutaneous sensory information when applied to the back of an individual. Riso and Ignani (1985) have demonstrated the efficacy of electrocutaneous feedback of shoulder position to assist individuals in producing more precise and more consistent command signals. These command signals are generated by the shoulder position transducer system and result in more effective control of the FES hand prosthesis.

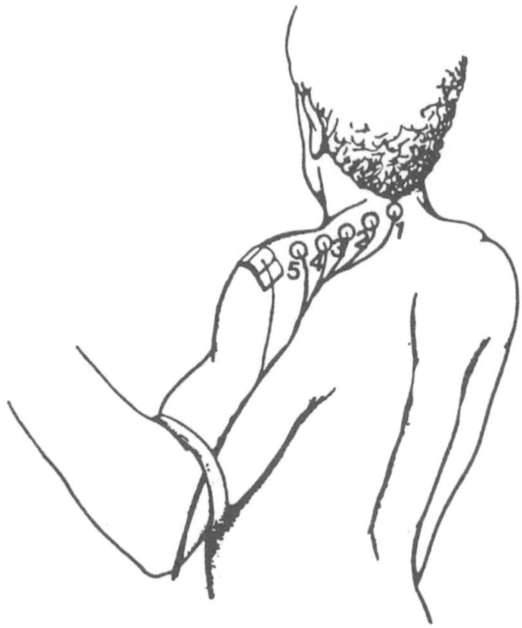

FIGURE 3.11. Five-electrode configuration on subject's shoulder used for sensory feedback for use with an FNS hand orthosis. (Reproduced from Riso and Ignagni, 1986, with permission of the Rehabilitation Engineering Society of North America.)

III. Lower Extremity Applications

Lower extremity applications of sensory feedback systems include both powered prostheses and FES systems. Limb load warning devices are representative of sensory feedback applications with powered prostheses. This section also reviews a specific FES–orthosis with a sensory feedback system for upright posture and balance.

A. Powered Prostheses

It is fully appreciated that there is a body of technical literature directed toward the lower extremity powered prosthesis for use by the amputee. However, only a few references could be found in the literature in which technological sensory feedback systems (similar to those used with upper extremity powered prostheses) had been applied to the lower extremity powered prosthesis. In general, these references describe limb load warning systems, which are used as an adjunct to rehabilitation.

Stolov et al. (1971) first indicated the need for an optimal progression of weight bearing. This progressive course was for patients fitted with immediate prostheses after below-knee amputation.

Endicott et al. (1974) developed a system that measured the load at the heel of an ambulatory patient. The system compared that load to a preset but adjustable required minimum or allowable maximum load. Such loads were prescribed by the physician, physiotherapist, or prosthetist. The device monitored the correct walking pattern by

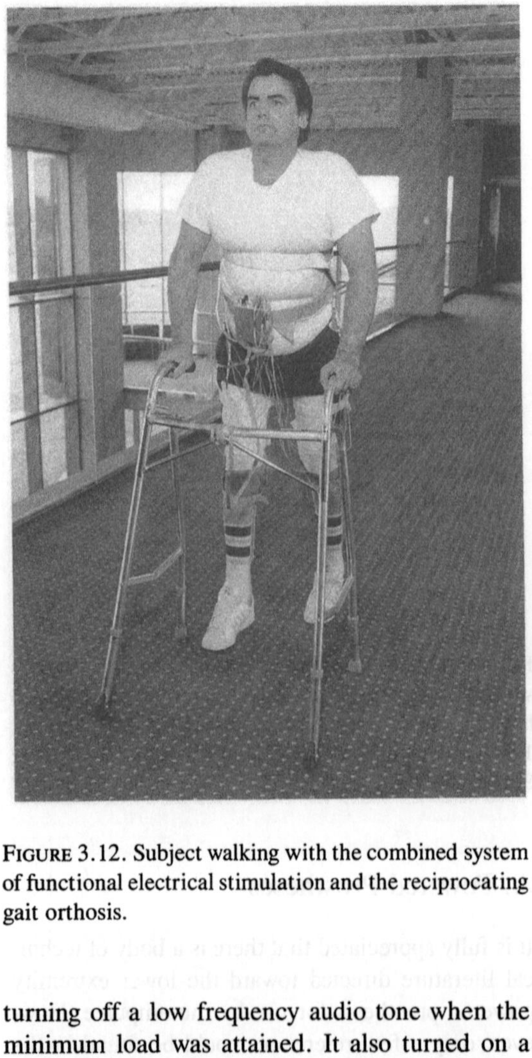

FIGURE 3.12. Subject walking with the combined system of functional electrical stimulation and the reciprocating gait orthosis.

turning off a low frequency audio tone when the minimum load was attained. It also turned on a high frequency tone when the maximum load was exceeded. Endicott et al. (1974) stated that the device would be useful to control the walking performance of lower extremity amputees during progressive weight bearing training, since it would help individuals to limit their weight bearing to a prescribed range.

Miyazaki and Iwakura (1978) also developed a portable electromechanical system for controlling the limb load exerted by an individual during partial weight bearing walking exercise. Their device measured the limb load with two removable load transducers affixed to the anterior and posterior aspect of the sole of the foot. The measured load was compared with two preset thresholds. It emitted a low frequency audio tone when the limb load was in the desired range. A high frequency audio tone occurred when the limb load exceeded the patient's maximum allowable load. Miyazaki and Iwakura (1978) provided performance data on four patients. The device displayed an accuracy of ±10% in all stages of the stance phase.

B. Neural (FES) Orthosis

During the past decade, considerable attention has been focused on the application of FES for standing and balancing, walking and sitting in the spinal-cord-injured individual. This type of neural prosthesis makes use of the fact that although central command has been lost, there is an intact neuromuscular system below the spinal cord lesion. With these lower extremity neural prostheses, the issues of both technological and natural, motor and sensory feedback control has received focused attention.

Basic pathophysiological changes occur in spinal-cord-injured individuals (who utilize neural prostheses). In addition to motor paresis below the level of injury, the complete spinal-cord-injured individual also has sensory anesthesia. Furthermore, the majority of spinal-cord-injured individuals have both legs completely affected. There is no proximal intact leg segment with sensate stump (and contralateral intact leg) to convey information, as with the amputee. Anyone who has had spinal anesthesia knows the feeling (or lack thereof).

Phillips and Petrofsky (1985a, 1986) have utilized vibrocutaneous stimulation in order to provide sensory feedback for the FES–orthosis (shown in Fig. 3.12). A four-channel stimulator system and complete carrier suppression are also features of the system (as described in Section III of Chapter 1). This combination of FES–orthosis and SFS (i.e., a total neural prosthesis) has resulted in the acquisition and maintenance of an upright posture and balance in a paraplegic subject (Phillips and Petrofsky, 1985b, 1986).

Phillips and Petrofsky (1986) reported that their sensory feedback system consisted of (a) foot load transducers (FLTs), which had strain gages mounted on lightweight portable force plates shaped to fit the bottom of the subjects' shoes, (b) the appropriate electronic signal conditioning, and (c) a vibrotacticle cutaneous interface with four spatially separated channels. These were located on the anterior chest wall (at T-2 to T-3 dermatomal levels). The two

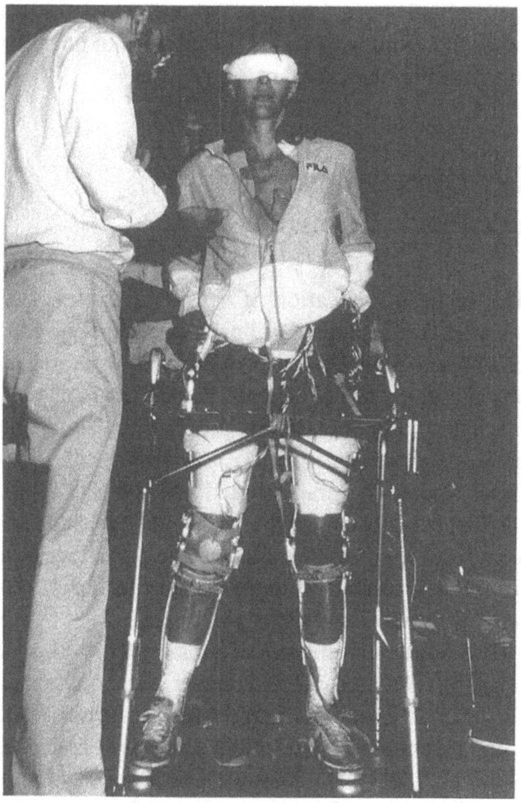

FIGURE 3.13. Paraplegic subject initially standing with the FES–orthosis and a sensory feedback system. (Reproduced from Phillips and Petrofsky, 1986, with permission of the American Academy of Neurological and Orthopedic Surgeons.)

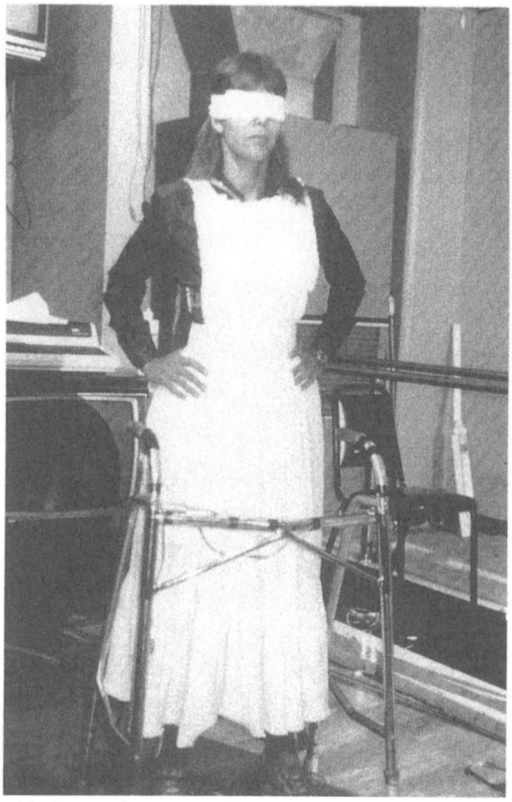

FIGURE 3.14. Paraplegic subject free-standing and balancing with FES–orthosis and a sensory feedback system. (Reproduced from Phillips and Petrofsky, 1986, with permission of the American Academy of Neurological and Orthopedic Surgeons.)

right-foot channels (one above the other) were located adjacent to the right sternal border. The two left-shoe channels (one above the other) were located adjacent to the left sternal border. The superiorally positioned channels conveyed information about force at the balls of the feet. The inferiorally positioned channels conveyed information about force at the heels of the feet.

Results are reported for a T4 level (complete) paraplegic subject. While upright and blindfolded (Fig. 3.13), there was an initial period averaging 30 seconds, during which time the subjects continuously adjusted their posture (and inclination at the hip varied to $\pm 5°$ peaks). This was followed by an average 5-minute period of prolonged standing (Fig. 3.14) (during which time hip inclination varied up to $\pm 1°$ peaks). The sensory feedback system was then deactivated while the subjects were still blindfolded. Within an average time of 5 seconds, the subjects lost their balance (at which time hip inclination exceeded $\pm 10°$) as shown in Figure 3.15.

Visual feedback has been utilized for the maintenance of upright posture, balance, and walking in all lower extremity neural (FES) prostheses to date (e.g., Brindley et al., 1978; Kralj et al., 1980; Vodovnik et al., 1981; Marsolais and Kobetic, 1983; Holle et al., 1984; Petrofsky et al., 1984a). The maintenance of upright posture and balance (in the presence of visual feedback) for the paraplegic subject who is using a neural (FES) lower extremity prosthesis has been mathematically modeled by Kralj (1986) and Jaeger (1986). These investigators have characterized the upright

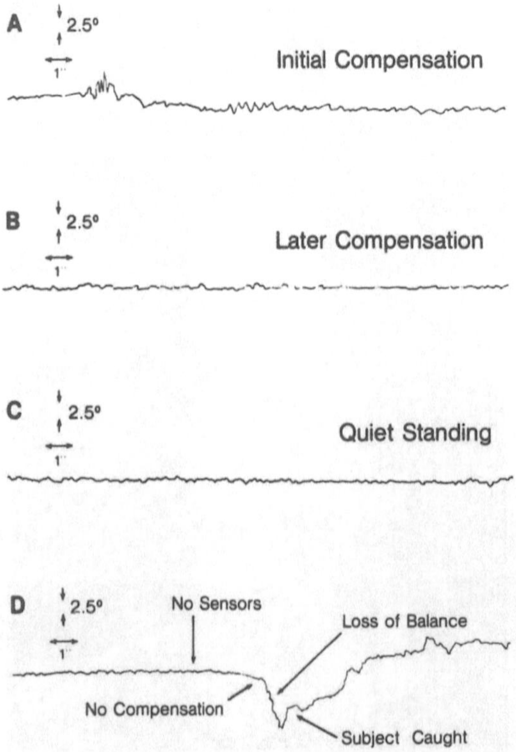

FIGURE 3.15. Sample of inclinometer data during an experimental trial (see text). (Reproduced from Phillips and Petrofsky, 1986, with permission of the American Academy of Neurological and Orthopedic Surgeons.)

posture and balance situation in terms of the familiar inverted pendulum model.

The most interesting aspect of the Phillips and Petrofsky (1985b, 1986) study was that upright body posture and balance can be maintained in a high level paraplegic without the use of visual feedback. This was considered an absolute test of the usefulness of the sensory feedback system, and it appears to have been satisfied. As stated earlier, upright posture and balance can be maintained with vestibular and tactile feedback alone. However, it is certainly not anticipated that future paraplegic subjects will rely exclusively on these two sensory modalities. Visual feedback will remain the most important sensation for maintaining upright posture and balance. However, the sensory feedback system (in combination with the FES–orthosis) will alleviate the continued necessity of having ambulating paraplegics look down at their legs and feet. Certainly, a large amount of their visual sense can now be redirected toward other more interesting and useful ways of interacting with the environment.

IV. Effective Sensory Feedback Control

The earliest application of sensory feedback systems was to "supplement" upper extremity powered prostheses. Sensory feedback systems were generally presented as an "add on" feature to an existing mechanical upper extremity prosthetic design. The basic objective was to make the mechanical prosthesis more "natural" (functional). Prior et al. (1976a) actually coined the term "supplemental sensory feedback," which reflected the prevailing attitude of a secondary role for sensory feedback systems with upper extremity prostheses.

The more central issue of sensory feedback systems for actual control of upper extremity prostheses was never formally addressed. However, some psychophysiologic studies have appeared (see Sections II and III). These studies have objectively described the usefulness of sensory feedback with respect to making upper extremity and lower extremity motor prostheses more functional.

There are four phases in the development of effective sensory feedback control of upper and lower extremity motor prostheses. Examples of recent work from this laboratory (Phillips and Petrofsky, 1986; Phillips and Buhrman, 1987; Buhrman and Phillips, 1989) will be used to illustrate specific phases.

Phase 1 requires a block diagram of the extremity prosthesis that includes both motor *and* sensory components. This phase is essential to understanding the complex interaction of sensory with motor components. Examples of this phase are given in Section I. In addition, the input and output variable for each functional block should be identified. As much as possible, the transfer function of each functional block should be specified, with particular attention to block gains and time delays.

Finally, the man–machine interface needs to be clearly identified for the system block diagram. This can be indicated by a broken line indicating the actual interface. This subtask of phase 1 must be completed as a prerequisite to the next phase.

Phase 2 requires a precise description and analysis of the interaction of the human operator and the prosthesis which occurs at the man–machine interface. This phase is exemplified by Phillips and Petrofsky (1986), who incorporate this procedure in their analysis of a total neural prosthesis. In order for the subject to maintain upright posture and balance, the person must interact with the FES-RGO and the sensory feedback system. This occurs at the man–machine interface.

In the upright and balanced position, the subjects initially adjust their balance by holding the handles of a standard walker. The intensity of tactile vibration of each cutaneous element is then increased (one channel at a time), using the four separate gain controls on the front of the signal-processing electronics package. The subjects will then sense four separate areas of tactile vibration on their anterior, superior chest wall corresponding to heel and ball pads of the right and left feet.

As mentioned in Section III of Chapter 1, the RGO has a pelvic band and thoracic support for better postural alignment. This aids in maintaining postural stability, especially for mid-and high level paraplegics. Also, the RGO has two cables attached to the anterior and posterior hip joints. A reciprocating action at the hips is produced by shifting upper body weight (Section III of Chapter 1).

The subject may begin to lose balance posteriorly (fall backward). This is detected by an increased tactile vibration from the heel (posterior) FLTs and a decreased tactile vibration from the anterior FLTs. The subject (with hands on the hips) then briefly displaces the shoulders and the elbows posteriorly so as to push anteriorly with the hands. This movement is transduced by the hip cables to cause a hip "tuck" (in effect, causing hip extension). Neck and head tilts are used to equilibrate the center of gravity for finer degrees of control (as necessary). This returns more of the body weight onto the anterior aspect of the foot and relieves some of the weight on the posterior (heel) of the foot. Thus, the subject corrects in order to maintain a constant, uniform level of vibration intensity on all four of the chest wall cutaneous elements.

The subject may begin to lose balance anteriorly (fall forward). This is detected by an increased tactile vibration from the anterior FLTs and a decreased tactile vibration from the posterior (heel) FLTs. The subject then displaces the shoulders anteriorly so as to push posteriorly with the hands. This movement is transduced by the hip cables to release the hip "tuck" (in effect, causing hip flexion). This returns more of the body weight onto the posterior aspect (i.e., heel) of the foot and relieves some of the weight on the anterior part of the foot. Again, the subject corrects in order to maintain a constant, uniform level of vibration intensity on all four of the chest wall cutaneous elements. Concurrent or subsequent head–neck tilt assists to reequilibriate the center of gravity.

Lateral instability is not a significant problem as the RGO provides rather rigid sideways bracing (along the medial and lateral sides of the legs). Also, the subject positions the two feet far enough apart so that the stance base is appropriately wide.

Phase 3 requires a physical model of the man–machine interaction. This model must be described mathematically and must be appropriately verified. The physical model will be suggested by the more qualitative analysis acquired during phase 2. However, it provides a more quantitative and controllable approach to further analysis of the man–machine interface. Phase 3 is exemplified by the work of Phillips and Buhrman (1987) and Buhrman and Phillips (1989). These investigators developed a model of the interaction of the FES–RGO and sensory feedback system for maintenance of upright posture and balance.

Sensory feedback studies by Phillips and Petrofsky (1986) had indicated that during standing and balance (in a blindfolded paraplegic) there was an upper body motion as the person attempted to compensate for variation in the center of gravity. This anterior–posterior swaying motion was measured with an inclinometer as an error signal of 0–5° (Fig. 3.15). This was suggestive of a type 0 servomechanism system for balance control, which could be described by the following second-order, linear differential equation (Fig. 3.16):

$$J\frac{d^2\theta_o}{dt^2} + C\frac{d\theta_o}{dt} + K(\theta_o - \theta_i) = 0 \qquad (3.1)$$

Buhrman and Phillips (1989) then employed the following methodology. Four steps were required.

1. A system transfer function for a step input perturbation was derived from the differential equation above as modified by Phillips and Buhrman (1987):

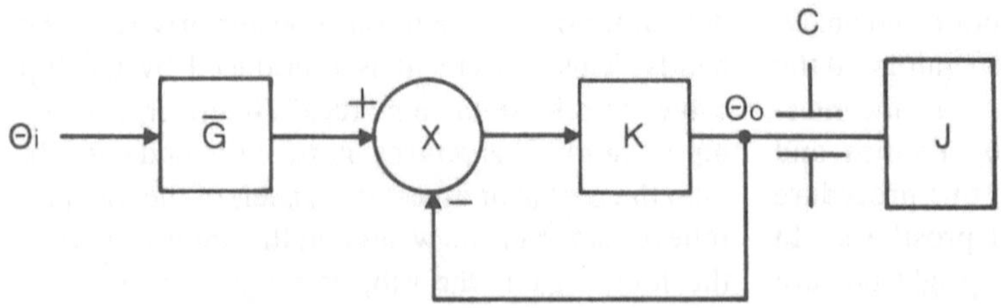

FIGURE 3.16. Type 0 servosystem with second-order differential equation.

$$J \frac{d^2\Theta_o}{dt^2} + C\frac{d\Theta_o}{dt} + K(\Theta_o - \bar{G}\Theta_i) = 0$$

$$\frac{\Theta_o}{\Theta_i} = 1 + \left(\frac{G}{\sqrt{1-z^2}}\right) e^{-zwt} \cdot \sin\left(w\sqrt{1-z^2}\, t - \phi\right) \quad (3.2)$$

where

$$\phi = \cos^{-1} z \quad (3.3)$$

2. A physical model of the inverted pendulum was constructed to simulate erect posture and balance (shown diagrammatically in Fig. 3.17). An eyebolt with a stiff shaft was spring-mounted to a platform adjustable from 0 to 5°. A transducer attached to the eyebolt head was interfaced to a paper chart recorder.
3. The platform was dropped from inclinations of 2.0, 3.5, and 4.5° (to the horizontal). The transducer gave oscillations in response to these step perturbations, which were plotted over time by the paper chart recorder.
4. The parameters z and w of the theoretical model were calculated from the experimental data. The theoretical model differential equation was then solved over time and superimposed on the recorded (experimental) data (Fig. 3.18). The scaling constant G was computed by comparing the peaks of the predicted and recorded values of θ_o and averaging the values.

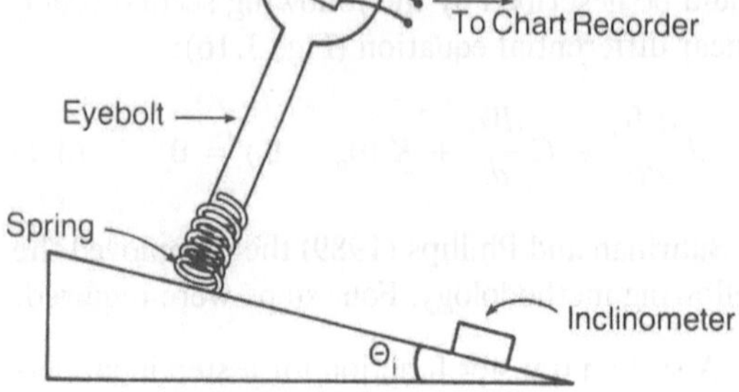

FIGURE 3.17. Physical model used to simulate a free-standing balance response. (Redrawn from Phillips and Buhrman, 1987.)

Buhrman and Phillips (1989) obtained these results: z was computed as 0.0904 ±0.006 and w was computed as 18.01 ±0.56 (rad) for 4.5° of perturbation respectively. The scaling constant G was calculated as an average value of 3.46 ±0.25.

The experimental data points were compared with those predicted by the mathematical model. Figure 3.18 shows a close approximation of the two curves with a slight magnitude underprediction at the first and second positive peaks.

The model has been tested in two ways. The first test was the superposition of the response calculated from the model upon the actual measured response. A close approximation of the measured to the calculated responses indicated a correct model. The second test was to achieve constant values of z and w. For a correct model, the values of z and w should remain constant for all values of θ_i.

Phase 4 requires the application of the physical model to the human operator–prosthetic system interaction. For example, the application of the Phillips–Buhrman model (1987) is that sensory feedback systems for lower extremity FES-orthoses can be mathematically analyzed. Specifically, the experiments of Phillips and Petrofsky (1986) can now be repeated at various degrees of tilt perturbation and mathematically analyzed with the model. Changes in system parameters (w and z) can be determined for upright standing and balance with and without the use of sensory feedback (visual and/or tactile). Optimal balancing strate-

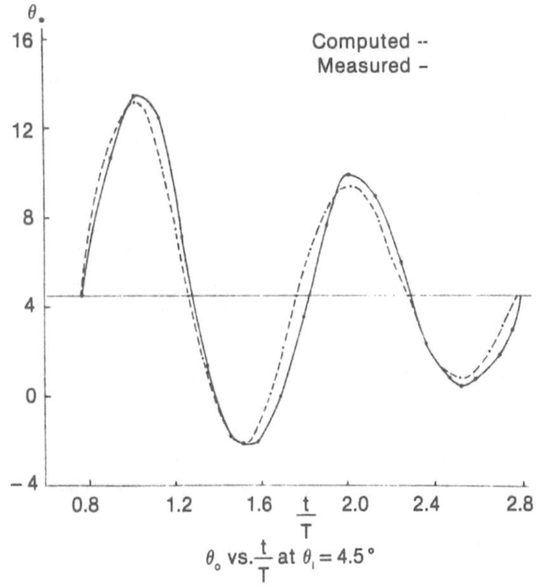

FIGURE 3.18. Plot of θ_o versus t/T at $\theta_i = 4.5°$ (see text).

gies (by voluntary movement of the upper body and FES movement of the paralyzed muscles) can now be developed by determining the appropriate changes in the *w* and *z* system parameters.

In summary, four phases are required in the development of effective sensory feedback control of upper and lower extremity motor prostheses. Phase 1 requires a block diagram of the extremity prosthesis that includes both the sensory and motor components. Phase 2 requires a precise description and analysis of the interaction of the human operator and the prosthesis which occurs at the man–machine interface. Phase 3 requires a physical model of the man–machine interaction. Phase 4 requires the application of the physical model to the human operator–prosthetic system interaction. The final result of this systematic approach will be the realization of more effective upper and lower extremity prostheses.

References

Beeker, T.W., During, J., and den Hertog, A. Artificial touch in a hand prosthesis. *Med. Biol. Eng.* 5:47–49, 1967.

Brindley, G.S., Polkey, C.E., and Rushton, D.N. Electrical splinting of the knee in paraplegia. *Paraplegia,* 16:428, 1978.

Buhrman, J.R., and Phillips, C.A. A mathematical model to analyze control of upright posture and balance. *Automedica* (London) 11:111–122, 1989.

Childress, D.S. Powered limb prostheses: Their clinical significance. *IEEE Trans. Biomed. Eng.* BME-20: 200–207, 1973.

Clippinger, F.W., Avery, R., and Titus, B.R. A sensory feedback system for an upper-limb amputation prosthesis. *Bull. Prosthes. Res.* 10-22:247–258, 1974.

Endicott, D., Roemer, R., Brooks, S., and Meisel, H. Leg load warning system for the orthopaedically handicapped. *Med. Biol. Eng.* 12:318–321, 1974.

Holle, J., Frey, M., Gruber, H., Kern, H., Stohr, H., and Thoma, H. Functional electrical stimulation of paraplegics. *Orthopedics,* 7:1145–1155, 1984.

Jaeger, R.J. Design and simulation of closed-loop electrical stimulation orthoses for restoration of quiet standing in paraplegia. *J. Biomech.* 19:825–835, 1986.

Kralj, A. Electrical stimulation of lower extremities in spinal cord injury. In *Spinal Cord Injury Medical Engineering,* D.N. Ghista and H.L. Frankel, Eds. Charles C. Thomas, Springfield, Ill, pp. 439–509, 1986.

Kralj, A., Bajd, T., and Turk, R. Electrical stimulation providing functional use of paraplegic patient muscles. *Med. Prog. Technol.* 7:3–9, 1980.

Lewis, E.A., Sheredos, C.R., Sowell, T.T., and Houston, V.L. Clinical application study of externally powered upper-limb prosthetic systems: The VA elbow, the VA hand, and the VA/NU myoelectric hand systems. *Bull. Prosthes. Res.* 10-24:51–136, 1975.

Mann, R.W. Efferent and afferent control of an electromyographic proportional-rate, force sensing artificial elbow with cutaneous display of joint angle. *Proc. Inst. Mech. Eng.* 183 (Part 3J):86–92, 1968.

Mann, R.W., and Reimers, S.D. Kinesthetic sensing for the EMG-controlled "Boston arm." *IEEE Trans. Man-Machine Syst.* MMS-11:110–115, 1970.

Marsolais, E.B., and Kobetic, R. Functional walking in paralyzed patients by means of electrical stimulation. *Clin. Orthop. Relat. Res.* 175:30–36, 1983.

Miyazaki, S., and Iwakura, H. Limb-load alarm device for partial-weight-bearing walking exercise. *Med. Biol. Eng. Comput.* 16:500–505, 1978.

Peckham, P.H., Mortimer, J.T., and Marsolais, E.B. Controlled prehension and release in the C5 quadriplegic elicited by functional electrical stimulation of the paralyzed forearm musculature. *Ann. Biomed. Eng.* 8:369–388, 1980.

Petrofsky, J.S., Phillips, C.A., and Heaton, H.H. Feedback control system for walking in man. *Comput. Biol. Med.* 14:135–149, 1984a.

Petrofsky, J.S., Phillips, C.A., Larson, P., and Douglas, R. Computer synthesized walking—An application of orthosis and functional electrical stimulation (FES). *J. Neurol. Orthop. Med. Surg.* 6:219–230, 1985.

Petrofsky, J.S., Phillips, C.A., and Stafford, D.E. Closed-loop control for restoration of movement in paralyzed muscle. *Orthopedics*, 7:1289–1302, 1984b.

Pfeiffer, E.A., Rhode, C.M., and Fabric, S.I. An experimental device to provide substitute tactile sensation from the anaesthetic hand. *Med. Biol. Eng*. 7:191–199, 1969.

Phillips, C.A., and Buhrman, J.R. A mathematical model to analyze control of upright posture and balance: Verification and implications for sensory feedback systems with an FES–orthosis. *Proceedings of the Ninth Annual Conference of the IEEE Eng. Med. Biol. Soc.* 87:609–610, 1987.

Phillips, C.A., and Petrofsky, J.S. Cognitive feedback as a sensory adjunct to functional electrical stimulation (FES) neural prosthesis. *J. Neurol. Orthop. Med. Surg*. 6:231–238, 1985a.

Phillips, C.A., and Petrofsky, J.S. Preliminary report on balance/stance in paraplegia with and without a cognitive feedback system. *Physiologist*, 28:302, 1985b.

Phillips, C.A., and Petrofsky, J.S. A total neural prosthesis for spinal cord injury rehabilitation: The cognitive feedback system with a functional electrical stimulation (FES) orthosis. *J. Neurol. Orthop. Med. Surg*. 7:225–234, 1986.

Prior, R.E., and Lyman, J. Electrocutaneous feedback for artificial limbs: Summary progress report, February 1, 1974 through July 31, 1975. *Bull. Prosthes. Res.* BPR 10–24:3–37, 1975.

Prior, R.E., Case, P.A., Scott, C.M., and Lyman, J. Supplemental sensory feedback for the VA/NU myoelectric hand: Background and preliminary designs. *Bull. Prosthes. Res.* BPR 10–26:170–191, 1976a.

Prior, R.E., Case, P.A., Scott, C.M., and Lyman, J. A myoelectric hand with supplementary sensory feedback. *Proc. San Diego Biomed. Symp.* 15:91, 1976b.

Prior, R.E., Case, P.A., and Lyman, J. Hook opening feedback for the below elbow amputee. *Proc. San Diego Biomed. Symp.* 16:357–368, 1977.

Riso, R.R., and Ignagni, A.R. Electrocutaneous sensory augmentation affords more precise shoulder position command generation for FNS orthoses. *Eighth Annual RESNA Conference on Rehabilitative Engineering*, pp. 228–230, 1985.

Riso, R.R., and Ignagni, A.R. Sensory feedback of machine state and shoulder position command information for use with FNS hand orthosis. *Proceedings of the Ninth RESNA Conference*, pp. 325-327, 1986.

Schmidl, H. The importance of information feedback in prostheses for the upper limbs. *Prosthet. Orthop. Interface*, 1:21–24, 1977.

Scott, R.N., Brittain, R.H., Caldwell, R.R., Cameron, A.B., and Dunfield, V.A. Sensory feedback system compatible with myoelectric control. *Med. Biol. Eng. Comput*. 18:65–69, 1980.

Shannon, G.F. A myoelectrically controlled prosthesis with sensory feedback. *Med. Biol. Eng. Comput*. 17:73–80, 1979.

Stolov, W.C., Burgess, E.M., and Romano, R.L. Progression of weight bearing after immediate prosthesis fitting following below-knee amputation. *Arch. Phys. Med. Rehabil*. 52:491–502, 1971.

Vodovnik, L., Bajd, T., Kralj, A., Gracanin, F., and Strojnik, P. Functional electrical stimulation for control of locomotor systems. *CRC Crit. Rev. Bioeng*. 6:63–131, 1981.

Part 2
Stationary Exercise Rehabilitation

4
Stationary Exercise Rehabilitation: Introduction and Overview

I. Introduction to Stationary Exercise Rehabilitation

A. Statement of the Problem

Most researchers in the area of spinal cord injury accept the idea that there are 300,000 to 500,000 Americans who have been permanently paralyzed by SCI resulting from trauma or disease. Every year, 10,000 to 15,000 new cases are added to this patient load. There are several times this many patients who are paralyzed by stroke, battle wounds, muscular dystrophy, and other conditions. Until recently, this enormous patient group has had little hope of ever regaining a quality of life that can be useful and satisfying.

To the paralyzed individual, the most dramatic effects of being paralyzed are the changes associated with lack of use of the paralyzed limbs. The unused muscles become smaller and weaker, and bone dissolution occurs. The cardiovascular and pulmonary systems can become deconditioned. The patient is depressed by this degradation of his physical condition and appearance and by his inability to function in a normal manner. There is a significant loss of dignity associated with not standing and always being looked down upon in a wheelchair.

Until recently, there was little hope for functional rehabilitation of paralyzed persons. Being inactive, they were continually suffering pressure sores, bone fractures (due to osteoporosis), bladder disorders, and kidney infections. They suffered from lowered immunity to general infection, and they were continually vulnerable to cardiopulmonary diseases. The combination of the personal loss to the paralyzed person himself, the loss of this large group of people to the needs of society, and the great economic cost to the people of America makes rehabilitation of these patients one of our most important problems.

A consideration equal to that posed by human/emotional loss is the health care financial burden. The costs of the care of paralyzed persons are enormous. Initial hospitalization costs for spinal cord injuries are estimated to be about $200,000, and continuing care adds tens of thousands of dollars with each year of life of the paralyzed person. In 1975 it was estimated that the lifetime medical costs of each person paralyzed early in life was about $1.5 million. In 1985 such costs had at least doubled, indicating that for each paralyzed person, the total cost of care may equal $3 million. Thus, caring for the hundreds of thousands of persons paralyzed by spinal cord injury alone may cost American society as much as $5 billion per year, over the next few decades. The cost of care of persons paralyzed for reasons other than spinal cord injury may multiply this amount several times. This adds appreciably to the medical cost burden of this country.

These two critical areas—the human suffering associated with the condition of paralysis and the enormous cost of medical care for the very large group of people who are paralyzed—more than justify a major effort toward obtaining a better life quality for these people. It is significant that we live in an era of high technology, and there is every reason to believe that technology can and must make an important contribution to helping paralyzed people become functional in the mainstream of life again.

B. General Introduction

The relationship between exercise and general physical conditioning is well documented. Alam and Smirk (1938) observed an increase in blood pressure with voluntary isometric muscle contractions in man. Lind et al. (1968), Petrofsky et al. (1981b), Miyazawa et al. (1977), Krayenbuehl (1973), and Helfant et al. (1971) have extended these observations and made studies of the effects of contraction of muscles on the body condition. McCloskey and Mitchell (1972) and Coote et al. (1971), who worked with electrical stimulation of cat muscles, observed an increase in arterial pressure with such stimulation. Lind et al. (1964) and Petrofsky et al. (1981a) made the same observations in animal research. Corbett et al. (1971) reported the same increase in arterial blood pressure with electrical stimulation of the paralyzed muscles of a quadriplegic patient. There have been numerous studies indicating that forced inactivity of the paralyzed individual results in a degradation of cardiovascular function. This work has been thoroughly reviewed by Davis et al. (1981). Nilsson et al. (1975), and Gass and Camp (1979), describing general cardiovascular degeneration in paralyzed persons.

Hjeltnes (1977) observed a large increase in atrioventricular oxygen difference in lower limb blood flow in paralytics and considered this to be an indication of impaired circulation associated with lower body blood pooling. Zwiren and Bar-Or (1975) compared paralytics and able-bodied controls and noted poor circulatory function in the paralytics; they were able to demonstrate a difference in the cardiovascular condition of active and sedentary paraplegics. In our experience we have noted large variations between physically active (e.g., handicapped athletes) and inactive paralytics.

Clausen (1977), in reviewing the cardiovascular effects of exercise in general, noted that the literature supports the concept that circulatory function is degraded in paralyzed patients. Shephard (1977) pointed out that normal methods of assessment of physical condition (e.g., treadmill stress testing) are not feasible in paralytics. Glaser et al. (1978, 1979) demonstrated that wheelchair ergometry and arm ergometry could be used for physical assessment maneuvers and for effective exercise of upper body muscles in paralytics. Davis et al. (1981) cited the writings of a number of researchers who have studied the use of upper body exercise to enhance total body fitness in paralyzed patients.

Shephard (1965) observed increases in fitness estimated at 10–30% in exercising normal humans (as opposed to sedentary normal humans), and Knutsson et al. (1973) reported an increase in work capacity and a decrease in heart rate response to exercise in paraplegics performing in an exercise program.

In well-designed experiments, Coutts et al. (1983) studied cardiopulmonary function and oxygen uptake in paraplegics and quadriplegics using wheelchair ergometry. They noted a distinct improvement after completion of an exercise program.

Major changes of pulmonary function have been noted in paralyzed persons after wheelchair ergometry. These changes have been primarily in those parameters indicating stiffness of the breathing apparatus. There is the combination of respiratory muscle dysfunction and impaired pulmonary function in paralyzed persons. Bergofsky (1964) noted a diminished thoracic compliance in patients with high level SCI and hypothesized that this might be due to loss of intercostal muscle function. Axen (1984) studied the responses of paralyzed patients to imposed pressure loading of the lungs and was not able to demonstrate significant differences between normal subjects and paraplegics.

C. Specific Introduction

In their work with FES, the Wright State University group has emphasized the development of systems and devices that rely primarily on surface electrodes. Their experience has led them to believe that surface stimulation is a perfectly reasonable approach when there is (a) incorporation of electrode garment systems (Granek and Granek, 1985) and (b) the intelligent application of the stimuli to produce muscle contractions. Much of the work over the past several years has been in the development of such FES systems and their testing with SCI patients.

Another important facet of the Wright State work has been the use of closed-loop control of the pattern of stimulation (Petrofsky and Phillips, 1979; Phillips and Petrofsky, 1980). Without such

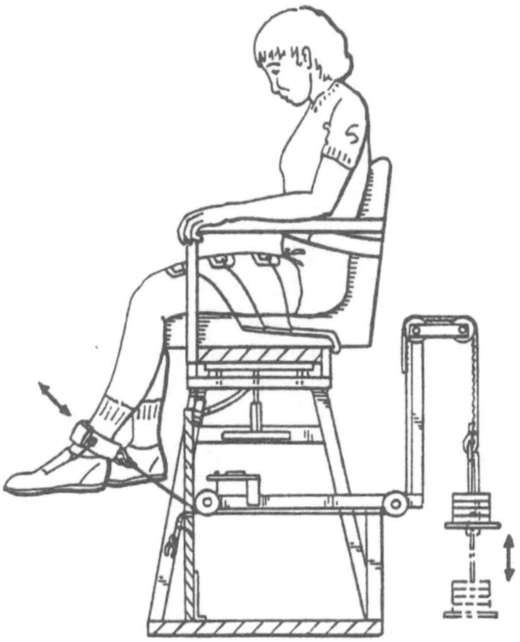

FIGURE 4.1. The leg trainer system.

control (i.e., in an "open-loop" system), smooth, coordinated movements are more difficult to produce. The resulting movements are often more jerky. It is significant that in the 1980s, high technology has produced electronic devices that bring computer control of stimuli for FES completely within the realm of possibility. The Wright State program has placed much of its effort on application of high tech devices for the production of computer-controlled, closed-loop, adjusted electrical stimuli for the production of smooth and coordinated movements in paralyzed muscles.

The Wright State FES techniques have been applied to the design and construction of a number of practical systems for exercising paralyzed individuals. These include a "leg trainer" (Petrofsky et al., 1984a) (Fig. 4.1) for the application of isokinetic exercise to the paralyzed muscles and a "bicycle ergometer" (Petrofsky et al., 1984b) (Fig. 4.2) for the production of aerobic exercise. Isokinetic exercise is accepted as the mode of choice for building size and strength of muscles. Aerobic exercise is known to exert a primary effect on general physical conditioning, especially of the cardiovascular and pulmonary systems. Both types of exercise must be involved in the rehabilitation of SCI-paralyzed individuals.

FIGURE 4.2. The exercise bicycle ergometer.

Significant training (conditioning) effects have been demonstrated by the use of these systems. Petrofsky and Phillips (1983) have reported a progressive increase in quadriceps muscle strength (over a 4-week period) using an isokinetic FES leg exerciser. Increase in quadriceps strength with FES leg lifting exercises has also been reported by Robinson et al. (1986). The cardiopulmonary conditioning has demonstrated significant changes in heart rate, cardiac output, and P-R interval as reported by Danopulos et al. (1986).

These devices for the exercising of paralyzed people are in use at several major rehabilitation centers in this country and abroad, and they have been used at Wright State University in the treatment of a large number of paralyzed patients. It has been amply demonstrated (Phillips et al., 1984; Hendershot et al., 1985; Petrofsky et al., 1985) that significant acute physiological effects are realized from their use.

The demonstration that closed-loop FES produces useful movements in paralyzed muscles provided the original basis for functional electrical rehabilitation. The Wright State effort in this area has been reviewed by Petrofsky and Phillips (1985).

In comparing data from Phillips et al. (1984), it appears that FES-induced exercise of the legs results in appreciably greater physical conditioning than exercise limited to the upper torso. This might be expected, since lower body exercise reverses the effect of blood pooling, and because the muscle mass involved in leg exercise is greater than that in upper torso exercise.

A set of preliminary studies recently reported by Phillips et al. (1989) identified a number of physiological parameters that do respond to long-term aerobic exercise of the paralyzed muscles. These parameters have been identified, although (in many cases) the specific biochemical–physiological mechanisms are not known.

II. Cardiovascular Responses and Thermoregulation*

A. Introduction

Historically, little research was done on the cardiovascular responses, trainability of muscle, or any other research associated with permanent spinal cord injury. Prior to World War II, as noted earlier, most individuals died soon after spinal cord injury as a result of kidney and bladder infections (Sunderland, 1968; Guttmann, 1976a). With the advent of sulfa drugs and better treatment of spinal shock, however, spinal-cord-injured patients have had a significant increase in longevity, allowing them to live a fairly normal life span (Cox and Grubb, 1974; Burke and Murray, 1975; Smart and Sanders, 1976). Technology and legislation have enabled wheelchair-bound patients to have greater mobility, and this, of course, has enormous psychological benefits (Burke and Murray, 1975). However, merely increasing mobility does little to alleviate the physiological problems that accompany life in a wheelchair.

A number of research areas are currently being pursued with the aim of curing patients with either acute or chronic spinal cord injury. These experiments range from omentum transplants (Goldsmith et al., 1983) and use of drugs (Faden and Holaday, 1979) to spinal cord reconstruction, neural regeneration, and spinal cord grafting. Another approach for restoration of mobility is to replace the lost central command to the alpha motor neurons with functional electrical stimulation. Electrical stimulation to paralyzed muscles has been used since 1840 to reduce the spasticity often found in spinal-cord-injured people (McNeal et al., 1969). However, it was not until the past 30 years that studies have appeared in the literature on restoring movement by electrical stimulation (Liberson et al., 1961; Crochetiere et al., 1967; Rebersek and Vodovnik, 1973; Zealear and Dedo, 1977; Brindley et al., 1978; Petrofsky and Phillips, 1979). Before any cure or other aid for chronic spinal cord injury can be attempted, the critical question to be answered is the reversibility of muscular atrophy and cardiovascular deconditioning that occurs with long-term spinal-cord-injured patients. As with any body system, lack of use of skeletal muscle and a reduction in the stress on the cardiovascular system results in considerable atrophy and deconditioning. Spinal cord injury also causes disruption of some of the autonomic pathways and may alter much of the normal control mechanisms for the cardiovascular system. If atrophy cannot be reversed and exercise done safely again (even with an impaired autonomic nervous system), a successful cure for spinal cord injury will never be widely applied. Finally, even while research continues on a cure for spinal cord injury, it is important to physically recondition the body in order to improve the well-being of the patient and reduce subsequent medical costs.

This section summarizes some of the work being done at Wright State University on the development of techniques for functional electrical rehabilitation. Our work was presented at the American Paralysis Association symposium in San Francisco (June 1984), and the work on active physical therapy is reviewed in this section.

*From J.S. Petrofsky and C.A. Phillips, The use of functional electrical stimulation for rehabilitation of spinal cord injured patients. Reprinted with permission from *Central Nervous System Trauma*, 1:57–74, 1984. Copyright 1984, Mary Ann Liebert, Inc.

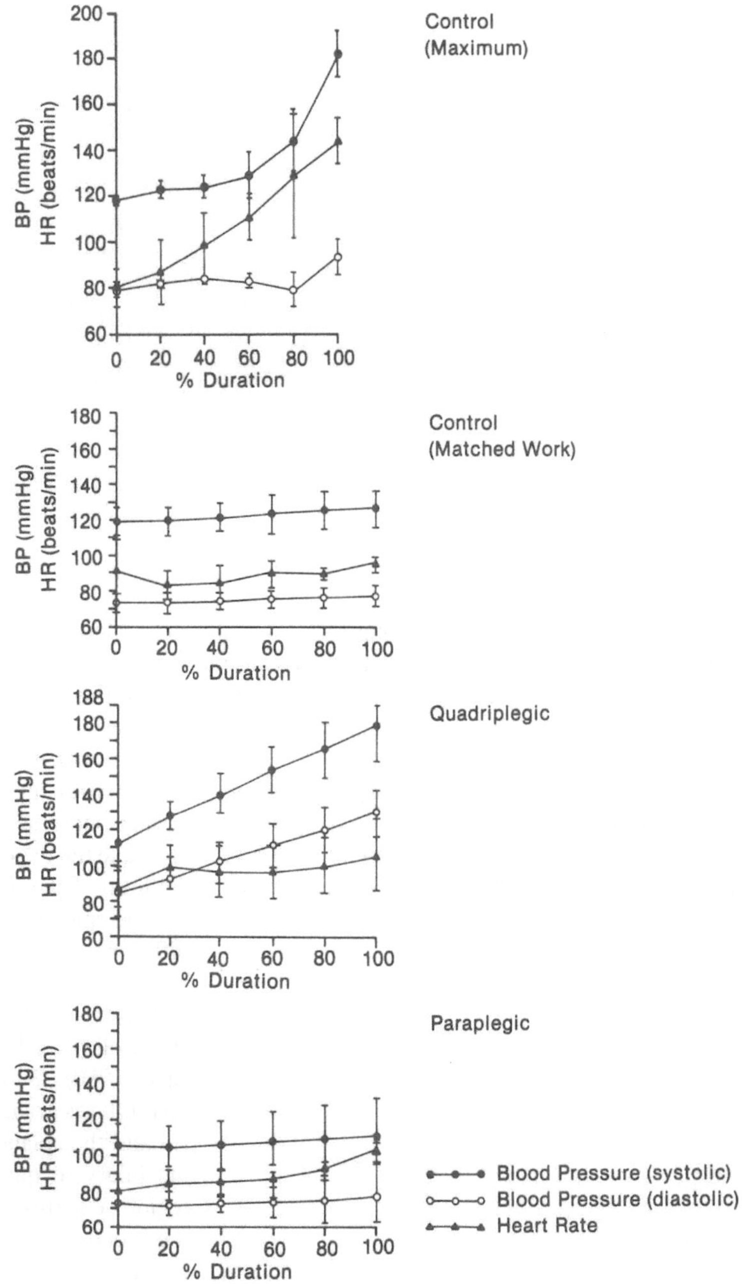

FIGURE 4.3. Average blood pressure and heart rate changes.

B. Cardiovascular Response to Electrically Induced Exercise

Functional electrical exercise places stress on skeletal muscle, soft tissue, and bone, but such electrically induced exercise stresses other body systems as well. Figure 4.3 shows the typical blood pressure and heart rate responses of the control, quadriplegic, and paraplegic subjects with this type of exertion. During a typical experiment (where subjects lifted weights of 9.1 kg for a period of 4 minutes), increased blood pressure and heart rate responses for different subjects are reviewed. In the control subjects during a voluntary effort (no electrical stimulation), blood pressure and heart rate increased during the exercise, but not very much. However, for the paraplegic and quadriplegic subjects the exercise was maximal (so that

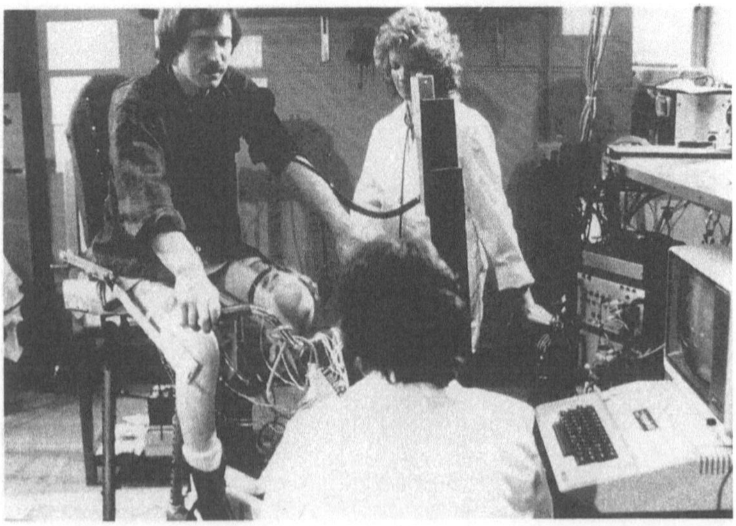

FIGURE 4.4. The bicycle ergometer used by paralyzed persons.

the exercise was fatiguing). Consequently, when the control subjects lifted a weight that would cause them to also fatigue in the 4 minutes, blood pressure and heart rate increased significantly, as shown in the top panel of Figure 4.3.

For the control subjects, both heart rate and blood pressure increased during the exercise. However, for the paraplegic subjects, blood pressure increased very little, but heart rate increased during the exercise. In contrast, for quadriplegic subjects, blood pressure increased significantly, but heart rate increased very little. Further examination of this phenomenon showed that the smaller increase in blood pressure occurred in paraplegic subjects with lower level injuries, whereas paraplegics with higher level injuries had a slightly greater increase in blood pressure. This would correlate the degree of autonomic nerve damage to the level of the injury in these groups of subjects. For paraplegic subjects, the normal autonomic reflex pathways arising from active skeletal muscle in the leg (which normally cause blood pressure to increase) would be disrupted (Coote et al., 1971; McCloskey and Mitchell, 1972; Petrofsky and Lind, 1980). With quadriplegic subjects, the normal reflexes arising between skeletal muscle in the leg and the splanchnic bed would be intact, since the injury to quadriplegics is found in the neck region. For this reason, it can be anticipated that the blood pressure response would be normal in quadriplegics, but damage to the pathways innervating the heart may reduce changes in heart rate during exercise.

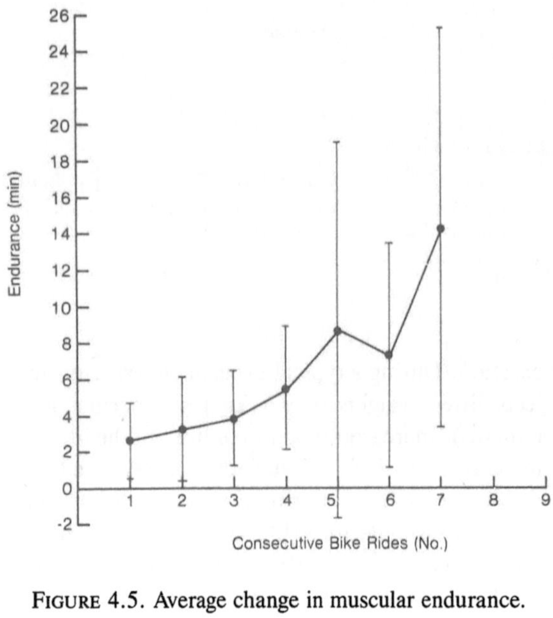

FIGURE 4.5. Average change in muscular endurance.

While stressing the cardiovascular system to a certain extent, isokinetic exercise is not usually used to build endurance and condition the cardiovascular system. To stress the cardiovascular system, dynamic exercise is typically used. In this form of exercise, cardiac output is usually increased to a sufficient extent that training of the cardiovascular system can result. A typical form of dynamic exer-

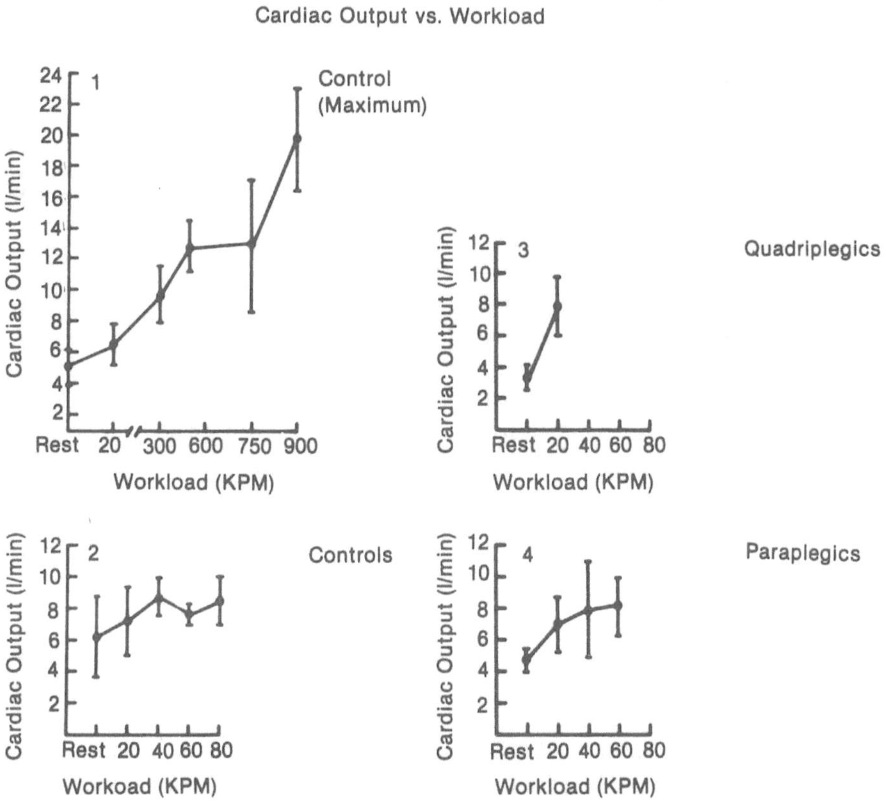

FIGURE 4.6. Cardiac output during bicycle ergometry.

cise is bicycling. To allow paralyzed subjects to bicycle, a closed-loop, electrically stimulated bicycle ergometer system was developed (Phillips and Petrofsky, 1983; Petrofsky et al., 1984b, 1985). Subjects in our laboratory have been stimulated to ride a Monark bicycle ergometer, which has been specifically modified for this purpose (Fig. 4.4). Stimulation involves the quadriceps and gluteus maximus muscles, and sensors provide information to the computer as to the position of the legs. Knee stabilizer bars are used to keep the knees in alignment, and a special high seat back with a shoulder harness and seatbelt are used to maintain the posture of the subject (Fig. 4.4). When subjects have been induced to bicycle for 15 minutes a day (3 days per week) at a workload set to fatigue the subject in that period of time, the results have been significant.

Subjects were first weight-trained to greater than 20 pounds of leg extension strength on the isokinetic leg trainer over a 6-week period. However, when they were first placed on the bicycle ergometer, endurance was very short (Fig. 4.5). The group of eight subjects demonstrated a significant increase in endurance with subsequent bike runs. When patients were exercised for just over 2 weeks, endurance was seen to increase by more than 500%. Further increases in endurance were noted as bicycling continued, and the workload could be progressively increased. Because of the dynamic nature of the exercise, cardiac output also increased. Fig. 4.6 illustrates the cardiac output of four paraplegic, four quadriplegic, and four control subjects during peddling of the exercise bicycle ergometer at workloads that ranged (for the controls) as high as 60 kilopond*·meters per minute (kp·m/min.). As the workload was increased, there was an increase in cardiac output; that is, cardiac output nearly doubled for the

*1 kilopond (kp) is the force acting on a mass of 1 kilogram at normal acceleration of gravity; 1 kp = 9.80665 Newtons, a unit of force.

average paraplegic and quadriplegic subject. With long-term exercise training and increase in the tolerance for work, cardiac output increased even further.

One interesting finding in these experiments involved the blood pressure stability of quadriplegic subjects. As described previously for the leg exerciser, blood pressure in quadriplegic and paraplegic subjects differed during exercise. The typical quadriplegic subject had a high blood pressure response during exercise, paralleling that of control subjects. A typical paraplegic subject had a very small increase in blood pressure during exercise. However, quadriplegic subjects had highly variable blood pressures during exercise on the bicycle ergometer. The quadriplegics, at the onset of these studies, were slightly hypotensive, with average resting blood pressures of 90/60 mmHg. Blood pressure increased to 150/100 mmHg at the end of fatiguing exercise in the first week of bicycling. At the end of the third week, blood pressure became more stable. Resting blood pressure increased to 100/70 mmHg, whereas exercising blood pressure decreased to about 135/90 mmHg at the end of the runs. This increase in the resting blood pressure and reduction in exercising blood pressure were accompanied by an increased feeling of well-being throughout the exercise and a reduction in post-exercise recovery times.

C. Thermoregulation

One problem associated with these experiments was that any type of exercise caused an increase in body heat load. Metabolic heat is generated during exercise. For example, bicycling (which is inherently one of the more efficient types of exercise) is at best 25% efficient, and lifting weights is only about 2% efficient: 98% of the metabolism used to lift the weights is converted to heat that must be dissipated by the body. When subjects worked at high levels (such as on the bicycle ergometer), it became apparent that their body temperatures were rising. Therefore, a series of experiments was performed to determine how active physical therapy techniques affected the thermoregulatory system in man.

A group of four resting paraplegic, four resting quadriplegic, and four resting control subjects were first exposed to environmental temperatures ranging form 30 to 40°C for 30 minutes. Tympanic temperature (central core temperature), specific sweat rates, and mean sweat rates for the entire body were measured. Following the conclusion of these experiments (which were conducted on separate days), subjects manually operated an arm crank ergometer at a workload of 25 W (at a speed of 50 rpm) for a period of 20 minutes. During this 20-minute exercise, the subjects were exposed to temperatures similar to those used in the earlier protocols. Central body temperature was again measured, as was sweat rate from five measuring sites. The measuring sites used were the chest, forehead, left arm, and both legs.

The results of these experiments are shown in Figures 4.7, 4.8, and 4.9. Core temperature at rest remains fairly stable over a 30-minute exposure in the environmental chamber for all three groups of subjects (Fig. 4.7). Mean sweat rate showed some small increase, which accounted for maintaining body temperature at rest. When work was added to heat exposure, the control subjects tolerated the heat very well, as indicated by measurements of core temperature (Fig. 4.8). In contrast, quadriplegic subjects at environmental temperatures of 30 and 35°C showed a linear increase in core temperature with workload as the room temperature was increased (Fig. 4.8). Quadriplegic subjects

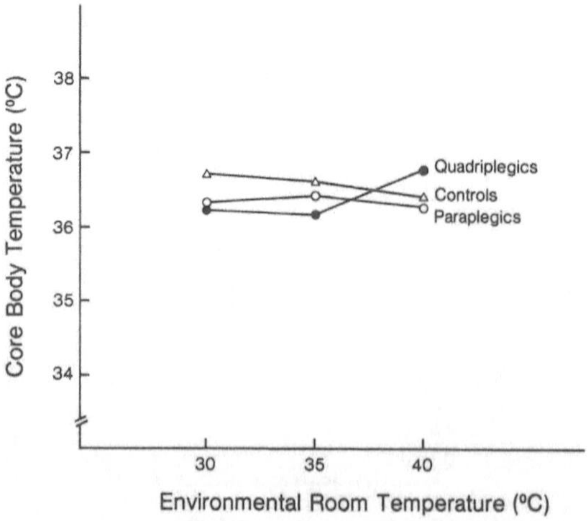

FIGURE 4.7. Resting core temperatures during heat exposure.

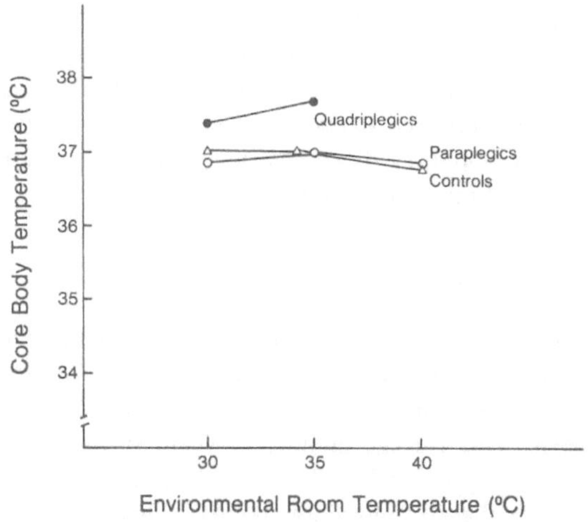

FIGURE 4.8. Core temperatures at the end of a 20-minute period of exercise.

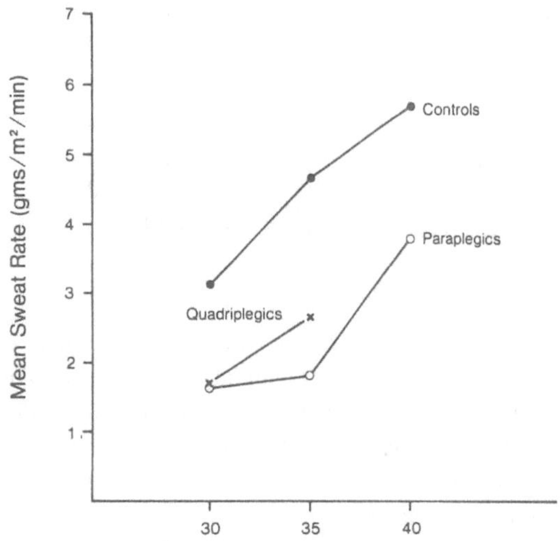

FIGURE 4.9. Mean sweat rates.

were not able to tolerate work in a 40°C room. Work was terminated when body temperature exceeded 37.8°C, and work could be tolerated only at 30 and 35°C for quadriplegics.

This intolerance for exercise can be explained with respect to sweat rate. The mean sweat rate of the body could be increased more with the controls than with the paraplegic and quadriplegic subjects (Fig. 4.9). Specific examination of regional sweat rates showed that no increase in sweating could be found (with the exception of transient reflex sweating) below the neck in the quadriplegic subjects or below the waist in the paraplegic subjects. Although hyperactivity of the sweat glands occurred in the forehead and upper chest areas of paralyzed subjects, this increase in sweat rate was not sufficient to cool the body enough to keep body temperature down. Therefore, even with a marginal increase in workload (such as 25 W of arm crank ergometry) in a warm environment, the quadriplegic subjects exhibited poor thermoregulation.

D. Discussion

This section has described the development of computer-controlled, closed-loop electrical stimulation exercise and its effect on the body. Any type of exercise modality used to train and maintain physical fitness must be evaluated not only regarding muscular strength and endurance but also regarding the effects on the cardiovascular system and the thermoregulatory system. This review of experiments has shown that the cardiovascular and muscular systems respond quite readily to closed-loop electrical exercise. By using a combination of strength training and endurance training programs, it appears that the body can be physically reconditioned safely. Additional studies need to be performed toward optimizing the training protocols. Areas that still need to be addressed include the development of optimum exercise protocols, optimally mixing the two forms of exercise for a balanced training program, and the long-term effects of the FES-induced exercise on the body.

However, a word of caution is in order. Thermoregulatory studies indicate that unless special heat-exchange clothing is used, subjects should not exercise to any large extent at a room temperature exceeding 80°F if paralysis involves a significant portion of the body. *Quadriplegics must be extremely careful with any type of exercise in a non-air-conditioned room.* These aspects of the research need to be thoroughly investigated before functional electrical exercise can be safely applied to a clinical setting. In the meantime, however, there are obvious benefits with these forms of exercise.

III. Cardiovascular Circulatory Dynamics with Quadriplegia*

A. Introduction

Cardiocirculatory changes have been known to occur in patients with spinal cord injury. The changes vary according to the level of injury and are different in the acute stage and the chronic posttraumatic stage as a result of the body's homeostatic adaptation.

During the acute stage (spinal shock), there is complete loss of motor and reflex activity as well as sensory input below the level of injury due to interruption of somatic motor and sensory nerve traffic (Ruch, 1974; Frankel and Mathias, 1976; Guttmann, 1976b). An autonomic nervous system crisis occurs whenever there is a complete cutoff of the sympathetic outflow below the spinal cord injury. In the chronic stage, adaptation of the autonomic nervous system and the body's homeostasis occurs. However, this adaptation only partially compensates for the autonomic nervous system deficit and the loss of function of a significant part of the body.

A number of studies of cardiovascular function in chronic quadriplegics have been performed, but only limited reports are available in the literature as to the nature of the changes in cardiac and circulatory dynamics and their mechanism. It has been known for some time that quadriplegics have a tendency to orthostatic hypotension. In these subjects, resting plasma catecholamines, as well as the response following head-up tilt, are lower than in matched control subjects. However, no comprehensive cardiac and circulatory measurements have been reported to evaluate the cardiocirculatory dynamics and the mechanism of the altered response in chronic quadriplegic or paraplegic subjects. Therefore, we undertook a study to compare young chronic quadriplegic subjects with matched normal subjects to noninvasively evaluate the cardiocirculatory dynamics in resting supine and 40° head-up tilt positions.

B. Methods

The subjects were nine young quadriplegics (eight males, one female) with the mean age of 25.7 years and five normal controls with the mean age of 24.6 years. The average number of years since spinal cord injury was 3.9. All subjects had closed cervical spinal cord transection at the levels of C4–C7. All lesions were clinically complete, with total loss of motor and sensory function below the lesion. The subjects volunteered to participate in the study as part of a long-term rehabilitation program including quadriceps training and bicycle exercise training using functional electrical stimulation. The reported results represent data obtained prior to beginning the training sessions. Each individual had two measurements on the same day. First, the measurements were obtained in supine position and subsequently in 40° head-up tilt position for 5 minutes using an electrically driven tilt table.

Systolic time intervals (STI) were obtained from simultaneously recording an electrocardiogram, phonocardiogram, and carotid pulse tracing on a multichannel oscilloscopic recorder (Electronics for Medicine, "E for M" model VR-12) using photographic paper at a speed of 100 mm/s as previously reported (Weissler et al., 1968, 1969). Heart sounds were recorded by a Cambridge transducer microphone placed on the left anterior chest wall at the level of the second or third intercostal space, assuring clear delineation of the initial high frequency vibration of the first and second heart sounds. The carotid pulse was recorded at the point of maximum pulsation with an E for M pressure pulse transducer. The transducer was adjusted to obtain a clear delineation of the onset of the rapid rise and the point of the nadir of the incisura (dicrotic notch) of the pulse tracing. The ECG lead was selected to record a clear onset of the QRS complex.

To obtain the total electromechanical systole (QS_2), the interval between the onset of the QRS complex of the ECG to the first high frequency

---

*From D.M. Danopulos et al., Cardiovascular circulatory dynamics in subjects with chronic cervical spinal cord injury in supine and head-up tilt position. Reprinted with permission from *Journal of Neurological and Orthopedic Medicine and Surgery*, 6:265–270, 1985. Copyright 1986, American Academy of Neurological and Orthopedic Surgeons.

vibration of the aortic component of the second sound in the phonocardiogram was measured. Left ventricular ejection time (LVET) was measured as the interval from the beginning of the rapid upstroke of the carotid pulse tracing to the incisura. The preejection period (PEP) was calculated by subtracting LVET from QS_2. The PEP represents the time required for the left ventricular electromechanical events to generate the pressure from the end-diastolic to the aortic level preceding systolic ejection.

The systolic time intervals were corrected for heart rate by linear regression equations as previously reported (Weissler et al., 1968, 1969). Indices were calculated as the product of the sum of the measured intervals and the appropriate normal regression slopes times the observed heart rates (index = observed interval + regression slope × heart rate). The regression slope for PEP was 0.4 for both male and female, for LVET 1.7 for male and 1.6 for female, and for QS_2 2.1 for male and 2.0 for female (Weissler et al., 1968, 1969). The PEP/LVET ratio was calculated for both supine and tilt positions.

Stroke volume (SV), cardiac output (CO), cardiac index (CI), and ejection fraction (EF) were calculated from echocardiographic measurements. Echocardiograms were recorded on E for M M-mode multichannel oscilloscopic recorder model VR-12 and two-dimensional and M-mode recorder (Hewlett-Packard, model Ekoline 5500). Care was taken to obtain recordings in the same individuals at the same level with identical transducer angulations both in supine and tilt positions.

Measurements obtained included left ventricular end-diastolic dimension (LVEDd) measured from the endocardial surfaces of the interventricular septum and left ventricular posterior wall (LVPW) at the onset of the QRS complex. The left ventricular end-systolic dimension (LVESd) was measured as the smallest left ventricular dimension without reference to the electrocardiogram (Sahn et al., 1978; D'Cruz, 1983).

End-diastolic and end-systolic LV volumes were calculated as the cube of the respective dimensions. Stroke volume was the difference between the end-diastolic and end-systolic volumes. Cardiac output was calculated by multiplying stroke volume with heart rate, and CI was the product of the CO divided by the body surface area. Ejection fraction was calculated by dividing SV by LV end-diastolic volume.

The experimental procedure was carefully explained to each subject and an informed consent approved by the Wright State University's Institutional Review Board was obtained. The study was performed in two stages: first in resting supine position, then after the participant had been secured at the legs and the knees with straps on the table, at 40° head-up tilt position. The subjects were fasting for at least 3 hours prior to the test, and the resting supine measurements were obtained 15 minutes after the participants were lying relaxed on the table. The measurements in the tilt position were obtained after 5 minutes of tilting.

To obtain the systolic time intervals, 10 consecutive cardiac cycles were measured and the data averaged. For the echocardiographic measurements, three cycles of the resting-supine and 40° head-up tilt recordings were analyzed and the data averaged for each participant. The results of these measurements were compared for quadriplegics and normals and the differences were statistically analyzed by the Student t-test. Differences in p value ≤ 0.05 were considered significant.

C. Results

Table 4.1 shows the complete data gathered from the study. Mean age, weight, and body surface area were not significantly different in normal controls and quadriplegics.

Supine resting measurements indicated that systolic and diastolic blood pressures were moderately but not significantly lower. The heart rate was significantly lower in quadriplegics. The LVEDd was similar in quadriplegics and controls and was within normal limits. The LVESd was smaller in quadriplegics, and this corresponded with the higher ejection fraction in quadriplegics. Cardiac output and cardiac index were higher in controls than in quadriplegics, and the difference was significant. All three systolic time intervals (PEP, LVET, QS_2) were longer in quadriplegics, but when corrected for heart rate the indices were similar in both groups and within normal range. The PEP/LVET ratios also were similar in resting-supine state in both groups.

Following 40° head-up tilt, there was a considerable fall both in SBP and DBP in quadriplegics. In

TABLE 4.1. Comparison of rest and 40° head-up tilt hemodynamics in quadriplegics and control subjects.

	Rest (Supine)		40° Head-up tilt	
Measurements	Quadriplegic	Normal	Quadriplegic	Normal
SBP (mmHg)	106 ± 10	127 ± 9	82 ± 11**	123 ± 13
DBP (mmHg)	69 ± 8	78 ± 2	54 ± 7**	85 ± 15
HR (beats/min)	53 ± 7	80 ± 9	76 ± 14*	86 ± 10
LVEDd (mm)	48 ± 4	49 ± 3	40 ± 3**	47 ± 2
LVESd (mm)	31 ± 4	36 ± 5	25 ± 3*	35 ± 6
SV (mL)	80 ± 14	71 ± 1	48 ± 13**	64 ± 1**
EF (%)	72 ± 5	61 ± 9	75 ± 7	59 ± 9
CO (L/min)	4.18 ± 0.7	5.7 ± 0.7	3.33 ± 0.6*	5.4 ± 0.7
CI (L/min/m²)	2.28 ± 0.2	3.0 ± 0.2	1.84 ± 0.3*	2.85 ± 0.2
PEP (ms)	100 ± 7	84 ± 3	112 ± 7*	89 ± 8
PEPI	122 ± 8	115 ± 3	143 ± 8**	116 ± 1
LVET (ms)	330 ± 28	270 ± 17	265 ± 24**	255 ± 19
LVETI	418 ± 27	406 ± 5	396 ± 19	404 ± 5
QS_2 (ms)	436 ± 34	357 ± 18	383 ± 31*	343 ± 26
QS_2I	544 ± 32	525 ± 13	543 ± 24	524 ± 7
PEP/LVET (%)	0.310 ± 0.03	0.316 ± 0.09	0.427 ± 0.06**	0.348 ± 0.09

SBP = systolic blood pressure; DBP = diastolic blood pressure; HR = heart rate; LVEDd = left ventricle end-diastolic dimension; LVESd = left ventricle end-systolic dimension; SV = stroke volume; EF = ejection fraction; CO = cardiac output; CI = cardiac index; PEP = preejection period; PEPI = PEP index; LVET = left ventricle ejection time; LVETI = LVET index; QS_2 = total electromechanical systole; QS_2I = QS_2 index.
$^*p \leq 0.05$.
$^{**}p \leq 0.001$.

controls, there was only a slight decrease of SBP while DPB moderately increased. There was also a considerable increase in heart rate, almost 50% greater than in quadriplegics, during head-up tilt. In controls the increase was small, less than 7%.

The LVEDd and LVESd changed only slightly in controls, paralleling the minimal change in CO and CI. In contrast, both these dimensions decreased significantly in quadriplegics, corresponding with the considerable decrease in CO and CI in quadriplegics.

Considerable difference was observed in the change of stroke volume during head-up tilt, decreasing in controls from 71 ± 1 mL to 64 ± 1 mL as compared to the decrease in quadriplegics from 80 ± 14 mL to 48 ± 13 mL.

The systolic time intervals (PEP, LVET, and QS_2) as well as the indices (PEPI, LVETI, SQ_2) showed almost no change in normal controls during head-up tilt. However, in quadriplegics, PEP as well as PEPI, was increased while LVET and LVETI decreased and QS_2I remained unchanged. The PEP/LVET ratio increased markedly in quadriplegics from 0.310 ± 0.03% to 0.427 ± 0.06%, as a result of the increase of PEP and decrease of LVET from supine to head-up tilt. This was in sharp contrast with the slight increase of the ratio in normal control subjects.

D. Discussion

In normal conditions, when the upright position is assumed following supine or passive tilting, the baroreceptor reflex immediately responds by enhancing sympathetic activity to restore blood pressure (Wagner, 1959; Kezdi and Geller, 1968; Koizumi and Brooks, 1980; Schatz, 1984). The afferent pathways of this reflex feedback loop are in the vagus and the glossopharyngeal nerves, which carry the impulses from the baroreceptors in the vascular walls to the nucleus tractus solitarius in the brain, and from there to the cardiovascular centers. The efferent pathways of the sympathetic course in the spinal cord join the paravertebral ganglia and innervate the heart and smooth muscles of the peripheral resistance vessels. A decrease in the systemic blood pressure is sensed immediately by the baroreceptors. They reduce the inhibition of the central sympathetic output, resulting in increased sympathetic drive. In turn, norepinephrine is released at the sympathetic alpha 1 receptors of the vascular smooth muscle. The resulting vasoconstriction and increased peripheral resistance returns blood pressure toward control. Also, secondary supporting hormonal mechanisms may enter in this reflex by mobilizing epinephrine from the adrenal medulla, activating the renin–angiotensin–aldosterone system, or enhancing vasopressin release.

In spinal cord injury at the cervical level, the sympathetic arm of the reflex loop is interrupted in varying degrees and the vasomotor center will lose control of the heart and vascular system in quadriplegics. Because of the decreased sympathetic drive, plasma norepinephrine and epinephrine levels were shown to be extremely low in quadriplegics (Guttman et al., 1963; Debarge et al., 1974; Mathias et al., 1975). Also during head-up tilt, the usual considerable rise of catecholamines in normals was absent in quadriplegics (Guttman et al., 1963; Mathias et al., 1975). The slower resting heart rate in quadriplegics was probably the result of predominant vagal tone to the heart not counterbalanced by the normal sympathetic drive. However, some investigators report no difference in heart rate in chronic quadriplegics when compared to normal subjects (Mathias and Frankel, 1983).

Resting stroke volume was higher in quadriplegics because the Starling mechanism compensated for the slower heart rate. However, during head-up tilt, the unloading of the baroreceptor reflex, while diminishing the vagal drive of the heart, fails to compensate with increased sympathetic drive of both the heart and the resistance vessels, resulting in a lack of increased inotropy and insufficient vasoconstriction. There is, then, excessive pooling of blood in the lower body and diminished venous return, which results in considerable fall in SBP and DBP. Quadriplegics often faint during head-up tilt or upright position because of the malfunction of the normal homeostatic mechanism. Also, the considerable increase in heart rate in quadriplegics during head-up tilt is probably due to the imbalance between vagal and sympathetic tones, with the greater fall in stroke volume leading to metabolically triggered acceleration of the rate of cardiac contraction.

Of the cardiac dimensions, LVEDd was similar and within normal limits in quadriplegics as in controls. However, LVESd was generally smaller than in controls, probably reflecting the higher ejection fraction observed in quadriplegics. The higher ejection fraction in supine resting condition in quadriplegics indicates normal myocardial performance adapting to the predominant vagal drive with bradycardia through the Starling mechanism, since the sympathetic drive does not counterbalance the vagal tones as in normals. The higher EF in quadriplegics with a higher SV maintains CI within the needs of the body metabolism during rest. During head-up tilt, EF further increases in an attempt to compensate for the lack of sufficient venous return.

It has been reported that the PEP/LVET ratio calculated from the systolic time intervals is a reliable index of left ventricular performance (Weissler et al., 1968, 1969; Stack et al., 1976). Since both PEP and LVET normally shorten proportionally to the increase of the heart rate, the PEP/LVET ration is not influenced by heart rate changes. The PEPI prolongs, LVETI shortens with LV dysfunction (Weissler et al., 1961; Diamont and Killip, 1970; Samson, 1970; Hodges et al., 1972), and the PEP/LVET ration increases in the presence of left ventricular dysfunction. Also, one may expect a close correlation between ejection fraction and PEP/LVET ratio as reported by different authors (Garrard et al., 1970; Ahmed et al., 1972; Lewis et al., 1976).

In our study, the high supine resting EF correlated well with the normal PEP/LVET ratio in quadriplegics, but during head-up tilt this correlation was absent, since EF remained high while PEP/LVET ratio increased from 0.310 ± 0.03 to 0.427 ± 0.06. Theoretically, this indicates impaired left ventricular function. The increased PEP/LVET ratio in quadriplegics during head-up tilt was due to both prolongation of PEP and shortening of LVET. Factors that prolong PEP are left ventricular failure, left ventricular conduction delay, negative inotropic agents, and diminished preload. The LVET may be shortened by decreased stroke volume, regardless of the cause, and increased rate of ejection due to use of positive inotropic agents. During head-up tilt, there is a substantial decrease in left ventricular preload in quadriplegics due to the lack of sympathetic drive, which normally prevents excessive pooling of blood in the lower part of the body. This is responsible for increased PEP and decreased LVET during head-up tilt in quadriplegics, and for the increased PEP/LVET ratio which is not related to impaired left ventricular function. On the other hand, in normal controls, the small increase in PEP/LVET ratio was within the normal range due to the same mechanism but less exaggerated conditions.

High resting ejection fraction, even with some increase during head-up tilt, and the high resting stroke volume exist in quadriplegics with normal systolic time intervals and normal echocardiographic measurements. These are all indicators of a normal myocardium and left ventricular function and adaptation of the cardiocirculatory system to the autonomic nervous system imbalance in chronic quadriplegics.

E. Concluding Statement

[The authors] have also completed a study of chronic adjustments the body may make in the circulatory dynamics. Long-term quadriceps and bicycle exercise training of chronic quadriplegic subjects using functional electrical stimulation are reported in Section II of Chapter 7.

References

Ahmed, S.S., Levinson, G.E., Schwartz, C.J., and Ettinger, P.O. Systolic time intervals as measures of the contractile state of the left ventricular myocardium in man. *Circulation*, 46:559, 1972.

Alam, M., and Smirk, F.H. Unilateral loss of blood pressure raising, pulse-accelerating reflex from voluntary muscle due to a lesion of the spinal cord. *Clin. Sci.* 3:247–252, 1938.

Axen, K. Adaptations of quadriplegic men to consecutively loaded breaths. *J. Appl. Physiol., Respir. Environ. Exercise Physiol.* 56(4):1099–1103, 1984.

Bergofsky, E.H. Mechanism for respiratory insufficiency after cervical cord injury. A source of alveolar hypoventilation. *Ann. Intern. Med.* 61(3):435–447, 1964.

Brindley, G.S., Polkey, C.E., and Rushton, D.N. Electrical splinting of the knee in paraplegia. *Paraplegia*, 16:428–435, 1978.

Burke, D.C., and Murray, D.D. *Handbook of Spinal Cord Medicine*. London, Macmillan, 1975.

Clausen, J.P. Effects of physical training on cardiovascular adjustments to exercise in man. *Physiol. Rev.* 57:779–815, 1977.

Coote, J.H., Hilton, S.M., and Perez-Gonzalez, J.F. The reflex nature of the pressor response to muscular exercise. *J. Physiol.* 215:789–804, 1971.

Corbett, J.L., Frankel, H.L., and Harris, P.J. Cardiovascular changes associated with skeletal muscle spasm in tetraplegic man. *J. Physiol.* 215:381–387, 1971.

Coutts, K.D., Rhodes, E.C., and McKenzie, D.C. Maximal exercise responses of tetraplegics and paraplegics. *J. Appl. Physiol., Respir. Environ. Exercise Physiol.* 55(2):479–482, 1983.

Cox, W., and Grubb, R. In *Trauma to the Central Nervous System in Neurological Pathophysiology*, A. Eliasson, A. Prensky, and W. Hardin, Eds. Oxford University Press, New York, 1974.

Crochetiere, W.J., Vodovnik, L., and Reswick, J.B. Electrical stimulation of skeletal muscle. A study of muscle as an actuator. *Med. Biol. Eng.* 5:111–125, 1967.

D'Cruz, L.A. *Echocardiographic Diagnosis*. Macmillan, New York, 1983.

Danopulos, D., Kezdi, P., Stanley, E.L., Petrofsky, J.S., Stacy, R.W., Phillips, C.A., and Meyer, L.E. Changes in cardiovascular circulatory dynamics after a twelve week active bicycle rehabilitation in young tetraplegics. *J. Neurol. Orthop. Med. Surg.* 7(2):179–184, 1986.

Davis, G.M., Kofsky, P.R., Kelsey, J.C., and Shephard, R.J. Cardiorespiratory fitness and muscular strength of wheelchair users. *CMA J.* 125:1317–1323, 1981.

Debarge, O., Christensen, N.J., Corbett, J.L., Eidelman, B.H., Frankel, H.L., and Mathias, C.J. Plasma catecholamines in tetraplegics. *Paraplegia*, 12:44, 1974.

Diamont, B., and Killip, T. Indirect assessment of left ventricular performance in acute myocardial infarction. *Circulation*, 42: 579, 1970.

Faden, A.I., and Holaday, J.W. Naloxone reversal of hypotension caused by spinal transection. In *Endogenous and Exogenous Opiate Agonists and Antagonists*, I.L. Way, Ed. Pergamon, New York, pp. 483–486, 1979.

Frankel, H.L., and Mathias, C.J. The cardiovascular system in tetraplegia and paraplegia. In *Handbook of Clinical Neurology*, Vol. 26, P.J. Vinken and G.W. Bruyn, Eds. American Elsevier, New York, pp. 313–328, 1976.

Garrard, C.L., Weissler, A.M., and Dodge, H.T. The relationship of alterations in systolic time intervals to ejection fraction in patients with cardiac disease. *Circulation*, 42:455, 1970.

Gass, C.G., and Camp, E.M. Physiological characteristics of trained paraplegic and tetraplegic subjects. *Med. Sci. Sports*, 11:256–259, 1979.

Glaser, R.M., Laubach, L.L., Sawka, M.N., and Suryaprasad, A.G. Exercise stress fitness evaluation and training of wheelchair users. In *Proceedings of the First International Conference on Lifestyle and Health*, A.S. Leon and G.J. Amundson, Eds. University of Minnesota Press, Minneapolis, pp. 167–170, 1978.

Glaser, R.M., Sawka, M.N., Laubach, L.L., and Suryaprasad, A.G. Metabolic and cardiopulmonary responses to wheelchair and bicycle ergometry. *J. Appl. Physiol.* 46:1066–1070, 1979.

Goldsmith, H., Steward, E., Chen, W.F., and Duckett, S. Application of intact omentum to the normal and traumatized spinal cord. In *Spinal Cord Reconstruction*, C. Kao, R. Bunge, and P. Reier, Eds. Raven Press, New York, 1983.

Granek, H., and Granek, M. Transcutaneous transducer garments: Current mammalian usage and future applications. *J. Neurol. Orthop. Med. Surg.* 6:271–278, 1985.

Guttmann, L., Munro, A.F., Robinson, R., and Walsh, J.J. Effect of tilting on the cardiovascular responses and plasma catecholamine levels in spinal man. *Paraplegia*, 1:4, 1963.

Guttmann, L. *Spinal Cord Injuries, Comprehensive Management and Research*. Blackwell Scientific Publications, Oxford, 1976a.

Guttmann, L. Spinal shock. In *Handbook of Clinical Neurology*, Vol. 26, P.J. Vinken and G.W. Bruyn, Eds. American Elsevier, New York, p. 243, 1976b.

Helfant, R.H., DeVilla, M.A., and Meister, S.G. Effect of sustained isometric handgrip exercise on left ventricular performance. *Circulation*, 44:982–993, 1971.

Hendershot, D., Petrofsky, J.S., Phillips, C.A., and Moore, M. Blood pressure and heart responses in paralyzed and nonparalyzed man during isokinetic leg training. *J. Neurol. Orthop. Med. Surg.* 6:259–264, 1985.

Hjeltnes, N. Oxygen uptake and cardiac output in graded arm exercise in paraplegics with low level spinal lesions. *Scand. J. Rehabil. Med.* 9:107–113, 1977.

Hodges, M., Halpern, B.L., Friesinger, G.C., and Dagenais, G.R. Left ventricular pre-ejection period and ejection time in patients with acute myocardial infarction. *Circulation*, 45:933, May 1972.

Kezdi, P., and Geller, E. Baroreceptor control of postganglionic sympathetic nerve discharge. *Am. J. Physiol.* 214:427, 1968.

Knutsson, E., Lewenhaupt-Olsson, E., and Thorson, M. Physical work capacity and physical conditioning in paraplegic patients. *Paraplegia*, 11:205–216, 1973.

Koizumi, K., and Brooks, C.M. The autonomic system and its role in controlling body functions. In *Medical Physiology*, Vol. I, 14th ed., V.B. Mouncastle, Ed. Mosby, St. Louis, p. 893, 1980.

Krayenbuehl, H.P., Rutishauser, W., Schoenbeck, M., and Amende, I. Evaluation of left ventricular function from isovolumic pressure measurements during isometric exercise. *Am. J. Cardiol.* 29:323–330, 1973.

Lewis, R.P., Boudoulas, H., Welch, T.G., and Forester, W.F. Usefulness of systolic time intervals in coronary artery disease. *Am. J. Cardiol.* 37:787, 1976.

Liberson, W.T., Holmquest, H.J., Scot, D., and Dow, M. Functional electrotherapy: Stimulation of the peroneal nerve synchronized with the swing phase of gait of hemiplegic patients. *Arch. Phys. Med Rehabil.* 42:101, 1961.

Lind, A.R., Taylor, S.H., Humphreys, P.W., Kennally, B.M., and Donald, K.W. The circulatory effects of sustained voluntary muscle contraction. *Clin. Sci.* 27:229–244, 1964.

Lind, A.R., McNichol, G.W., Bruce, R.A., and MacDonald, H.R., and Donald, K.W. The cardiovascular responses to sustained contractions of a patient with unilateral syringomyelia. *Clin. Sci.* 35:45–53, 1968.

Mathias, C.J., and Frankel, H.L. Autonomic failure in tetraplegia. In *Autonomic Failure*, Vol. 1, R. Bannister, Ed. Oxford University Press, Oxford, p. 453, 1983.

Mathias, C.J., Christensen, N.J., Corbett, J.L., Frankel, H.L., Goodwin, T.J., and Peart, W.S. Plasma catecholamines, plasma renin activity and plasma aldosterone in tetraplegic man, horizontal and tilted. *Clin. Sci. Mol. Med.* 49:291, 1975.

McCloskey, D.I., and Mitchell, J.H. Reflex cardiovascular and respiratory responses originating in exercising muscle. *J. Physiol.* 224:173–186, 1972.

McNeal, D.R., Wilemon, W.K., Money, V., Boggs, R., and Tumaki, J. The effect of peripheral nerve implanted electrical stimulation on motor control in stroke patients. In *World Congress of Neurological Science*, New York, 1969.

Miyazawa, K., Haneda, T., Shirato, K., Honna, T., and Takishima, T. Effects of isometric handgrip exercise on coronary sinus blood flow in idiopathic cardiomyopathy. *Tohoku J. Exp. Med.* 122:1–8, 1977.

Nilsson, S., Staff, P., and Pruett, E. Physical work capacity and the effect of training on subjects with long standing paraplegia. *Scand. J. Rehabil. Med.* 7:51–56, 1975.

Petrofsky, J.S., and Lind, A.R. The blood pressure response during isometric exercise in fast and slow twitch skeletal muscle in the cat. *Eur. J. Appl. Physiol.* 44:223–230, 1980.

Petrofsky, J.S., and Phillips, C.A. Constant velocity contractions in skeletal muscle by sequential stimulation of muscle efferents. *Med. Biol. Eng. Comput.* 17:583–592, 1979.

Petrofsky, J.S., and Phillips, C.A. Active physical therapy—A modern approach for rehabilitation of the disabled. *J. Neurol. Orthop. Med. Surg.* 4:165–173, 1983.

Petrofsky, J.S., and Phillips, C.A. Closed-loop control of movement of skeletal muscle. *CRC Crit. Rev. Biomed. Eng.* 13(1):35–96, 1985.

Petrofsky, J.S., Phillips, C.A., and Lind, A.R. The influence of fiber composition, recruitment order, and muscle temperature on the pressor response to isometric contractions in skeletal muscle of the cat. *Circ. Res.* 48:132–136, 1981a.

Petrofsky, J.S., Phillips, C.A., Sawka, M., Haupeter, D., and Stafford, D. Blood flow and metabolism during isometric contractions in cat skeletal muscles. *J. Appl. Physiol.* 50:493–502, 1981b.

Petrofsky, J.S., Heaton, H.H., and Phillips, C.A. Leg exerciser for training of paralyzed muscle by closed-loop control. *Med. Biol. Eng. Comput.* 22:298–303, 1984a.

Petrofsky, J.S., Phillips, C.A., Heaton, H.H., and Glaser, R.M. Bicycle ergometer for paralyzed muscle. *J. Clin. Eng.* 9:13–19, 1984b.

Petrofsky, J.S., Phillips, C.A., and Hendershot, D. The cardiorespiratory response during dynamic exercise in paraplegic and quadriplegic man. *J. Neurol. Orthop. Surg.* 6:252–258, 1985.

Phillips, C.A., and Petrofsky, J.S. Velocity of contraction of skeletal muscle as a function of electrical activation and fiber composition: A mathematical model. *J. Biomech.* 13:549–558, 1980.

Phillips, C.A., and Petrofsky, J.S. Computer controlled movement of paralyzed muscle: The medical perspective. *Proceedings of the First Vienna International Workshop on Functional Electrostimulation*, paper 5.2, 1983.

Phillips, C.A., Petrofsky, J.S., Hendershot, D., and Stafford, D. Functional electrical exercise: A comprehensive approach for physical conditioning for the spinal cord paralyzed individual. *Orthopedics*, 7:1112–1123, 1984.

Phillips, C.A., Danopulos, D., Kezdi, P., and Hendershot, D. Muscular, respiratory and cardiovascular responses of quadriplegic persons to an FES bicycle ergometer conditioning program. *Int. J. Rehabil. Res.* 12:147–157, 1989.

Rebersek, S., and Vodovnik, L. Proportionally controlled functional electrical stimulation of the hand. *Arch. Phys. Med. Rehabil.* 54:378–382, 1973.

Robinson, C.J., Bolam, J.M., Chinoy, M., et al. Response to surface electrical stimulation of the quadriceps in individuals with spinal cord injury. *Proc. RESNA*. 6:282–284, 1986.

Ruch, T.C. Transection of the spinal cord. In *Howell-Fulton's Physiology and Biophysics*, Vol. 1, 20th ed. Saunders, Philadelphia, pp. 1–11, 1974.

Sahn, D.J., Demaria, A., Kisslo, J., and Weyman, A. (The Committee on M-Mode Standardization of the American Society of Echocardiography) Recommendations regarding quantitation in M-mode echocardiography. Results of a survey of echocardiographic measurements. *Circulation*, 58:1072, 1978.

Samson, R. Changes in systolic time intervals in acute myocardial infarction. *Br. Heart J.* 32:839, 1970.

Schatz, I.J. Orthostatic hypotension. 1. Functional and neurogenic causes. *Arch. Intern. Med.* 144:773, 1984.

Shephard, R.J. The development of cardiorespiratory fitness. *Med. Serv. J. Can.* 21:533–544, 1965.

Shephard, R.J. *Endurance Fitness*. 2nd ed. University of Toronto Press, Toronto, 1977.

Smart, C.N., and Sanders, C.R. *The Costs of Motor Vehicle Related Spinal Cord Injuries*. Insurance Institute for Highway Safety, Washington, DC, 1976.

Stack, R.S., Lee, C.C., Reddy, B.P., Taylor, M.L., and Weissler, A.M. Left ventricular performance in coronary artery disease evaluated with systolic time intervals and echocardiography. *Am. J. Cardiol.* 37:331, 1976.

Sunderland, S. *Nerves and Nerve Injuries*. Williams & Wilkins, Baltimore, 1968.

Wagner, H.N., Jr. Orthostatic hypotension. *Bull. Johns Hopkins Hosp.* 105:322, 1959.

Weissler, A.M., Peeler, R.G., and Roehill, W.H. Relationships between let ventricular ejection time, stroke volume and heart rate in normal individuals and patients with cardiovascular disease. *Am. Heart J.* 62:367, 1961.

Weissler, A.M., Harris, W.S., and Schoenfeld, C.D. Systolic time intervals in heart failure in man. *Circulation*, 37:149–159, 1968.

Weissler, A.M., Harris, W.S., and Schoenfeld, C.D. Bedside technics for the evaluation of ventricular function in man. *Am. J. Cardiol.* 23:577–583, 1969.

Zealear, D.L., and Dedo, H.H. Control of paralyzed axial muscles by electrical stimulation. *Trans. Am. Acad. Ophthalmol. Otolaryngol.* 84:310, 1977.

Zwiren, L.D., and Bar-Or, O. Responses to exercise of paraplegics who differ in conditioning level. *Med. Sci. Sports*, 7:94–98, 1975.

5
Acute Effects with the Leg Exerciser System: The Physiological Process

I. Introduction to the Leg Exercise System

A. Background

During normal, voluntary activity, strength and endurance are trained by two different mechanisms. For example, progressive resistance exercise is used to train for muscular strength (Muller, 1932). This type of exercise involves lifting weights at a very slow rate up and down. The rate is one important variable, while the second is the amount of weight that is lifted. To optimize this type of training, the weight must be approximately 50% of the person's maximum voluntary strength. When this weight can be handled, strength builds quite rapidly. This type of training is used to cause muscular hypertrophy and to increase the strength of muscle for weight lifters. It does not, however, train for muscle endurance.

To train for endurance, the blood supply and the oxidative capacity of the muscle must be increased. Such training is called aerobic training because it involves the utilization of oxygen that is extracted from the blood by active muscle. Aerobic training, which is totally different from anaerobic training such as weight lifting, is typified by bicycling or running. With aerobic training, muscles work with a light load but at a high rate. When the rate is set to stress the muscle in the proper manner (the muscle must work near its anaerobic threshold), the muscle begins to develop increased circulation. As a secondary benefit, the stress on the circulation improves cardiovascular and respiratory fitness. A good balanced training program is a combination of both aerobic and anaerobic methods. Such a balanced training program is best used for normal physical conditioning of the body.

Over the last 20 years, a number of techniques have been used to train paralyzed muscle. These typically involve stimulating the muscle at various rates with an undetermined weight. Although the neural traffic is similar to that which occurs during voluntary activity, stress on the body is not. Basically, the muscle metabolism must be stressed in a precise manner as a stimulus for muscle growth (Vrbova, 1963). Over the past few years we have developed at Wright State a computer-controlled stimulation system to stress muscle in a manner similar to that which would occur during voluntary exercise training (see, e.g., Petrofsky et al., 1984a). This is termed "active physical therapy," (APT), and two different types of device were developed to implement it. The first type (the leg exercise system) is described in this chapter. The second type (the exercise bicycle ergometer) is described in Chapter 6.

B. The Leg Exerciser System

The first device was a weight trainer, similar in many respects to the weight-lifting apparatus available at any modern gymnasium. This equipment features tray, to and from which weights can be added and removed. The weights are connected through a cable system to a harness that is placed around the appropriate appendage to allow muscles to operate. When a given muscle (such as the quadriceps muscle of the leg) contracts, the weights are lifted up and down. A typical subject is seen in the

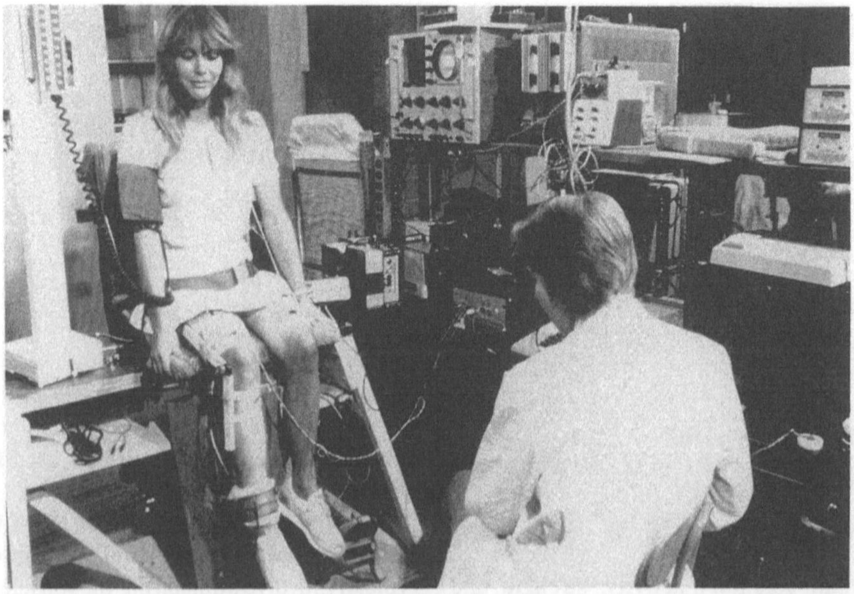

FIGURE 5.1. Patient using the isokinetic leg trainer.

isokinetic exercise trainer developed for strength training (Fig. 5.1). In this example, the muscle being exercised is the quadriceps muscle. Operationally, a strap is placed around the ankle of the subject and secured by Velcro fasteners. The leg is connected (through a steel cable connected to the strap) to the weight tray, which is under the back of the chair. The subject obviously cannot feel the

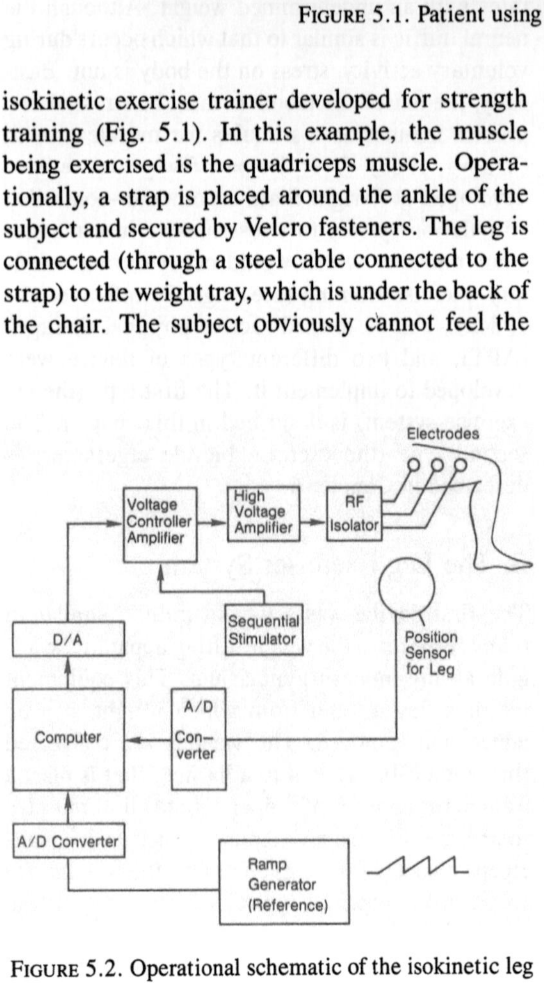

FIGURE 5.2. Operational schematic of the isokinetic leg trainer (see text).

limb, nor can the subject see the weights that are being lifted. Therefore, a sensor along the length of the cable provides information to a computer with respect to the position of the leg (Fig. 5.2). The computer used in these studies was an Apple II microprocessor (Apple Computer Company, Berkeley, California). To interface the computer to the sensor, an A/D converter was used (Petrofsky et al., 1983a).

The A/D converter used in the system was made by National Semiconductor, and was the ADC0808. The ADC0808 is an eight-channel, multiplexed A/D converter having a conversion time of 100 µs. This conversion time would normally be too slow for most applications involving programs written in machine code. The programs described below for this particular application were written in BASIC, however, and the interpreter time to process BASIC commands was long enough ($\approx$ 1 ms) to allow the A/D converter to operate at its own speed without impairing the speed of the operation of the program. The circuitry (Fig. 5.3) simply involves decoding an input/output select line from the Apple bus (into which the board is inserted) to select the A/D converter. Each I/O select line of the Apple selects 256 addresses. By connecting the three low order bits of the address bus to the A/D converter, a write instruction to this address latches the multiplexer to the A/D converter to receive

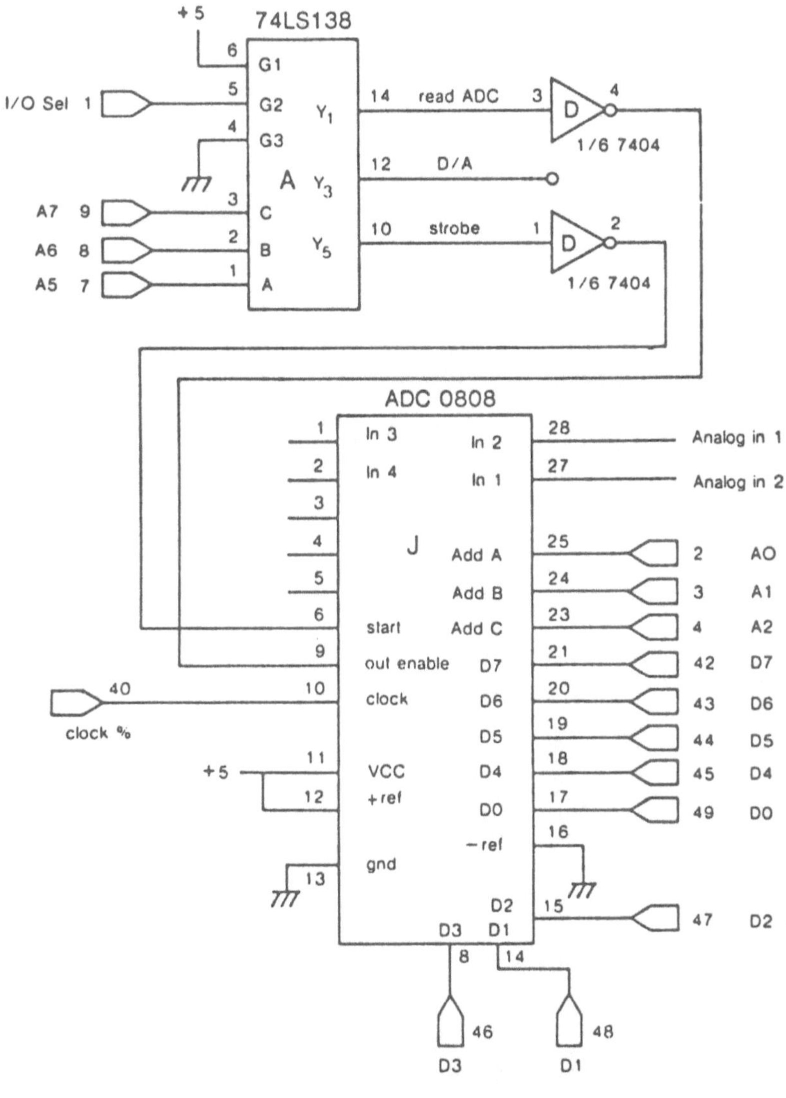

FIGURE 5.3. The Apple A/D converter (electrical schematic).

one of the eight inputs (which can be connected to the A/D converter chip). The same write cycle also strobes the A/D converter to begin processing the input data. A minimum of 100 μs later, a read to any one of the 256 memory locations results in reading of the A/D converter.

Digital-to-analog control is provided through a single D/A converter (Petrofsky et al., 1984b), a DAC0831, as shown in Figure 5.4. The DAC0831 is a double-buffered, eight-bit D/A converter. One buffer register is permanently disabled for this use. The DAC0831 is plugged into a second Apple board such that only the I/O select line is used to toggle data from the data bus onto the D/A converter chip.

The tracking program that operates the chair continually changes the stimulation voltage applied through the D/A converter and then senses the position of the leg to determine whether it is going up or down at the appropriate rate, as determined by a ramp generator (Fig. 5.5), which provides a triangular wave with a rise time of 3 seconds and a fall time of 3 seconds. The output of the ramp generator is set to match the gain of the sensor for the leg such that the computer program simply tries to match the target analog voltage to the actual input of the A/D converter. The computer program matches the synthetically generated ramp and tries to make the leg do the same thing. If the ramp voltage is going up

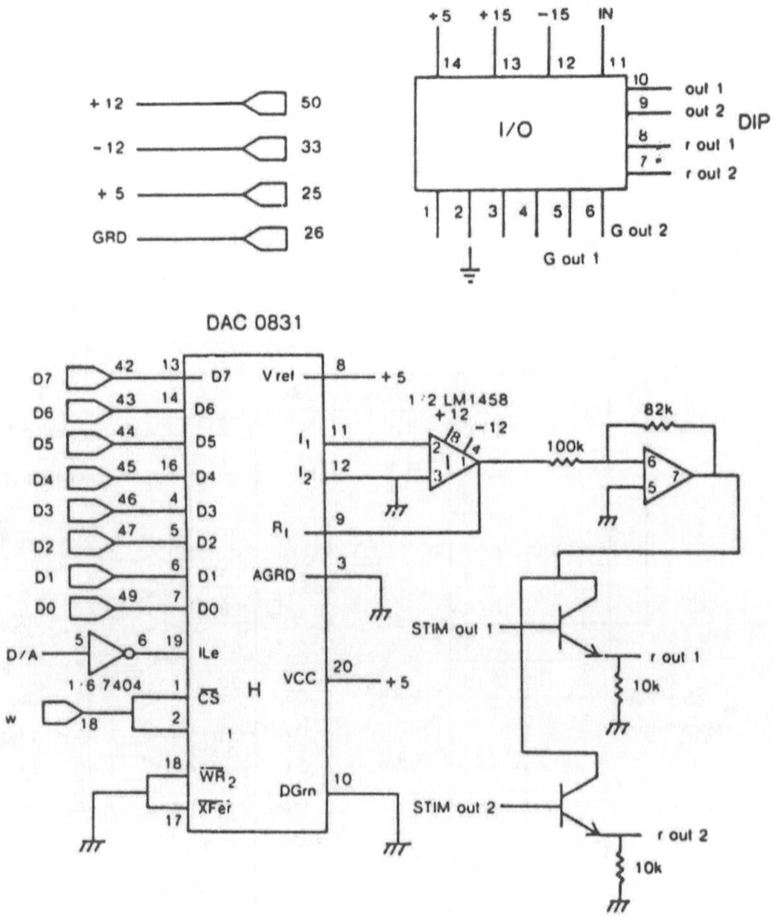

FIGURE 5.4. The Apple D/A converter (electrical schematic).

faster than the leg is moving, stimulation voltage is increased. If the ramp voltage is going up more slowly than the leg voltage, stimulation voltage is reduced.

The D/A converter is interfaced to the leg by a high voltage converter circuit. This involves using two 2N3904 transistors as voltage-controlled amplifiers so that a pair of asynchronously (180° out of phase) applied pulses are generated. The amplitude of these pulses is controlled by the D/A converter. This output is then buffered and inverted through a transformer to 100 V maximum, which is then applied, through high voltage transistors, to the leg. In this manner, then, the Apple computer interface controls a stimulation voltage that can vary from 0 to 300 V under digital control.

This section has provided a synoptic overview of the leg exerciser system. For a complete technical description, the interested reader is referred to our earlier work (Petrofsky et al., 1983a, 1984b).

II. Functional Electrical Exercise with the Leg Exercise System*

A. Introduction and Background

Active physical therapy (APT) can be defined as the use of paralyzed muscle as an actuator which actively participates in the physical reconditioning of a previously deconditioned individual. It differs

*From C.A. Phillips et al., Functional electrical exercise: A comprehensive approach for physical conditioning of the spinal cord paralyzed individual. Reprinted with permission from *Orthopedics*, 7:1112–1123, 1984. Copyright 1984, SLACK, Incorporated.

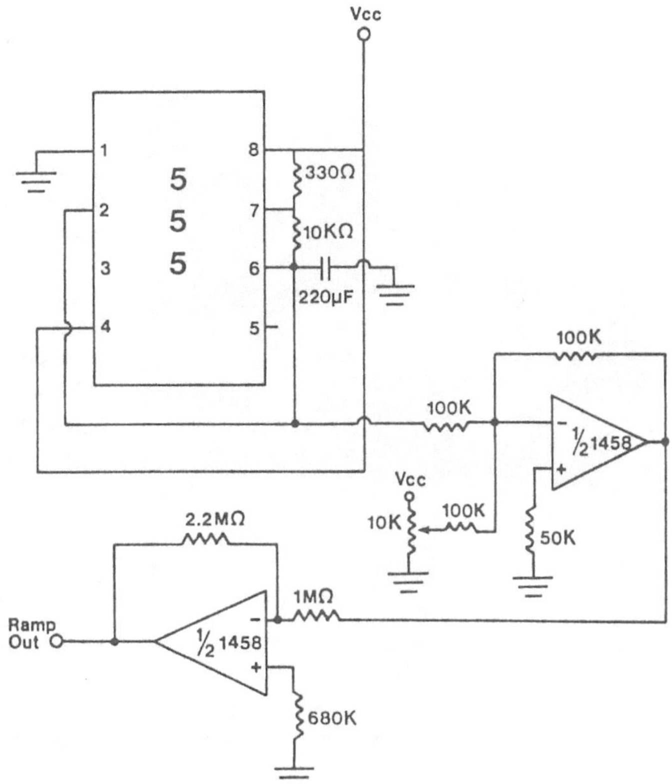

FIGURE 5.5. The Apple ramp generator (electrical schematic).

from conventional rehabilitation therapy, in which the paralyzed muscle is passive and is manipulated by various external influences. Active physical therapy requires that paralyzed muscle be *electrically stimulated* in a manner that is *functional* (i.e., the muscle is made to perform useful external work). An introduction to APT that reports preliminary results of physically reconditioning muscular strength and endurance is available (Petrofsky and Phillips, 1983b, Phillips and Petrofsky, 1983).

A number of investigators have looked at the effects of electrical stimulation on fast- and slow-twitch skeletal muscle (Salmons and Vrbova, 1969; Pette et al., 1972, 1975; Peckham et al., 1973, 1976; Riley and Allin, 1973; Brown et al., 1976; Hudlicka et al., 1977). However, a common factor in most of these studies was that electrical stimulation was applied at different frequencies and amplitudes with no control over the muscle force and/or velocity. The results of these studies clearly indicated the trophic influence of the stimulated nerve upon the muscle. However, there was no systematic control of muscle force and/or velocity as they affect return of muscular strength and/or endurance.

Training of skeletal muscle generally falls into two categories: training for strength and training for endurance. Training for strength is associated with progressive resistance exercise. Training for endurance is associated with such activities as bicycling or running, where muscles can work at or near their anaerobic threshold (Astrand and Rodahl, 1977). For example, Muller (1932) showed that lifting weights slowly up and down at approximately 50% of the muscle's maximum strength resulted in a rapid increase in muscular strength. This type of exercise is commonly used by weight lifters to increase muscular strength by causing the hypertrophy of fast-twitch muscle fibers (Simonson, 1971; Astrand and Rodahl, 1977).

Because progressive resistance exercise does not increase capillary density or oxidative enzymes, it does not improve endurance. Endurance is increased by working muscle close to the anaerobic threshold. Bicycling at the anaerobic threshold rapidly increases the concentration of enzymes

FIGURE 5.6. The isokinetic leg trainer.

involved in oxidative phosphorylation of glucose and fats and increases mitochondrial and capillary density. This type of exercise preferentially trains slow-twitch muscle fibers.

B. Leg Exercise System

The isokinetic leg trainer (Fig. 5.6) is similar in many respects to the types of weight lifting equipment available at a modern gymnasium. This equipment utilizes a weight tray where weights can be added and lifted. The weights are connected through a cable system to a harness that can be placed around the appropriate appendage and allow muscles to operate. When a given muscle (such as the quadriceps muscle of the leg) contracts, the weights can be lifted up and down. Technical details regarding the various subsystems of the leg exercise chair have been provided in Section I of this chapter.

The operation of the complete chair system simply involves a tracking algorithm, which continually changes the stimulation voltage applied through the D/A converter and then senses the position of the leg to determine if the leg is going up or down at the appropriate rate (Fig. 5.2). The rate is determined by a ramp generator to provide a triangular wave with a rise time of 3 seconds and a fall time of 3 seconds. The output of this ramp generator is set to match the gain of the sensor for the leg, so that the computer program tries to match the target analog voltage to the actual input of the A/D converter. The program matches the synthetically generated ramp and tries to have the leg do the same thing. If the ramp voltage is going up faster than the leg is moving, stimulation voltage is proportionately increased. If the ramp voltage is going up slower than the leg voltage, stimulation voltage is proportionately reduced.

C. Subjects and Protocols

Six spinal-cord-injured subjects were used in these studies. Four subjects were quadriplegic with lesions at C5, C6, or C7. Two subjects were paraplegic with lesions at T9/T10 or T11.

The subjects were all volunteer, undergraduate university students, divided equally between male and female with ages ranging from 18 to 35. All subjects were free of significant secondary medical diseases and underwent a complete medical history, physical examination, and laboratory testing (including an ECG and full X-rays of both lower legs and upper arms). All subjects were informed of all experimental procedures and signed a duly witnessed informed consent. Care was taken to ensure subject safety, privacy, and confidentiality. All protocols were fully approved by Wright State University's Institutional Review Board.

The cardiopulmonary response protocol (reported in this section) involved all six SCI patients (four quadriplegic and two paraplegic). These subjects were all untrained at the time the data were gathered. Preexercise data were acquired (see Section D, "Cardiopulmonary Measurements"). The subject then performed a 15-minute fatiguing contraction under voluntary control using a standard arm crank ergometer (Monarch model 881) at a rate of 50 rpm. Alternately, the subject would perform a 15-minute fatiguing exercise under FES control of the lower limbs using the leg exercise trainer. The load on each device would be appropriately set to ensure that a fatiguing exercise occurred within a 15-minute period. Postexercise cardiopulmonary measurements were collected immediately at the fatigue end point.

D. Cardiopulmonary Measurements

Blood pressure was measured by auscultation of an inactive arm while the subject was at rest, every 2 minutes during exercise (excluding arm cranking activity), immediately postexercise, and every 2 minutes for 6 minutes postexercise. The heart rate was measured by a continuous recording of the precordial ECG (modified lead II configuration) over the same time periods. The resting heart rate (HR) and blood pressure (BP), as well as the immediate postexercise HR and BP, are reported in this study.

At three different periods (resting, at the point of muscular fatigue, and 4 minutes postfatigue), a 100 microliter (μL) arterialized blood sample was drawn from each subject's fingertip. From the resting and fatigue end-point samples, arterial pO_2, pCO_2, and pH were measured by a microanalytical technique (Petrofsky et al., 1980, 1981).

The resting and 4-minute postfatigue 100 μL blood samples were used to determine the serum lactic acid (LA). Each blood sample was mixed with 200 μL of perchloric acid (8%). The mixture was spun down in a microfuge for 1 minute. Two NAD vials each (consisting of 1 mL glycine buffer, 2 mL of H_2O, and 0.05 mL of lactic dehydrogenase) were inverted to dissolve. Three tubes were then used for the measurement. First, into each we added 1.4 mL of 8% perchloric acid. Next, into the *standard tube* we added 0.1 mL of standard (0.40 mg/L lactic acid). Finally, into the *sample tube* we added 0.1 mL of the blood–perchloric acid mixture (using supernatant). The tubes were incubated for 30 minutes at 37°C in a water bath. The samples were read at 340 nm on a Spectronic 21 (Bausch & Lomb).

The ventilatory equivalent ($\dot{V}_E$) and oxygen consumption ($\dot{V}_{O_2}$) were determined as follows. Expired air (CO_2 and unused O_2) was collected via blood gas syringes and expelled into the Beckman Medical Gas Analyzer, where O_2 and O_2 percentage by volume were calculated for the number of breaths that were expired per 30-second unit of time ($\dot{V}_E$, in liters per half-minute). The correction factor was determined by a standard chart using barometric pressure (mmHg) and temperature (°C). The corrected $\dot{V}_E$ was obtained by multiplying the correction factor by 1.44 and by twice the $\dot{V}_E$ (so that the two 30-second time periods would equal one minute). The $\dot{V}_{O_2}$ was obtained by taking the corrected $\dot{V}_E$ times the percent true O_2 and dividing by 100; $\dot{V}_{O_2}$ was the maximum oxygen uptake rate.

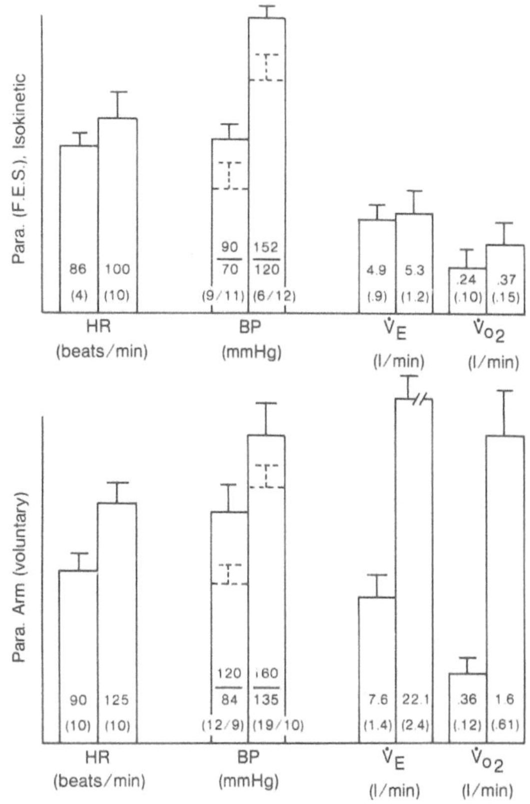

FIGURE 5.7. Change in HR, BP, $\dot{V}_E$, and $\dot{V}_{O_2}$ with FES (isokinetic leg trainer) versus voluntary (arm) activity in paraplegics.

$$\frac{(\text{cor } \dot{V}_E \ (\% \text{ true } O_2)}{100} \tag{5.1}$$

and

$$\text{cor. } \dot{V}_E = (2)\ (\dot{V}_E)\ (\text{cor. factor})\ (1.44) \tag{5.2}$$

where $\dot{V}_E$ is in liters per half-minute.

E. Results

For the two paraplegic subjects, using FES and the leg trainer versus voluntary arm cranking, BP increased markedly in both cases between resting values and the fatigue end point; HR, $\dot{V}_E$ and $\dot{V}_{O_2}$ increased only minimally for FES and the leg trainer, but increased markedly for voluntary arm cranking activity (between resting values and the fatigue end point) (Fig. 5.7).

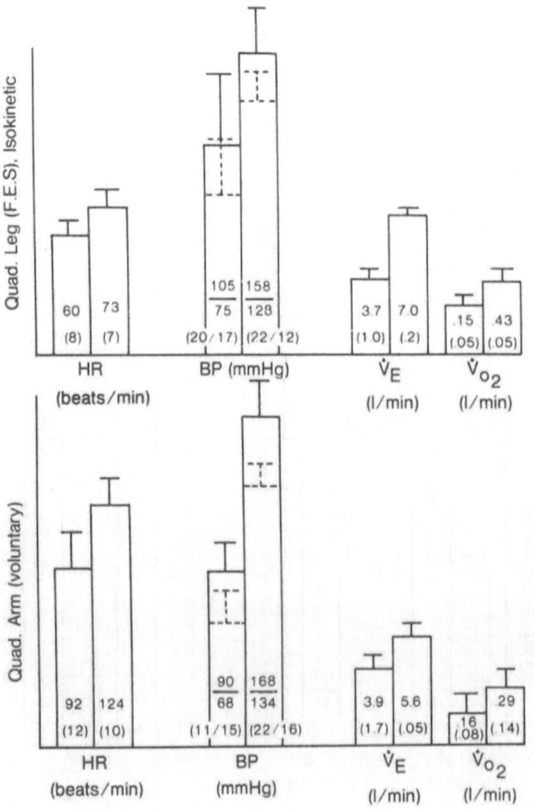

FIGURE 5.8. Change in HR, BP, $\dot{V}_E$, and $\dot{V}_{O_2}$ with FES (isokinetic leg trainer) versus voluntary (arm) activity in quadriplegics.

With the four quadriplegic subjects, BP and HR showed the same patterns as for the paraplegic subjects; $\dot{V}_E$ and $\dot{V}_{O_2}$ more than doubled for FES and the leg trainer (between resting values and fatigue end-point values). However, for voluntary arm cranking activity, $\dot{V}_E$ and $\dot{V}_{O_2}$ rose only moderately for the four quadriplegic subjects (Fig. 5.8).

F. Concluding Statement

This section has reported on functional electrical exercise with the leg exerciser system. Section III of Chapter 6 further reports on functional electrical exercise with respect to the bicycle ergometer system. The two systems are discussed collectively at the conclusion of Chapter 6.

III. Blood Pressure and Heart Rate Responses with Leg Trainer Exercise*

A. Introduction

Following spinal cord injury, muscle atrophies rapidly and bone demineralizes. There is a gradual reduction in the dynamics of the cardiorespiratory system due to disuse and also from damage to the autonomic nervous system (Guttmann, 1976). Paraplegics can obtain some exercise simply by moving their wheelchairs, but for the quadriplegics whose injuries can prevent even this level of exercise, the extent of cardiovascular deconditioning can be significant. Blood pressure is often low, and orthostatic hypotension is common (Guttmann, 1976). Throughout life, cardiovascular deconditioning progresses in quadriplegics and, when combined with osteoporosis and atrophy of muscle, can present many secondary clinical problems.

To increase the range of motion and to reverse atrophy in skeletal muscle in the paralyzed portions of the body, electrical stimulation has been commonly used in the clinical setting. By applying electrical impulses either through the skin (Petrofsky and Phillips, 1983a; Petrofsky et al., 1983b, 1984c; Malezic et al., 1984) or with deep muscle (Peckham et al., 1976; Marsolais and Kobetic, 1982) or nerve electrodes (Holle et al., 1984), electrical stimulation has been used to restore some degree of movement in the paralyzed. By sequentially activating muscle through multiple channels (Rack and Westbury, 1969; Petrofsky et al., 1976) smooth contractions have been developed. These stimulation frequencies are well within the physiological range, and modifying the amplitude of the stimulus may be used to alter the extent of recruitment. A number of studies have been done to quantify the effects of stimulation on long-term training of paralyzed skeletal muscle (Salmons and Vrbova, 1969; Peckham

*From D.M. Hendershot et al., Blood pressure and heart responses in paralyzed and non-paralyzed man during isokinetic leg training. Copyright 1985, American Academy of Neurological and Orthopedic Surgeons. Reprinted with permission from *Journal of Neurologic and Orthopedic Medicine and Surgery*, 6:259–265, 1985.

et al., 1973; Pette et al., 1975; Brown et al., 1976; Hudlicka et al., 1977). However, little has been done to quantify the effect of electrical stimulation on other body systems such as the cardiorespiratory system. In previous papers, it has been shown that blood pressure and heart rate do change in quadriplegics during electrically induced exercise (Corbett et al., 1971; Phillips and Petrofsky, 1983; Phillips et al., 1984) because of deconditioning of the heart and cardiovascular system. The increases in blood pressure observed previously may be dangerous for individuals with undiagnosed cardiovascular disease.

The origin of the blood pressure reflex during exercise has been attributed to a local reflex in the lower spinal cord associated with the accumulation of metabolites in fatiguing muscle (Lind et al., 1966; Coote et al, 1971; McCloskey and Mitchell, 1972). These studies on the mechanism of the blood pressure raising reflex were accomplished using isometric exercise, a form of exercise where the muscle contracts continually. In contrast, with phasic types of exercise (e.g., isokinetic exercise, as might be used in the clinical setting), the muscle contracts and relaxes briefly with rest intervals between the contractions. It is possible that with sequential contracting and relaxing of muscle, the postexercise blood flow (which perfuses the muscle between the contractions) might remove a significant amount of metabolites so that muscle fatigue might occur without increasing blood pressure (Barcroft and Millen, 1939). Therefore, the present series of experiments was conducted to better quantify the extent of the changes in heart rate and blood pressure during electrically induced exercise as might be encountered in a clinical setting.

Furthermore, we investigated whether the interval between the periods of exercise had an effect on the magnitude of the observed heart rate and blood pressure changes. Finally, because of the different level of damage in the spinal cord between paraplegics and quadriplegics, the present series of experiments also was accomplished to quantify the exact relationship between the magnitude of the blood pressure and heart rate response with the level of spinal cord injury.

B. Subjects

Nine subjects participated in these experiments, and their general characteristics are listed in Table 5.1. Subjects were divided into control, paraplegic, and quadriplegic groups. All were free of any major medical problems, including cardiovascular and renal disease, and were not taking any type of medication at the time of the studies. For both paraplegic and quadriplegic groups, subjects were between 1 and 10 years postinjury. The ages of the subjects ranged from 20 to 30 years. For the paraplegic subjects, it was also required that the level of injury be no lower than T11 because of the possibility of lower motor neuron damage with T12, L1, and L2 injuries. The subjects were informed of all experimental procedures and signed a consent form approved by Wright State University's Institutional Review Board prior to any experimental runs.

TABLE 5.1. General characteristics of subjects.

Subjects	Height (cm)	Weight (kg)
Control		
1	175	82
2	177	70
3	183	67
Quadriplegic		
4	180	65
5	191	95
6	185	77
Paraplegic		
7	173	65
8	173	65
9	188	77

C. Methods

Isokinetic exercise was accomplished on a specially designed leg trainer as shown in Figure 5.6. The device consisted of a seat that was specially padded to prevent pressure sores in the paraplegic and quadriplegic subjects. This seat was made movable with a series of roller bearings and locks so that the subject could be moved back and forth. The subject was positioned so that the angle of the knee was at 90° and the dependent leg (to be exercised) was directly over a pulley located at the bottom of the system. An ankle bracelet was then connected around the lower ankle (at the level of the malleoli) with Velcro. When the leg extended (Fig. 5.9), the ankle bracelet allowed a weight pan to be lifted by means of a steel cable and a series of pulleys.

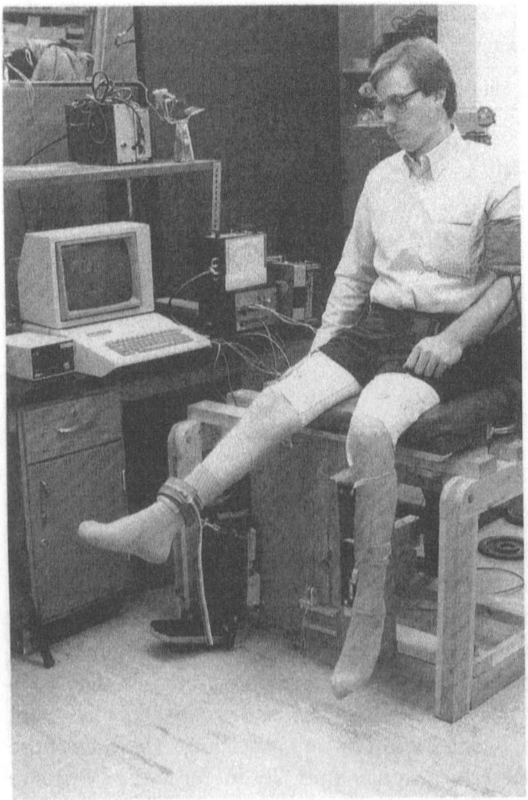

FIGURE 5.9. The experimental setup for isokinetic leg exercise.

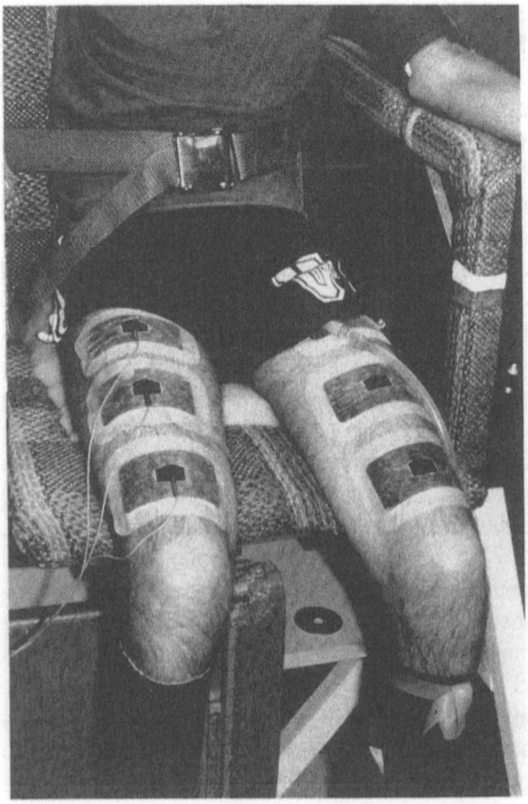

FIGURE 5.10. Electrode configuration for sequential stimulation.

For the nonparalyzed subjects, exercise was simply a matter of extending the leg by voluntary contraction of the quadriceps muscle through a 60° arc very evenly over a 3-second period. The leg was then relaxed to the 90° knee-dependent position over another 3-second period. Since voluntary control was obviously impossible for the paralyzed subject, the quadriceps muscle was stimulated by sequential electrical stimulation. Three carbonized rubber electrodes (Medtronics, Inc.) were placed diagonally across the quadriceps muscle of the exercising leg. The two outside electrodes were active electrodes, and the inside electrode was the ground (see Fig. 5.10). By alternately stimulating the two sides of the muscle, sequential stimulation was achieved. Sequential electrical stimulation has an advantage over a single pair of electrodes in that it allows the muscles to contract very smoothly at frequencies that are within the physiological range (Rack and Westbury, 1969; Peckham et al., 1976; Petrofsky et al., 1976; Petrofsky and Phillips, 1976). Typically, smooth contractions can be elicited at stimulation frequencies far below 40 Hertz. In the present set of experiments, therefore, a stimulation frequency of 40 Hz was chosen to fully tetanize the muscle. Stimulation was delivered under constant current control. Biphasic pulses were delivered alternately to the two active electrodes with the current level being controlled by an Apple II microprocessor. A sensor (placed in series with the chain connected to the ankle) translated the position of the limb during the contraction. The positional data then were used to modify the level of stimulation (current) to allow the muscle to contract smoothly and at the proper velocity. A more detailed description of the leg exerciser is given in Section I of this chapter.

1. Blood Pressure and Heart Rate

Heart rate was obtained from a continuous recording of the ECG. Heart rate was determined over a 15-second interval just before, at the onset of, and at 20, 40, 60, 80, and 100% of the duration of the exercise. Blood pressure was measured by auscultation of the arm. Blood pressure was taken before the exercise, as often as possible during the exercise, and every 30 seconds postexercise. Blood pressure was then interpolated or extrapolated to obtain blood pressure at 20, 40, 60, 80 and 100% of the duration of the exercise.

2. Statistics

Statistical analysis involved calculations of means, standard deviations, and related and unrelated t-tests. The level of significance was chosen at $p < 0.05$.

D. Procedures

For both the paralyzed and nonparalyzed subjects, a 2-week training period was first conducted. The subjects exercised for 15 minutes a day, three times a week, with progressively increasing weights using the leg trainer and exercising the quadriceps muscle.

The posttraining protocols were fairly simple. For the two paralyzed experimental groups, subjects were stimulated to lift their legs over a 6-second exercise cycle (3 seconds up and 3 seconds down) to extend the knee over a 60° arc. The workload was adjusted so that all subjects would fatigue at the end of a 15-minute period when there were no rest intervals between lifts. This workload was held constant for all other experimental procedures. Subjects exercised for 15 minutes, during which the interval between the exercise cycles was zero seconds, and/or for an additional 15 minutes with rest intervals of 3, 5, 10, 30, or 60 seconds. During the exercise periods, blood pressure and heart rate were continuously measured as described above. Two series of experiments were accomplished. In one, protocols were accomplished as given above. In a second set of experiments, an arterial occlusion cuff was placed around the leg. It was inflated to 250 mmHg throughout the duration of the exercise and for 2 minutes after. Occlusion was done to see how limiting blood flow during and after the exercise might alter the cardiovascular response.

E. Results

As might be expected, the control subjects showed an increase in blood pressure and heart rate throughout the duration of the exercise. However, the increase in blood pressure associated with the exercise was found to be proportional to the recovery interval.

Figure 5.11 shows the systolic and diastolic blood pressures for the control, paraplegic, and quadriplegic subjects for 6 seconds of active exercise with no rest between the periods of exercise, and with 3, 5, and 60 seconds of rest. For the control subjects, there was a sharp increase in both the systolic and diastolic blood pressure with no rest. The increase in blood pressure was gradually reduced when rest intervals were allowed, so that if greater than 5-second rest intervals were used, there was no significant increase in blood pressure over the 15-minute exercise period. The blood pressure at the end of the exercise period exponentially decayed as a function of the rest interval between the exercise cycles (Fig. 5.12). The same phenomenon occurred with the quadriplegic subjects (Figures 5.11 and 5.12). Paraplegic subjects had only a small increase in blood pressure during any experimental protocol.

To determine if the increase in blood pressure was due to a metabolic buildup within the muscle, a second series of experiments was accomplished. The subjects again contracted their muscles for a 6-second exercise cycle with no rest between the contractions. During these experiments, the circulation to the limb was occluded by an arterial blood pressure cuff inflated to 250 mmHg. After the exercise was terminated, the arterial occlusion cuff remained inflated for 2 minutes. The results of these experiments are shown in Figure 5.13.

Blood pressure in the control and quadriplegic subjects increased from a resting level of 100/64 mmHg to a final value of 158/78 mmHg at the end of the exercise. Once the exercise had been terminated, the blood pressure remained elevated at this level until the blood pressure cuff was removed. Paraplegics, again, had no significant change in blood pressure throughout the duration of the

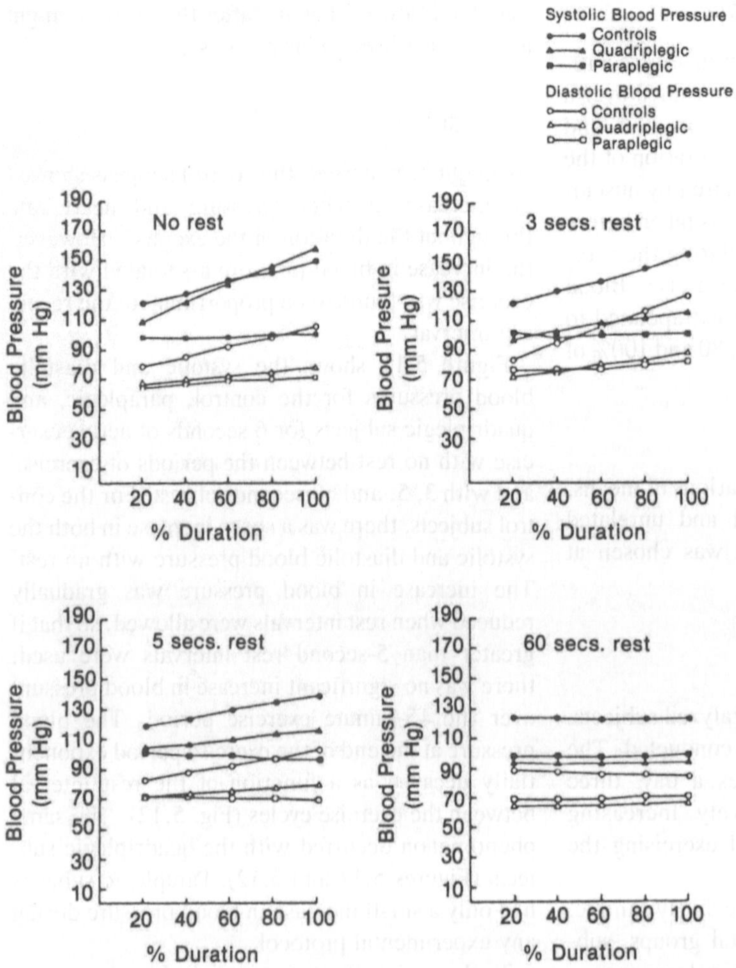

FIGURE 5.11. The blood pressure response throughout 15 minutes of fatiguing contractions in 3 paraplegic, 3 quadriplegic, and 3 control patients, illustrating average results for the groups of patients with interconnection intervals of 0, 3, 5, and 60 seconds. The abscissa has been normalized in this and subsequent figures.

exercise, and there was no additional or lasting effect of arterial occlusion.

The heart rate changed very little and in no specific pattern throughout the duration of exercise in any of the subjects examined (Fig. 5.14). For example, the resting heart rate averaged 72 beats per minute for the quadriplegic subjects. Final exercising heart rate with no rest interval between exercise cycles was 66 beats per minute for that group.

F. Discussion

The blood pressure and heart rate responses were examined in control, paraplegic, and quadriplegic subjects during intermittent or fast phasic isokinetic contractions. The overall result was little change in heart rate, but a significant change in blood pressure when rest intervals between contractions were kept short. The reason for the higher increase in blood pressure associated with short intervals between contractions is probably twofold. First of all, with short intervals between contractions, the muscle would obviously fatigue more than with longer intervals between contractions. Rohmert (1968) showed that allowing rest intervals between contractions caused an exponential reduction in fatigue. The time course and magnitude of these changes certainly approaches that shown for blood pressure in the present investigation.

Second, part of the differences in blood pressure may be related to a flow phenomenon in which the blood perfuses the muscle between the contractions. Simonson (1971) showed that the blood pres-

FIGURE 5.12. Peak systolic and diastolic blood pressures (blood pressures recorded at 100% duration of the contractions) as a function of intercontraction interval in the paraplegic, quadriplegic, and control patients.

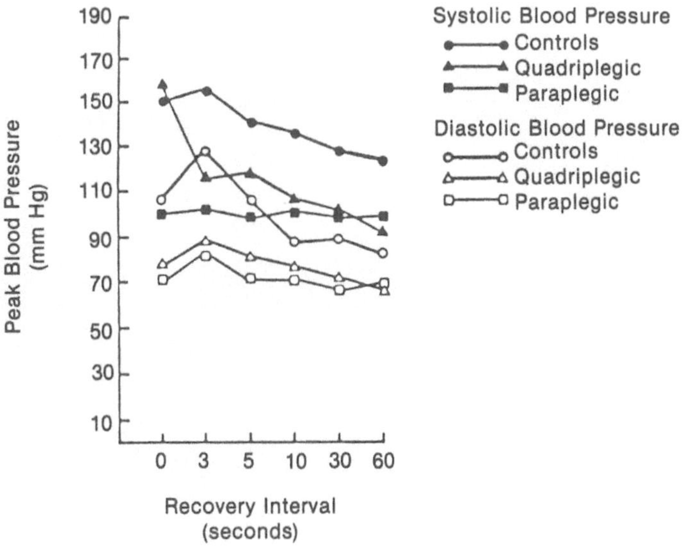

sure response during isometric exercise in man could be maintained after the exercise by occluding the circulation, hence trapping the metabolites in the muscle. Therefore, part of the difference in blood pressure with longer rest intervals might be due not only to the fatigue process, but also to a greater washout between contractions. Barcroft and Millen, (1939) found that washout is due to the postexercise hyperemia usually associated with intermittent contractions (Barcroft and Millen, 1939). Postexercise hyperemia (immediately after the exercise) elicited an increase in muscle blood flow as much as 40- to 50-fold higher than the blood flow during exercise.

The fact that paraplegics showed no increase in blood pressure supports the commonly held view that the blood pressure raising reflex is peripheral in origin for this type of exercise. There is significant damage to the lower spinal cord in paraplegics. In these cases, the reflex pathways associated with the response would be disrupted. Coote et al. (1971), McCloskey and Mitchell (1972), and Lind et al. (1966) have all pointed to the reflex arising in active muscle from the accumulation of a metabolite. That reflex causes a powerful splanchnic

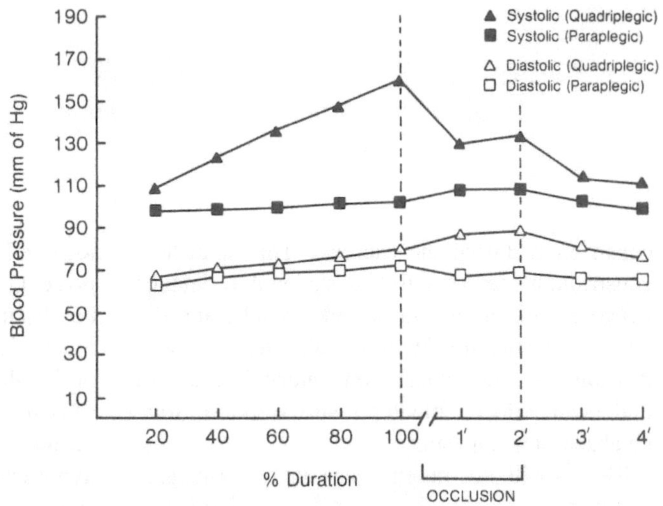

FIGURE 5.13. The blood pressure response in quadriplegic and paraplegic patients with no rest interval during experiments, where the circulation of the limb was restricted by arterial occlusion. The occlusion was maintained following the exercise for 2 minutes to observe the blood pressure and heart rate changes.

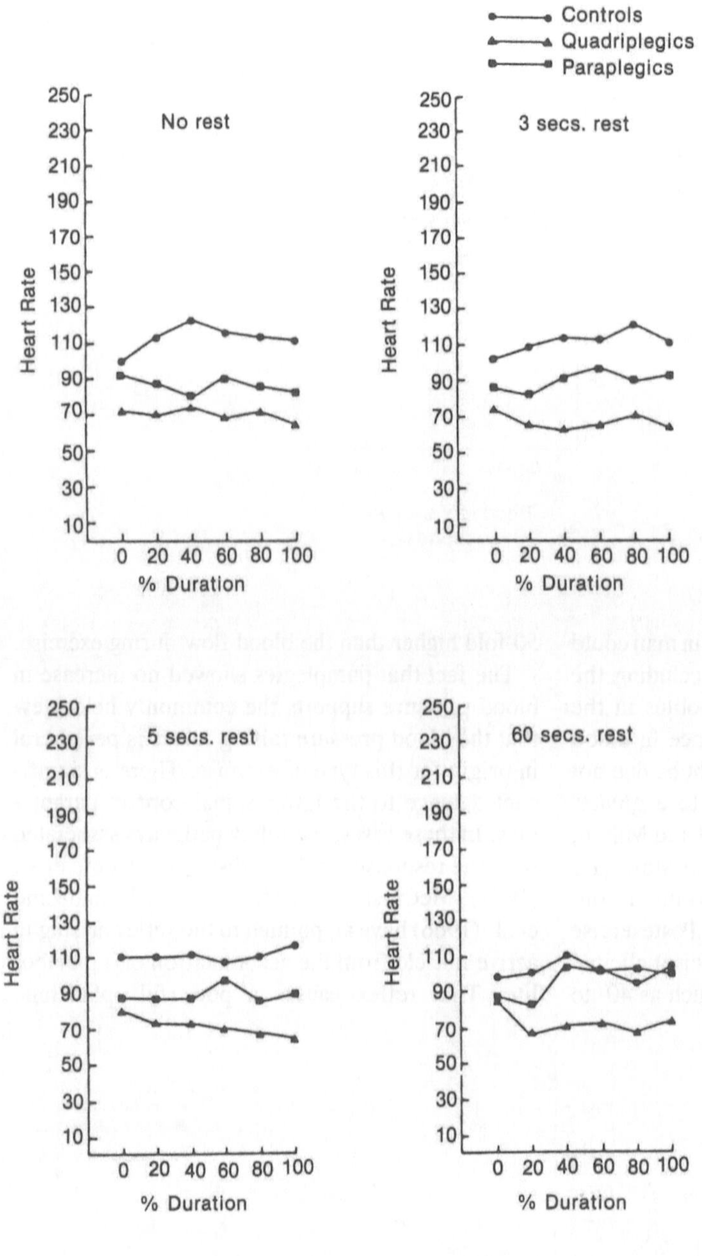

FIGURE 5.14. The heart rate response throughout 15 minutes of fatiguing contractions in 3 paraplegic, 3 quadriplegic, and 3 control patients. These are the average results for the groups of patients with intervals of 0, 3, 5, and 60 seconds.

constriction during the exercise. This splanchnic constriction increases total peripheral resistance, hence causes an increase in both systolic and diastolic blood pressure. With the splanchnic pathways disrupted by lower spinal cord injury, it is easy to understand why the blood pressure response would be absent in paraplegics.

This would not be the case in quadriplegics. Since the damage is higher in the quadriplegics, local reflexes in the lower spinal cord would still be present, while centrally mediated reflexes would be absent. Freyschuss (1970) has suggested that much of the heart rate response and certainly some of the blood pressure response, during exercise, is caused by central mediation from higher brain centers. The present study would argue against this hypothesis. The blood pressure increase in quadriplegics, where no central outflow is possible, is equal to that of the control subjects. Further, there is no blood pressure raising reflex in the paraplegics where central command is possible.

Paraplegic and quadriplegic patients might have a compromised cardiovascular system due to years of disuse. There is also the potential complication of cardiovascular disease. Therefore, this study indicates that a rest interval of at least 10 seconds between contractions would minimize the blood pressure changes during exercise. However, the associated reduction of workload on the heart will also have the disadvantage of reducing any cardiovascular training. A safe approach might be to use 3-second intervals between contractions. This will result in some increase of blood pressure to stress the heart, but not the marked increase seen with shorter intercontraction intervals.

References

Astrand, P.O., and Rodahl, K.: *Textbook of Work Physiology*, 2nd ed. McGraw-Hill, New York, 1977.

Barcroft, H., and Millen, J.L.E. The blood flow through muscle during sustained contractions. *J. Physiol.* 97:17–31, 1939.

Brown, M.D., Cotter, M.A., Hudlicka, O., and Vrbova, G. The effects of different patterns of muscle activity on capillary density, mechanical properties and structure of slow and fast rabbit muscles. *Pfluegers Arch.* 361:241–248, 1976.

Coote, J.H., Hilton, S.M., and Perez-Gonzalez, J.F. The reflex nature of the pressor response to muscular exercise. *J. Physiol.* 215:789–804, 1971.

Corbett, J.L., Frankel, H.L., and Harris, P.J. Cardiovascular changes associated with skeletal muscle spasms in tetraplegic man. *J. Physiol.* 215:381–387, 1971.

Freyschuss, U. Elicitation of heart rate and blood pressure increase on muscle contraction. *J. Appl. Physiol.* 28:758–761, 1970.

Guttmann, L. *Spinal Cord Injuries: Comprehensive Management and Research*, 2nd ed. Blackwell Scientific Publications, Oxford, 1976.

Holle, J., Frey, M., Gruber, H., Kern, H., Stohr, H., and Thoma, H. Functional electrostimulation of paraplegics. *Orthopedics*, 7:1145–1155, 1984.

Hudlicka, O., Brown, M., Cotter, M., Smith, M., and Vrbova, G. The effect of long-term stimulation of fast muscles on their blood flow, metabolism and ability to withstand fatigue. *Pfluegers Arch.* 369:141–147, 1977.

Lind, A.R., McNicol, G.W., and Donald, K.W. Circulatory adjustments to sustained muscular activity. *Physical Activity in Health and Disease.* Williams & Wilkins, Baltimore, 1966.

Malezic, M., Stanic, U., Kljajic, M., Acimovic, R., Krajnikj, Gros, N., and Stopac, M. Multichannel electrical stimulation of gait in motor disabled patients. *Orthopedics*, 7:1187–1195, 1984.

Marsolais, E.B., and Kobetic, R. Functional walking of paralyzed patients by means of electrical stimulation. In *Proceedings of the Fifth Annual Conference on Rehabilitation Engineering*, Houston, 1982.

McCloskey, D.I., and Mitchell, J.H. Reflex cardiovascular and respiratory responses originating in exercising muscle. *J. Physiol.* 224:173–186, 1972.

Muller, E.A. Das Arbeitsmaximum bei statischer Haltearbeit. Arbeitsphysiologie, 5:605, 1932.

Peckham, P.H., Mortimer, J.T., and Van Der Muelen, J.P. Physiologic and metabolic changes in white muscle of cat following induced exercise. *Brain Res.* 50:424, 1973.

Peckham, P.H., Mortimer, J.T., and Marsolais, E.B. Alteration in the force and fatigability of skeletal muscle in quadriplegic humans following exercise induced by chronic electrical stimulation. *Clin. Orthop.* 114:326, 1976.

Petrofsky, J.S., and Phillips, C.A. Constant velocity contractions in skeletal muscle of the cat. *Med. Biol. Eng. Comput.* 17:583–592, 1979.

Petrofsky, J.S., and Phillips, C.A. Computer controlled walking in the neurological paralyzed individual. *J. Neurol. Orthop. Surg.* 4:153–164, 1983a.

Petrofsky, J.S., and Phillips, C.A. Active physical therapy—a modern approach to rehabilitation therapy. *J. Neurol. Orthop. Surg.* 4:165–173, 1983b.

Petrofsky, J.S., Rinehart, J.S., and Lind, A.R. Isometric strength and endurance in slow and fast muscles in the cat. *Fed. Proc.* 35:291, 1976.

Petrofsky, J.S., Phillips, C.A., Sawka, M., Hanpeter, D., and Weber, C. Mechanical, electrical and biochemical correlates of isometric fatigue in the cat. *Adv. Physiol. Sci.* 18:229–236, 1980.

Petrofsky, J.S., Phillips, C.A., Sawka, M., Hanpeter D., and Stafford, D. Blood flow and metabolism during isometric contractions in cat skeletal muscles. *J. Appl. Physiol.* 50:493–502, 1981.

Petrofsky, J.S., Heaton, H.H., Phillips, C.A., and Glaser, R.M. Modification of the Apple computer as an exercise device for paralyzed muscle. *Coll. Microcomput.* 1:97–104, 1983a.

Petrofsky, J.S., Phillips, C.A., and Heaton, H.H. Computer control of walking in paraplegic subjects. *IEEE NAECON Record*, 2:1162–1171, 1983b.

Petrofsky, J.S., Danset, P., and Phillips, C.A. Closed-loop control of skeletal muscle. *Proceedings of the Third International Conference on System Engineering*, paper WA3.2, pp. 213–221, 1984a.

Petrofsky, J.S., Heaton, H.H., and Phillips, C.A. Leg exerciser for training paralyzed muscle by closed-loop control. *Med Biol Eng. Comput.* 22:298–303, 1984b.

Petrofsky, J.S., Phillips, C.A., and Heaton, H.H.: Feedback control system for walking in man. *Comput. Biol. Med*. 14:135–149, 1984c.

Pette, D., Smith, M.E., Staudte, H.W., and Vrbova, G. Effects of long-term electrical stimulation on some contractile and metabolic characteristics of fast rabbit muscles. *Pfluegers Arch*. 338:257–272, 1972.

Pette, D., Ramirez, B. and Mueller, W. et al. Influence of the intermittent long-term stimulation on contractile, histochemical and metabolic properties of fibre population in fast and slow rabbit muscles. *Pfluegers Arch*. 361:1–7, 1975.

Phillips, C.A., and Petrofsky, J.S. Computer controlled movement of paralyzed muscle: The medical perspective. *Proceedings of the First Vienna Workshop on Functional Electrostimulation*, paper 5.2, 1983.

Phillips, C.A., Petrofsky, J.S., Hendershot, D.M., and Stafford, D. Functional electrical exercise: A comprehensive approach for physical conditioning for the spinal cord paralyzed individual. *Orthopedics*, 7:1112–1123, 1984.

Rack, P.M.H., and Westbury, D.R. The effects of length and stimulus rate on tension in the isometric cat soleus muscle. *J. Physiol*. 204:443–460, 1969.

Riley, D.A., and Allin, E.F. The effects of inactivity, programmed stimulation, and denervation of the histochemistry of skeletal muscle fiber types. *Exp. Neurol*. 40:391, 1973.

Rohmert, W. Rechts–Links Vergleich bei isometrischem Arm Muskeltraining. *Int. Z. Angew, Physiol*. 26:363–370, 1968.

Salmons, S., and Vrbova, G. The influence of activity on some contractile characteristics of mammalian fast and slow muscles. *J. Physiol*. 201:535, 1969.

Simonson, E. *Physiology of Work Capacity and Fatigue*. Charles C. Thomas, Springfield, IL, 1971.

Vrbova, G. The effect of motor neurone activity on the speed of contraction of striate muscle. *J. Physiol*. 169:513–526, 1963.

6
Acute Effects with the Exercise Bicycle System: The Physiological Process

I. Introduction to the Exercise Bicycle System*

The clinical application of FES exercise cycling recently has become commercially available through a company, T.T.I.† This section describes the basic apparatus which may be used for both isokinetic leg training (phase I) and exercise bicycle ergometry (phase II). The clinical procedures for each phase (as currently recommended by the manufacturer) are also presented.

A. Apparatus

The computer-controlled FES exercise ergometer is the REGYS I Clinical Rehabilitation System (Fig. 6.1). It is comprised of three primary subsystems: the lower extremities ergometer, the stimulus control unit, and the patient chair. The ergometer is a standard Monarch ergometer used for nonparalyzed individuals, extensively modified to reduce resistance levels and to provide resistance and position sensors for patient safety. The stimulus control unit is a computer that controls and monitors the electrical stimulation according to prescribed parameters entered into a microprocessor by a remote control keyboard.

Six separate channels for sequential surface muscle stimulation are used during ergometry with a computer-controlled closed-loop system. Each channel supplies a monophasic current pulse to each of two active electrodes at a 30 Hz rate. Because the stimulus delivered to the active electrodes is sequential and 180° out of phase, the effective stimulation frequency at the single reference electrode is 60 Hz. Each stimulus pulse has a duration of 350 μs and an amplitude that is determined by the system's microprocessor.

The amplitude determination is based on performance inputs that are read from various sensors. The amplitude commanded by the microprocessor can range from zero to 132 mA. Constant current output circuitry maintains the desired amplitude unless a voltage of 220 V or an impedance of 1666 Ω is reached. At that point, the computer automatically disables the stimulation, based on the assumption that an electrode or wire has pulled loose.

Carbon-filled silastic electrodes are used to conduct the current to the skin area over the muscle group to be stimulated. Three electrodes (two active and one reference) are applied over each muscle group. On the quadriceps, electrodes measuring 4.5 by 10 cm are used. On the hamstring and gluteal muscles, electrodes measuring 4.5 by 4 cm are used. Each electrode is prepared by coating with a buffered electrode gel that provides a conductive interface between the electrode and skin.

A pedal position sensor is used to continuously monitor computed average velocity and instantaneous position, in order for the computer to control

---

*From K.T. Ragnarsson et al., Clinical evaluation of computerized functional electrical stimulation after spinal cord injury: A multicenter pilot study. Reprinted with permission from *Archives of Physical and Medical Rehabilitation*, 69:672–677, 1988. Copyright 1988, Editorial Board.

†Therapeutic Technologies Incorporated, 5709 Johns Road, Tampa, Florida 33634.

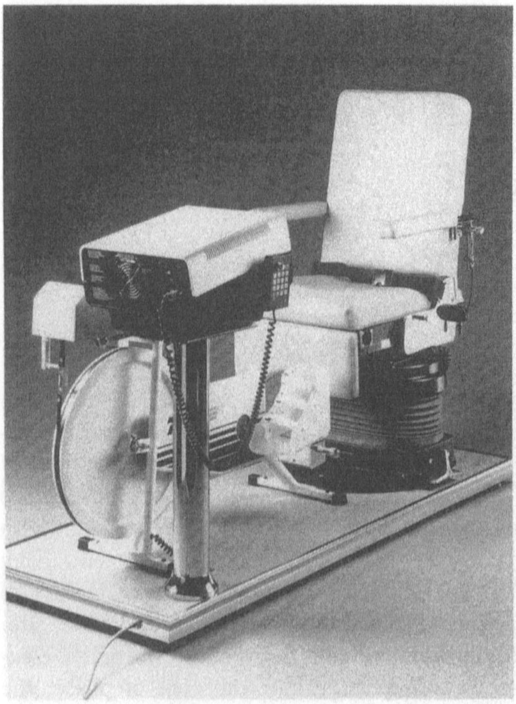

FIGURE 6.1. Computer-controlled FES exercise ergometer (REGYS I Clinical Rehabilitation System).

the exact stimulus amplitude required for each of the six muscle groups to allow the patient to maintain a constant speed of 50 rpm. Continuous feedback to the computer of pedal position allows the control of the rate of pedaling and stimulation sequence to achieve a smooth, rhythmic motion.

The resistance sensor is used to guarantee that the patient is subjected only to the prescribed resistance. Pedaling resistance levels for ergometers are typically quantified in units known as kiloponds (kp), a measure of the drag force applied tangential to the flywheel [1 kp is the force acting on the mass of 1 kg (2.2 lb) at normal acceleration of gravity; 1 kp = 9.80665 newtons, a unit of force]. Whereas conventional exercise bikes usually offer resistances of up to 10 kp, much lower resistances are used in therapy for SCI patients (i.e., from zero to 7/8 kp, in 1/8 kp increments).

The patient chair has a high back and is adjustable in height and seat depth in order to provide optimal firing angles, that is, the pedal position at which the leg can apply a smooth, efficient force to the crank as a result of stimulation applied to that muscle group. It also provides for rapid reclination of the seat to the horizontal (level) position. The chair is equipped with shoulder and lap belts to prevent sliding. Leg restraints are attached to the patient chair to allow pedaling only in a planar motion.

A computer-controlled FES home exercise bicycle ergometer is also available (ERGYS Home Rehabilitation System). This is shown in Figure 6.2.

B. Procedures

Phase I consists of 4 weeks of quadriceps training three times a week. Three surface electrodes were attached to the anterior thigh. Each session was terminated after 45 completed lifts, or when the patient became fatigued. Fatigue was defined as the inability to complete at least 50% of the lifts in a 45° arc of motion (from 90° of knee flexion to 45°). Each leg lift lasted no more than 8 seconds. There was a 14-second rest period between lifts.* Each subject began by attempting to lift a one-pound ankle cuff weight. Resistance was increased by the incremental application of two-pound ankle cuff weights once the subject had been able to successfully complete 45 lifts. A subject's inability to complete at least 30 lifts resulted in a decrease in ankle cuff weights by one pound. Immediately before and after each session, heart rate (HR), blood pressure (BP), and temperature (T) were measured. The number of completed lifts and resistance applied were documented for each session.

Phase II consisted of 12 weeks of lower extremity ergometry. Superficial electrodes were attached over the quadriceps, hamstring, and gluteus maximus muscles. There were three sessions per week, with each session consisting of one or two pedaling runs of up to 15 minutes each. If a subject successfully completed a 15-minute run, a second run was not given on that day. However, the resistance was increased by 1/8 kp after three consecutive sessions of 15-minute runs. If a subject could not complete a 15-minute run, a 5-minute rest period was given,

Author's (C.A.P.) note: For example, the right leg would perform an 8-second lift cycle. After a 3-second pause, the left leg would perform an 8-second lift cycle. After another 3-second pause, the right leg would start another lift cycle, etc. Therefore, each leg had a 14-second rest period between lifts.

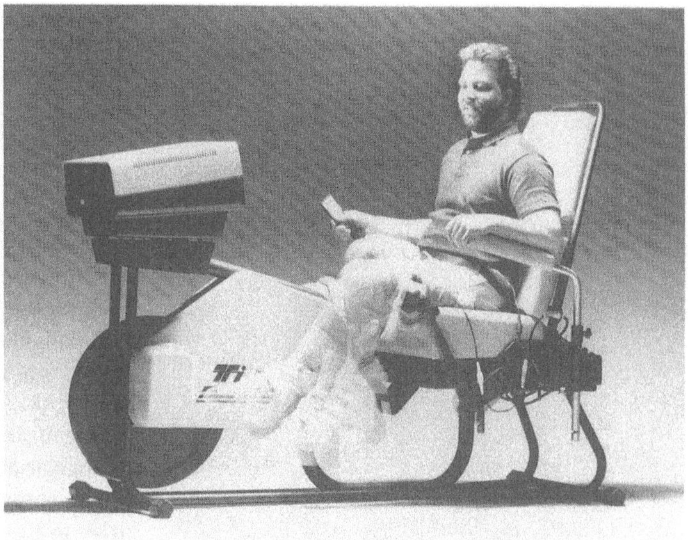

FIGURE 6.2. Computer-controlled FES exercise ergometer (ERGYS Home Rehabilitation System).

the resistance decreased if possible, and a second run of 15 minutes (or less) was given. Each run was halted if fatigue (inability to maintain pedaling speed of 35 rpm) was noted. Resistance was variable, depending on an individual subject's completed run time. Immediately before, during, and immediately after each session, HR, BP, and T were recorded. Length of each run and resistance were documented for each session.

II. The Cardiorespiratory Stress Effects During Dynamic Exercise*

A. Introduction

Following spinal cord injury, muscle atrophies rapidly, bone demineralizes, and there is a gradual reduction in the dynamics of the cardiorespiratory system due to disuse as well as from damage to the autonomic nervous system caused by the injury (Guttmann, 1976). For many quadriplegics, blood pressure at rest is low and orthostatic intolerance is commonly found (Guttmann, 1976). Throughout life, cardiovascular deconditioning progresses in many wheelchair-bound individuals. This problem, when accompanied by osteoporosis and atrophy of muscle, presents secondary medical complications. For wheelchair-bound individuals this is as important and probably more clinically significant than the paralysis itself. Upper body exercise has been used in paraplegics to recondition the cardiovascular system, but because of the degree of paralysis in most quadriplegics, voluntary exercise is difficult.

FES is defined as the use of electrical stimulation to move paralyzed limbs for practical tasks. These electrical impulses can be applied through the skin (Petrofsky et al., 1983a; Petrofsky and Phillips, 1984) with deep muscle (Peckham et al., 1976; Marsolais and Kobetic, 1982) or nerve (Holle et al., 1984) electrodes. Whatever the source, electrical stimulation can be used to return some degree of movement to paralyzed muscle. With the development of closed-loop control technology, movement in muscle can now be controlled more precisely (Petrofsky, 1978, 1979; Petrofsky and Phillips, 1979a).

By sequentially activating muscle (Rack and Westbury, 1969; Peckham et al., 1976; Petrofsky et al., 1976; Petrofsky, 1978) smooth contractions have been developed. Stimulation frequencies well within the physiological range have been used to minimize fatigue of the muscles (as seen in voluntary recruitment). With both open-loop and

*From J.S. Petrofsky et al: Cardiorespiratory stresses which occur during dynamic exercise in paraplegics and quadriplegics. Reprinted with permission from *Journal of Neurological and Orthopedic Surgery*, 6:252–258, 1985. Copyright 1985, American Academy of Neurologic and Orthopedic Surgeons.

FIGURE 6.3. The bicycle ergometer.

closed-loop control, electrical stimulation has been used to train skeletal muscle (Salmons and Vrbova, 1969; Peckham et al., 1973; Pette et al., 1975; Brown et al., 1976; Hudlicka et al., 1977). However, little has been done to quantify the effect of electrical stimulation on other body systems such as the cardiorespiratory system. Therefore, quantification of the cardiorespiratory responses of paralyzed muscle to aerobic exercise is very important. The purpose of the present investigation was to accomplish this objective.

B. Subjects

The subjects of this study were four quadriplegic (C5–C7), four paraplegic (T8–T11), and four control subjects without spinal cord injury. Except for the paralysis of the paraplegic and quadriplegic subjects, all subjects were diagnosed as being free of any major medical problems, including cardiovascular or renal diseases, and were not taking any medication throughout the experiments. All subjects were informed of all experimental procedures and signed a consent form. All procedures were approved by Wright State University's Institutional Review Board.

C. Methods

Dynamic exercise was accomplished on a modified Monark bicycle ergometer, which had been modified and adapted to be used with a computer-controlled stimulator (Fig. 6.3). A special high-back seat with seat belt and shoulder harness was designed to provide postural support for the paraplegic and quadriplegic subjects. In addition, a linear potentiometer was connected by a chain drive to the pedals of the bicycle so the computer would be provided with a means of measuring the pedal position. The details of this bicycle and computer control system are described elsewhere (Petrofsky and Phillips, 1983; Petrofsky et al., 1984b).

Using closed-loop control, the computer stimulated the quadriceps and iliacus muscles to cause subjects to pedal the bicycle ergometer smoothly at a speed of 40 rpm. The frequency of sequential stimulation of the muscle was 40 Hz, while the pulse width was always 300 μs. Stimulation at a frequency of 40 Hz has been shown to be sufficient to tetanized paralyzed muscle with sequential stimulation (Rack and Westbury, 1969; Petrofsky, 1978, 1979). Three carbonized rubber electrodes (Medtronics Corp.) were placed diagonally across the muscles. Sequential stimulation involved alternately stimulating the outside two electrodes with a biphasic square wave using the inside electrode as a reference. Two alternating channels are the minimum number of channels of stimulation required to elicit the development of smooth contractions in skeletal muscle by electric stimulation at stimulation frequencies of 40 Hz (Rack and Westbury, 1969). This technique also minimizes muscle fatigue (Petrofsky and Phillips, 1979b).

1. Blood Pressure and Heart Rate

Heart rate was measured from a continuous recording of the ECG. QRS complexes were counted for 15-second intervals just before, at the onset of, and at 20, 40, 60, 80, and 100% of the duration of the exercise. Blood pressure was measured by auscultation of the arm. Blood pressure was taken as often as possible during the exercise and at 30-second intervals during the postexercise period. Blood pressure was then interpolated or extrapolated to obtain blood pressure at 20, 40, 60, 80, and 100% of the duration of the exercise.

2. *Cardiorespiratory Measurements*

Oxygen uptake, ventilation, and carbon dioxide production were all measured on a Gould Pulmonary Function Analyzer (model 9000 IV). The analyzer was calibrated each day and checked against standardized gases before each run. Cardiac output was measured by CO_2 rebreathing on the same analyzer.

3. *Arterialized Blood Gases*

The pO_2, pCO_2, and pH were measured from arterialized fingertip blood samples on a Radiometer Copenhagen automated blood gas analyzer (model ABL3). The subject's hands were placed in warm water (40°C) for 10 minutes prior to the sampling to ensure that the arterioles in the fingertip were fully dilated. Fingertip blood samples (100 µL) were then taken from free-flowing arterialized blood to determine pO_2, pCO_2, and pH.

4. *Procedures*

Three groups of subjects participated in these experiments. For two of the experimental groups (the paralyzed groups), subjects were stimulated by a computer to elicit pedaling of the bicycle ergometer at a speed of 50 rpm during a pretraining period. This period was necessary, since the muscles in the paralyzed subjects were so weak from years of disuse atrophy that even 4 minutes of work at 0 kp was impossible. The workload was adjusted so that the subjects could be induced to pedal for more than 10, but less than 15 minutes at a speed of 50 rpm before their muscles fatigued. For the first experiments, this required a load of 0 kp on the Monark bicycle ergometer. Subjects were exercised 3 days per week (Monday, Wednesday and Friday) for a period of 6 successive weeks. Throughout this time, if subjects were able to complete the 15 minutes of exercise, the workload was then increased for the next experimental day by 1/8 kp.

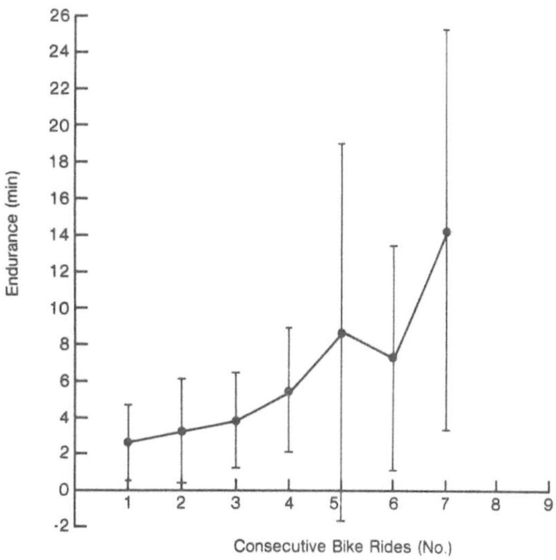

FIGURE 6.4. Endurance of the paralyzed patients.

When the 6-week training period was over, nonparalyzed and paralyzed subjects pedaled at similar workloads, during which time cardiorespiratory measurements were made. For these posttraining measurements, a modified Balke test was performed. All paralyzed subjects were initially stimulated to pedal the bicycle ergometer during a 4-minute period of work at 0 kp. The workload was then increased in 1/8 kp steps in successive 4-minute periods until a 4-minute period of exercise could not be completed; 15 minutes was allowed between successive work periods. Nonparalyzed subjects worked at these same loads with the same modified Balke procedure (Petrofsky et al., 1976) but pedaled voluntarily. At each load, during the third and fourth minute, oxygen uptake, cardiac output, stroke volume, and ventilation were measured. At the end of the 4-minute period, a fingertip arterialized blood sample was taken for the determination of pO_2, pCO_2, and pH.

D. Results

Endurance increased progressively throughout the training period. Figure 6.4 shows the increase in endurance for the eight subjects pedaling the bicycle at 0 kp. This figure shows that the endurance increased rapidly after just seven sessions (over 2 weeks). Generally, the eight paralyzed subjects could pedal the bicycle ergometer for only a few

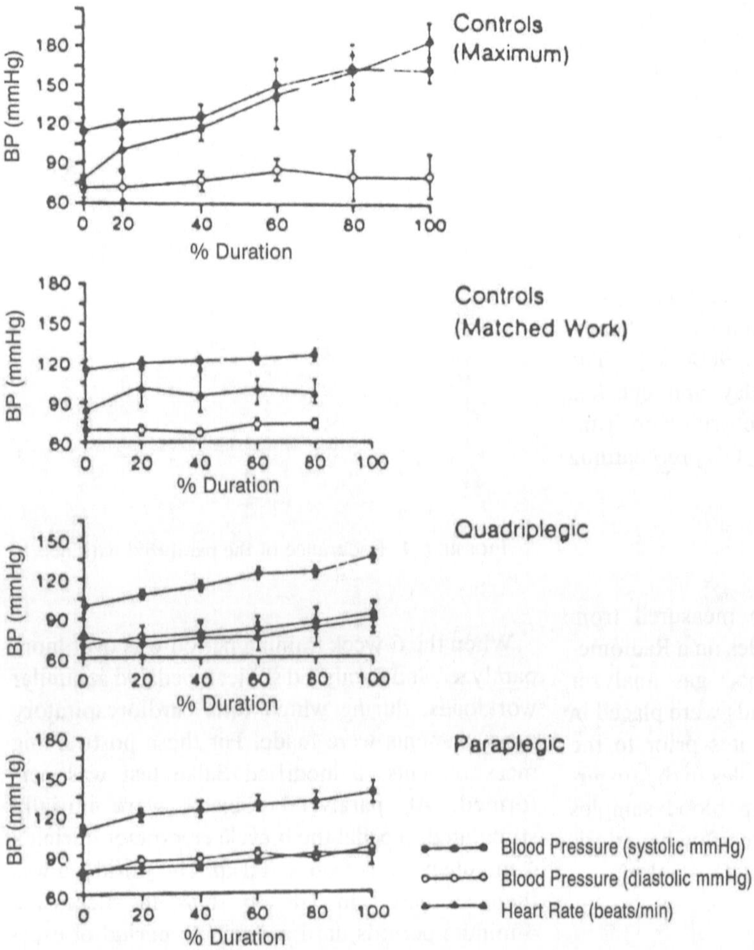

FIGURE 6.5. Heart rate and blood pressure responses.

seconds during the first training session at a load of 0 kp. However, after the 6-week training period, the endurance increased more than 10 times in initial value, and subjects were able to pedal the bicycle ergometer easily at loads above 1/8 kp. At the end of the 6-week training period, the subjects participated in a modified Balke stress test to measure their cardiorespiratory responses during dynamic exercise and to compare it to that of the control.

Heart rate and blood pressure increased in all subjects during exercise. The increases in blood pressure and heart rate were roughly proportional to the workload for each of the 4-minute periods. There were, however, noticeable differences between groups of subjects in the magnitude of the responses. Figure 6.5 shows the average increase in heart rate and blood pressure during the most fatiguing period of work. This figure shows the upper limit of these responses. Although blood pressure and heart rate increased in all subjects throughout the duration of the exercise, blood pressure showed a much smaller increase in the paraplegic and quadriplegic subjects than in the control subjects ($p < 0.01$). When control subjects did maximal work (the same relative workload as the paralyzed groups), both heart rate and blood pressure rose substantially. Maximum blood pressure and heart rate of controls compared to the other two groups were significantly different ($p < 0.01$) (Fig. 6.5).

Cardiac output increased in all subjects during the exercise (Fig. 6.6). The increase in cardiac output was similar for both groups of paralyzed subjects and was roughly proportional to workload. Further, the increase in cardiac output was not

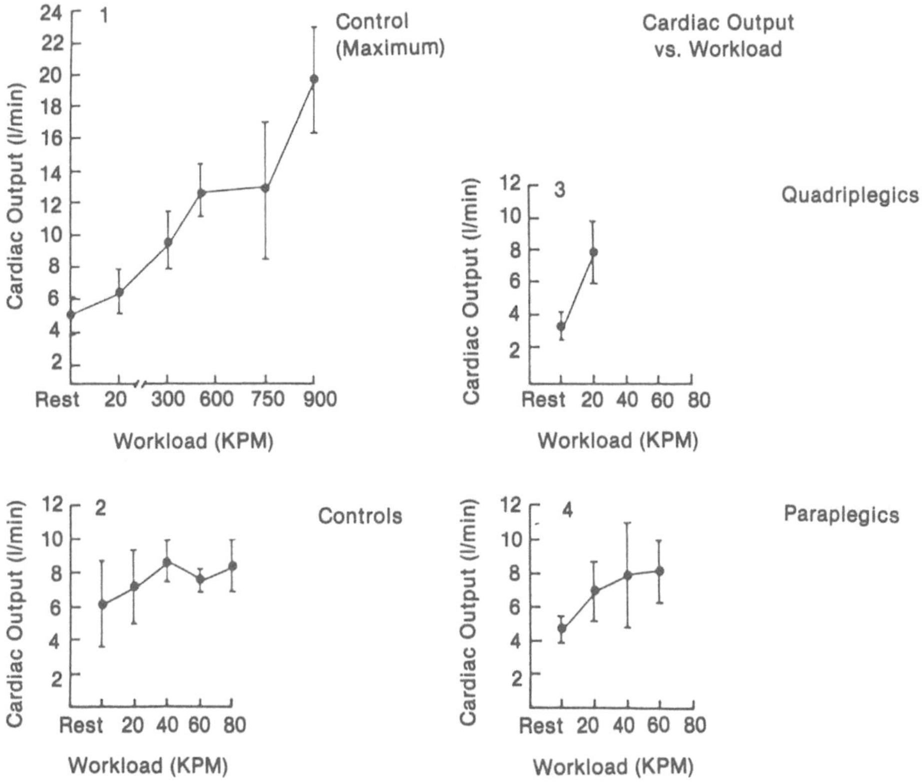

FIGURE 6.6. Cardiac output in the paraplegics and quadriplegics.

statistically different from that of the control group during exercise at the same load. Therefore, for a given absolute level of exercise, there was no significant difference in cardiac output in all subjects ($p > 0.05$). However, when the increases in cardiac output due to fully fatiguing workloads were compared, the paraplegic and quadriplegic subjects showed a significantly smaller increase than control subjects ($p < 0.01$). The mechanical efficiency of the exercise remained fairly constant in all groups of subjects, averaging 23% with no statistical difference among the groups ($p > 0.05$).

Ventilation and oxygen uptake increased in proportion to the workload for all groups of subjects (Fig. 6.7). For a given workload, there was no statistical difference between the groups of subjects in terms of oxygen uptake or ventilation, but there was a significant difference between the two paralyzed groups of subjects in the manner in which the increase in ventilation was achieved. In the paraplegic subjects, ventilation was increased by increasing tidal volume, while for the quadriplegic subjects, it was achieved by an increase in respiratory rate. For example, at the end of the 4-minute period of heaviest exercise, the respiratory rate and tidal volumes of the paraplegic subjects were averaged 32 breaths per minute with 111 mL of air per breath, while the quadriplegics averaged 42 and 85, respectively.

Arterialized blood gases showed little change throughout the exercise. At various workloads, arterialized blood gases remained fairly constant throughout exercise, even during maximal work (for the paralyzed subjects).

E. Discussion

In previous papers by ourselves (Petrofsky et al., 1983b, 1984a; Phillips et al., 1984), it has been shown that electrically induced exercise involving weight training can result in a large increase in muscle strength. Here those earlier experiments

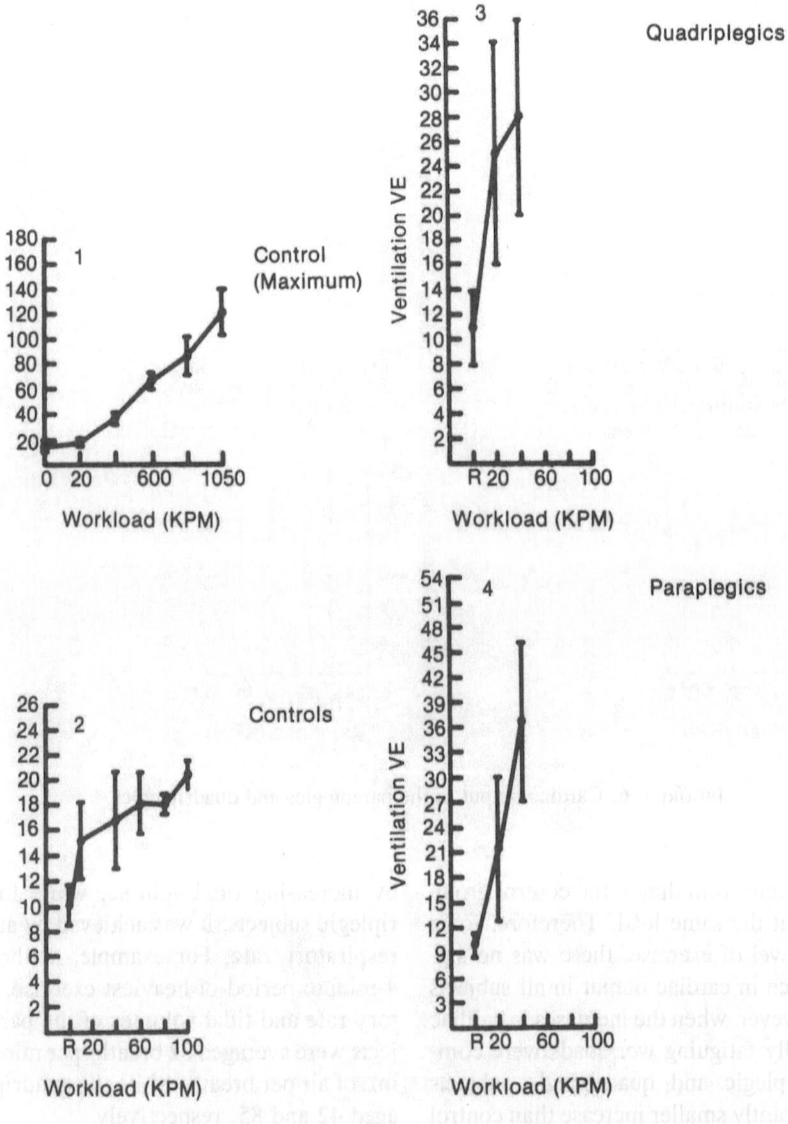

FIGURE 6.7. Ventilatory response in the paraplegics and quadriplegics.

have been expanded to look at the effect of aerobic exercise on the cardiorespiratory system during dynamic exercise. Generally, the responses of paraplegic and quadriplegic subjects compared quite favorably with those of controls, both in the magnitude and the direction of the response. However, some differences must be noted.

At a given absolute workload, the blood pressure and heart rate responses were similar in paraplegic and quadriplegic subjects. The paralyzed group of subjects had a significantly lower heart rate and blood pressure response at a given workload than did the control subjects. The heart rate and blood pressure responses of the paralyzed subjects were also less than those of control subjects doing maximal work. Therefore, it would appear that the relative stress at a given workload might be higher for paralyzed than for nonparalyzed subjects. Certainly some evidence supporting this theory is presented, since there was a large increase in ventilation associated with exercise for a given workload in the paralyzed subjects.

Cardiac output, a measure of the severity of the work, appeared to be largely dependent upon the degree of the work. Cardiac output was not significantly different in paraplegic, quadriplegic, and control subjects at a given level of work. This evidence (combined with the arterial blood gases during maximum work) indicates that the energy cost to the body was about the same with respect to cardiac output and mechanical efficiency of the exercise (calculated from the oxygen uptake).

These physiological stresses, however, may be related not so much to the level of work but to the degree of fatigue of the muscle. It has been well established that muscle fatigue, at least during isometric exercise, causes an increase in blood pressure which is independent of the level of the exercise (Lind et al., 1964). For example, isometric contractions sustained to fatigue at 20% of the maximum voluntary strength have been shown to elicit the same blood pressure responses as isometric contractions sustained at 70% of the muscle's maximum strength. The important issue is muscle fatigue. For the paralyzed subjects, 4 minutes of work at a maximal dynamic workload totally fatigued the muscle. For the control subjects, this workload was not fatiguing.

If, in fact, the paralyzed muscles were contracting or exercising at a higher relative workload, the production of metabolites would be greater than that of the control subjects working at the same absolute workload. Coote et al. (1971), and later McCloskey and Mitchell (1972), showed that a local blood pressure raising reflex arises from the dorsal tracts of the spinal cord associated with contraction of the skeletal muscle.

The pathways associated with this reflex appear to arise from metabolites accumulating in the fatiguing muscle and to elicit trains of impulses along the small, unmyelinated nerve fibers in the dorsal tracts returning to the spinal cord. These afferent signals then cause a reflex increase in splanchnic constriction, which, in turn, increases both systolic and diastolic blood pressure. While this mechanism may work during static or isokinetic exercise, it does not appear to be that important in dynamic exercise. During dynamic exercise, the increase in systolic blood pressure is usually attributed to an increase in cardiac output, while an increase in diastolic pressure is attributed to an increase in peripheral resistance (Astrand and Rodahl, 1977). Therefore, it is not surprising that blood pressure was shown to increase during exercise in paraplegics in the present investigation. In previous studies (Petrofsky and Phillips, 1984; Phillips et al., 1984), blood pressure did not increase in paraplegics during isokinetic exercise, presumably due to disruption of the peripherally mediated reflexes caused by the spinal cord injury. But here, where blood pressure was driven by a more centrally mediated mechanism, blood pressure response to exercise was similar in all three groups of subjects.

The experiments presented here have demonstrated the feasibility of using electrically induced dynamic exercise for physical training. The changes in endurance as seen during the 6-week training period were significant. However, very little was done to optimize the exercise protocols for building strength and endurance. The present protocol resulted in a large increase in endurance, and it is certainly possible that less exercise would result in a similar increase. Varying the exercise protocols (i.e., the length or intensity of the work) may result in more efficient protocols for building strength and endurance.

However, muscle endurance may not be the most significant factor in determining an optimal exercise protocol. Other parameters, such as the effect of this type of exercise on the heart, lungs, and bone, may be equally important in any study of exercise on a computer-controlled bicycle ergometer. The increase in cardiac output seen with this type of exertion and its relative safety makes this form of exercise worth exploring further. Certainly, there is a potential for reconditioning muscle and bone and for restoring cardiovascular fitness with this type of technology. However, these studies must be conducted on a substantially larger group of subjects before any definitive conclusions can be reached. The present series of experiments must be reviewed as only a pointer for what may be a practical and useful tool for rehabilitation.

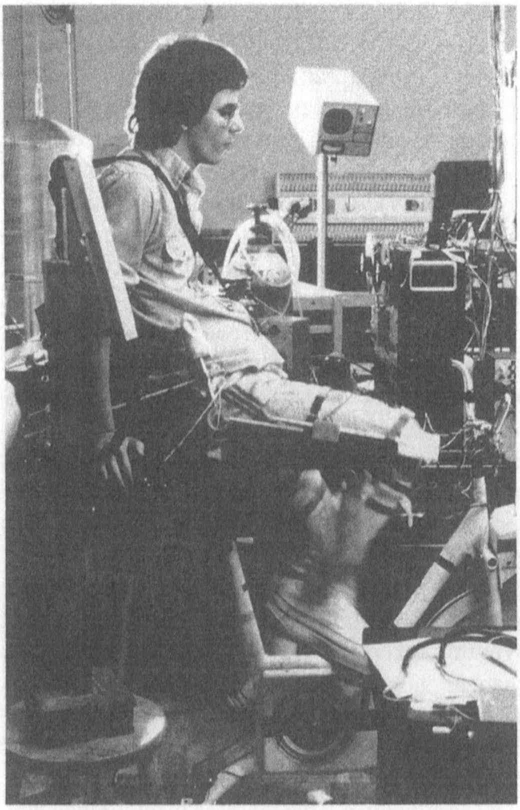

FIGURE 6.8. Subject and the exercise bicycle ergometer.

III. Functional Electrical Exercise with the Exercise Bicycle System*

A. Exercise Bicycle Ergometer

One common form of dynamic exercise used in physical fitness training is bicycling. For the paraplegic or quadriplegic subject, bicycling offers the advantage of supporting the body weight while allowing the muscles to exercise in a manner similar to what would occur during normal voluntary walking. Computer-controlled bicycling is accomplished by a modified Monark exercise bicycle ergometer system (Petrofsky et al., 1984b) shown in Figure 6.8. The bicycle ergometer was modified by a pedal sensor that indicated the position of the pedals. The bicycle ergometer was also modified by a high-back seat, shoulder harness, and seat belt to allow subjects to maintain balance. This is especially critical for quadriplegic subjects, who have no discernible upper body balance. To operate the system, an Apple II microcomputer was again utilized and modified with a hardware interface. This required that four channels of stimulus control be built similar to the leg trainer (Fig. 6.9). These four channels were controlled through four D/A converters (i.e., a DAC0831 solid-state, eight-bit D/A converter chip). The interfaces were identical except that four discrete addresses were decoded from the Apple address bus. Data inputs were provided from a single A/D converter. Additionally, Apple game paddles were used as analog inputs. Because of the use of multiple muscles, it was possible to develop a high level of force on the legs and joints during pedaling. In order to protect both muscle and bone, only the quadriceps and iliacus muscles of each leg were used. The timing was set assuming that the very top rotation of the pedal was a 0° rotation. For the first 90° of rotation of each leg, the appropriate quadriceps muscle was activated. For the last 180° of rotation, the appropriate iliacus muscle was activated. By adjusting the stimulus voltage on these two muscles (on each leg), pedaling could be sustained on a standard Monark exercise bicycle. The exercise bicycle has recently been modified to serve as a recreational device as part of a comprehensive APT program (Petrofsky et al., 1983a).

B. Subjects and Protocols

Six spinal-cord-injured subjects were used in these studies. Four subjects were quadriplegic with lesions at C5, C6, or C7. Two subjects were paraplegic with lesions at T9, T10, or T11. The subjects were all volunteer, undergraduate university students, divided equally between male and female, with ages ranging from 18 to 35 years. All subjects were free of significant secondary medical diseases and underwent a complete medical history, physical examination, and laboratory testing (including an ECG and full X-rays of both lower legs and upper arms). All subjects were informed of all experimental procedures and signed a duly witnessed informed consent. Care was taken to ensure

*From C.A. Phillips et al., Functional electrical exercise: A comprehensive approach for physical conditioning of the spinal cord paralyzed individual. Reprinted with permission from *Orthopedics*, 7:1112–1123, 1984. Copyright 1984, SLACK, Incorporated.

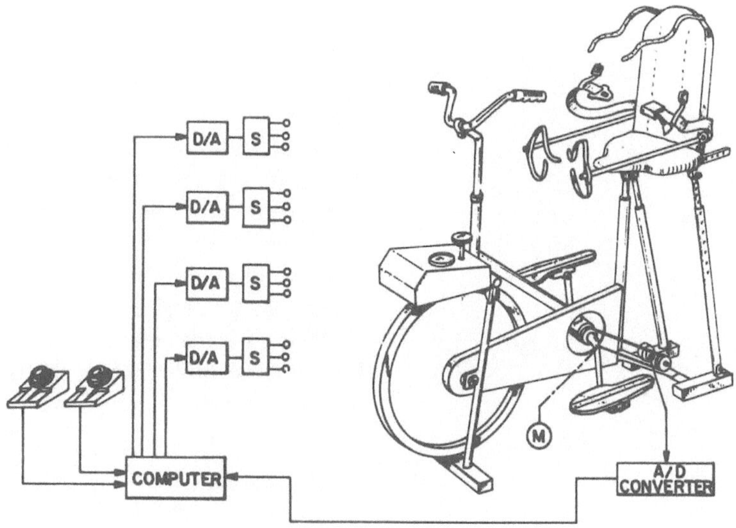

FIGURE 6.9. Operational schematic of the exercise bicycle ergometer (see text).

subject safety, privacy, and confidentiality. All protocols were fully approved by Wright State University's Institutional Review Board.

The cardiopulmonary response protocol involved six SCI patients (four quadriplegic and two paraplegic). These subjects were not trained. Preexercise data were acquired (see Section II, D, "Cardiopulmonary Measurements," of Chapter 5). The subject then performed a 15-minute fatiguing contraction under voluntary control using a standard arm crank ergometer (Monark model 881) at a rate of 50 rpm. Alternately, the subject would perform a 15-minute fatiguing exercise under FES control of the lower limbs using the exercise bicycle ergometer. The load on each device would be appropriately set to ensure that a fatiguing exercise occurred within a 15-minute period. Postexercise cardiopulmonary measurements were then collected immediately at the fatigue end point.

The lactic acid–blood gas protocols involved six SCI patients (four quadriplegics and two paraplegics). The subjects were not trained, and all quadriplegic data were acquired at the beginning of their 4-week training period. Preexercise data were acquired as before. The subject then performed a 15-minute fatiguing exercise under FES control of the lower limbs using the exercise bicycle ergometer. Appropriate postexercise cardiopulmonary measurements were then collected immediately at the fatigue end point.

C. Results

1. *Exercise Bicycle Ergometer*

For the two paraplegic subjects, using FES and the exercise bicycle versus voluntary arm cranking, BP and HR increased moderately with the bicycle but markedly with arm cranking between resting values and the fatigue end point. Ventilatory equivalent ($\dot{V}_E$) increased markedly with FES and the exercise bicycle, but increased only moderately for voluntary arm cranking activity (between resting values and the fatigue end point. Oxygen consumption ($\dot{V}_{O_2}$) more than quadrupled for both types of exercise in the two paraplegic subjects (Fig. 6.10).

With the four quadriplegic subjects, BP increased markedly for both types of exercise, but HR increased more markedly for voluntary arm cranking activity compared to its increase for FES and bicycle ergometer activity (between resting and fatigue values). $\dot{V}_E$ and $\dot{V}_{O_2}$ tripled for FES and bicycle ergometer activity, but only doubled for voluntary arm cranking (between resting and fatigue values) for the four quadriplegic subjects (Fig. 6.11).

Physical training increased the workload that could be tolerated by the subjects, hence the $\dot{V}_E$ and oxygen uptake, but had no effect on the pattern of response. For all these responses, the common denominator was variability. The average responses

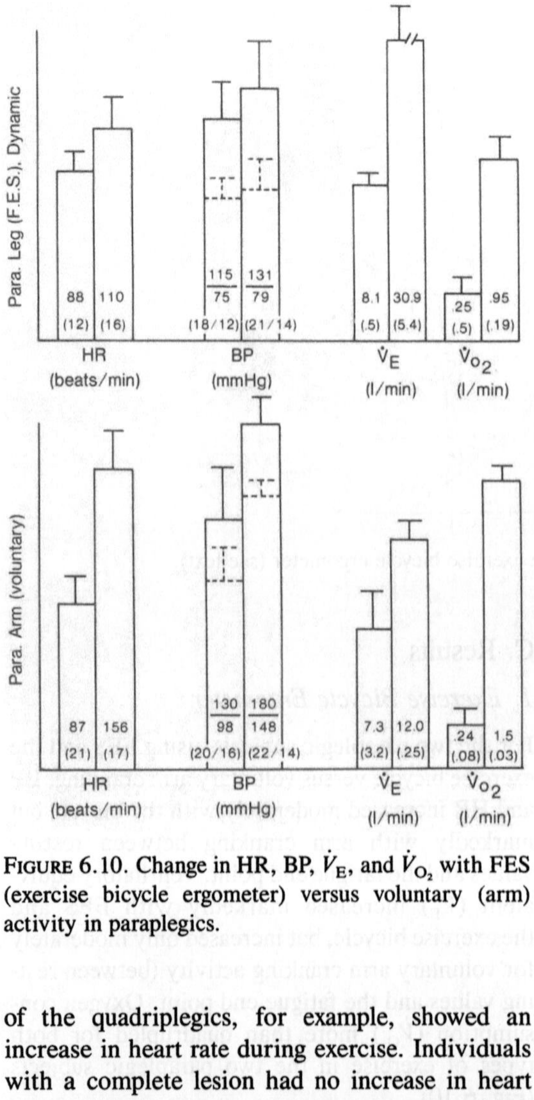

FIGURE 6.10. Change in HR, BP, $\dot{V}_E$, and $\dot{V}_{O_2}$ with FES (exercise bicycle ergometer) versus voluntary (arm) activity in paraplegics.

of the quadriplegics, for example, showed an increase in heart rate during exercise. Individuals with a complete lesion had no increase in heart rate, while the largest response was in the subject with the most incomplete injury. In a similar manner, an increase in blood pressure during leg exercise was not found in complete paraplegics, but was evident in incomplete paraplegics and quadriplegics. Ventilation and arterial blood gases showed the same variability.

2. *Lactate and Blood Gases*

For the two paraplegic untrained subjects using the exercise bicycle ergometer, lactic acid rose fivefold between rest and the fatigue end point. This was matched by an associated decrease in the blood pH. Arterialized pO_2 increased somewhat and pCO_2 remained relatively unchanged between rest and end exercise for the two paraplegic subjects (Fig. 6.12).

With the four quadriplegic subjects using FES and the exercise bicycle ergometer, LA and arterialized pH followed the same pattern as for the paraplegic subjects. Arterialized pO_2 actually fell somewhat and pCO_2 remained relatively unchanged between resting and postexercise values for the four quadriplegic subjects (Figure 6.12).

Physical training increased the workload that could be tolerated before lactic acid was produced. However, no difference in the overall pattern of response was noted.

D. Discussion

This section (and Section II of Chapter 5) indicates that APT stresses the cardiopulmonary system. It does this by different degrees, depending upon the

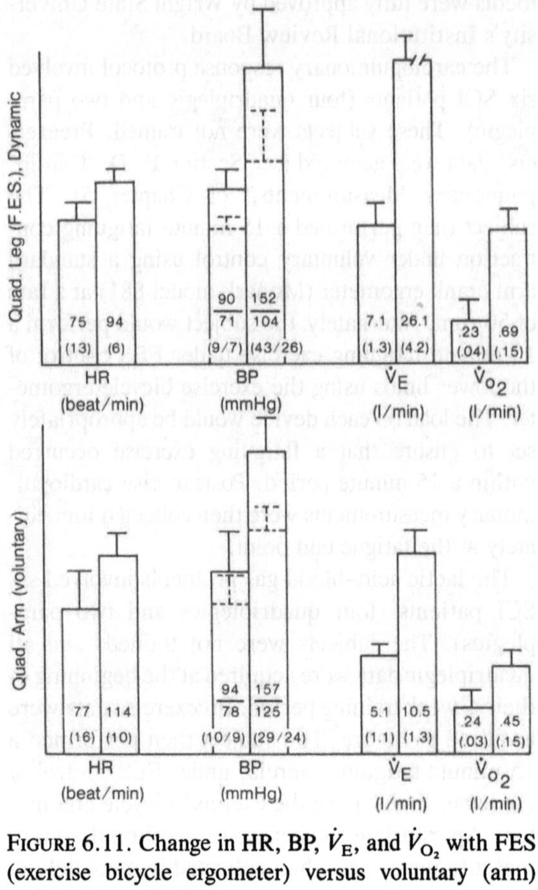

FIGURE 6.11. Change in HR, BP, $\dot{V}_E$, and $\dot{V}_{O_2}$ with FES (exercise bicycle ergometer) versus voluntary (arm) activity in quadriplegics.

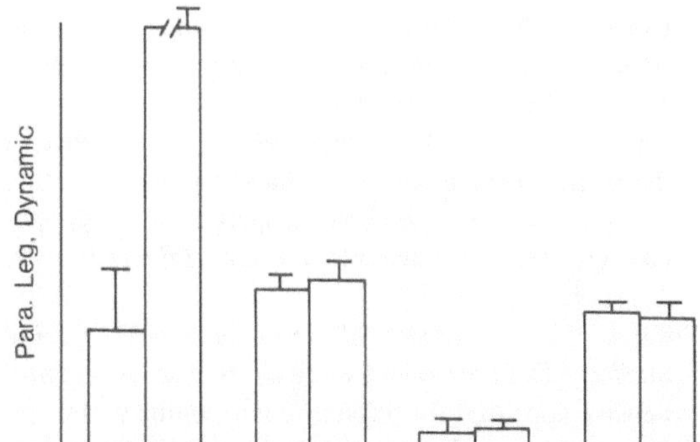

FIGURE 6.12. Changes in LA, pO_2, pCO_2, and pH with FES (exercise bicycle ergometer) in paraplegic and quadriplegic patients.

location and extent of the spinal cord injury. Further studies are now indicated in a number of directions. First, various populations need to be studied, each with a specific lesion location and extent in order to characterize lesion-specific physiological responses to APT. Second, definitive physiological mechanisms for the lesion-specific responses need to be postulated and critically tested. Third, the training effect must be evaluated. Specifically, the cardiopulmonary responses in this study have been reported for "untrained" subjects. Chapter 7 describes how cardiopulmonary responses change with long-term APT when applied to spinal-cord-injured patients.

In conclusion, a new area of scientific investigation has been identified: exercise physiology of the SCI individual. Here we have identified only the dimensions of the problem to be addressed.

References

Astrand, P.O., and Rodahl, K. *Textbook of Work Physiology*, 2nd ed. McGraw-Hill, New York, 1977.

Brown, M.D., Cotter, M.A., Hudlicka, O., and Vrbova, G. The effects of different patterns of muscle activity on capillary density, mechanical properties and structure of slow and fast rabbit muscles. *Pfluegers Arch.* 361:241–247, 1976.

Coote, J.H., Hilton, S.M., and Perez-Gonzalez, J.F. The reflex nature of the pressor response to muscular exercise. *J. Physiol.* 215:789–804, 1971.

Guttmann, L. *Spinal Cord Injuries: Comprehensive Management and Research*, 2nd ed. Blackwell Scientific Publications, Oxford, 1976.

Holle, J., Frey, M., Gruber, H., Kern, H., Stohr, H., and Thomas, H. Functional electrostimulation of paraplegics. *Orthopedics*, 7:1145–1155, 1984.

Hudlicka, O., Brown, M., Smith, M., and Vrbova, G. The effect of long-term stimulation of fast muscles on their blood flow, metabolism and ability to withstand fatigue. *Pfluegers Arch.* 369:141–144, 1977.

Lind, A.R., Taylor, S.H., Humphreys, P.W., Kennelly, B.M., and Donald, K.W. The circulatory effects of sustained voluntary muscle contraction. *Clin. Sci.* 27: 229–244, 1964.

Marsolais, E.B., and Kobetic, R. Functional walking of paralyzed patients by means of electrical stimulation. *Proceedings of the Fifth Annual Conference on Rehabilitation Engineering*, Houston, p. 61, 1982.

McCloskey, D.I., and Mitchell, J.H. Reflex cardiovascular and respiratory responses originating in exercising muscle. *J. Physiol.* 224:173–186, 1972.

Peckham, P.H., Mortimer, J.T., and Van Der Muelen, J.P. Physiological and metabolic changes in white muscle of cat following induced exercise. *Brain Res.* 50:424–429, 1973.

Peckham, P.H., Mortimer, J.T., and Marsolais, E.B. Alteration in the force and fatigability of skeletal muscle in quadriplegic humans following exercise induced by chronic electrical stimulation. *Clin. Orthop.* 114: 326–334, 1976.

Petrofsky, J.S. Control of the recruitment and firing frequencies of motor units in electrically stimulated muscles in cat. *Med. Biol. Eng. Comput.* 16:302–308, 1978.

Petrofsky, J.S. Sequential motor unit stimulation through peripheral motor nerves in the cat. *Med. Biol. Eng. Comput.* 17:87–93, 1979.

Petrofsky, J.S., and Phillips, C.A. Constant velocity contractions in skeletal muscle by sequential stimulation of muscle efferents. *Med. Biol. Eng. Comput.* 17: 583–592, 1979a.

Petrofsky, J.S., and Phillips, C.A. Microprocessor controlled stimulation in paralyzed muscle. *IEEE NAECON Record*, 79:198–210, 1979b.

Petrofsky, J.S., and Phillips, C.A. Active physical therapy–A modern approach to rehabilitation therapy. *J. Neurol. Orthop. Surg*. 4:165–173, 1983.

Petrofsky, J.S., and Phillips, C.A. The use of functional electrical stimulation for rehabilitation of spinal cord injured patients. *CNS Trauma J*. 1:57–74, 1984.

Petrofsky, J.S., Rinehart, J.S., and Lind, A.R. Isometric strength and endurance in slow and fast muscles in the cat. *Fed. Proc*. 35:291, 1976.

Petrofsky, J.S., Heaton, H.H., and Phillips, C.A. Outdoor bicycle for exercise in paraplegics and quadriplegics. *J. Biomed. Eng*. 5:292–296, 1983a.

Petrofsky, J.S., Heaton, H.H., Phillips, C.A., and Glaser, R.M. Application of the Apple as a microprocessor controlled stimulator. *Coll. Microcomput*. 1: 97–104, 1983b.

Petrofsky, J.S., Heaton, H.H., and Phillips, C.A. Leg exerciser for training of paralyzed muscle by closed-loop control. *Med. Biol. Eng*. 22:298–303, 1984a.

Petrofsky, J.S., Phillips, C.A., Heaton, H.H., and Glaser, R.M. Bicycle ergometer for paralyzed muscle. *J. Clin. Eng*. 9:13–19, 1984b.

Pette, D., Ramirez, B., Mueller, W., et al. Influence of the intermittent long-term stimulation on contractile, histochemical and metabolic properties of fibre population in fast and slow rabbit muscles. *Pfluegers Arch*. 361:1–7, 1975.

Phillips, C.A., Petrofsky, J.S., Hendershot, D.M., and Stafford, D. Functional electrical exercise: A comprehensive approach for physical conditioning of the spinal cord paralyzed individual. *Orthopedics*, 7:1112–1123, 1984.

Rack, P.M.H., and Westbury, D.R. The effects of length and stimulus rate on tension in the isometric cat soleus muscle. *J. Physiol*. 204:443–460, 1969.

Salmons, S., and Vrbova, G. The influence of activity on some contractile characteristics of mammalian fast and slow muscles. *J. Physiol*. 201:535, 1969.

7 Chronic Exercise Effects: The Therapeutic Outcome

I. Muscular Response with the Leg Exercise System

A. Introduction

Electrical stimulation causes muscle to contract by the generation of action potentials on nerve and muscle following the application of electrical current to the skin, motor nerve, or the muscle directly (see, e.g. Peckham, 1976; Petrofsky et al., 1976; Solomonow et al., 1978; Thoma et al., 1978). Current design of microprocessors allows multichannel control through standard feedback control techniques for movement in muscle (Petrofsky and Phillips, 1979, 1980, 1981; Phillips and Petrofsky, 1980). Using these techniques, it is now possible to control movement in muscle. However, one problem that may not be apparent is the deconditioning of the body that follows spinal cord injury. After SCI, muscle is not used and atrophies. This so-called disuse atrophy is similar to the atrophy that occurs in muscles after prolonged hospital stays or when a leg or arm is immobilized in a cast for a few months. Associated with disuse atrophy is demineralization of bones. This is analogous to the demineralization that occurs with prolonged bed rest and with prolonged stays outside the earth's gravity. When bone is not stressed in a normal manner, calcium leaves the bone and is excreted by the body. This makes the bone fragile and predisposes it to injury. The fragility of bones creates a long-term medical problem for people who have been in wheelchairs for many years. The lack of use of part of the body also results in deconditioning of the heart and the cardiovascular and pulmonary systems. This predisposes SCI individuals to cardiovascular disease, renal disease, and pulmonary disease (such as pneumonia). Consequently, before wide-scale computer-controlled technology can be applied to wheelchair-bound individuals or before any successful reversal of SCI can be attempted, the patient needs to be reconditioned to normal physical fitness or as near to normal as possible.

During normal, voluntary activity, strength and endurance are trained for by two different mechanisms. Training for muscular strength calls for isokinetic exercise (Muller, 1932), in which weights are very slowly lifted up and set down. The rate of lifting is one important variable, while the second is the amount of weight lifted. To optimize isokinetic training, the weight should be approximately 25–50% of the person's maximum voluntary strength. When this level has been reached, strength builds very rapidly. This type of training is used to cause muscular hypertrophy and to increase the strength of muscle (e.g., by weight lifters).

Over the past 20 years, a number of techniques have been used to exercise paralyzed muscle. These typically involved stimulating muscle at various rates with an undetermined weight. Although the neural traffic was similar to what occurred during voluntary activity, stress on the body was not. Basically, to be a stimulus for muscle growth, muscle metabolism must be stressed in a precise manner (Vrbova, 1963). Over the past few years we have developed in our laboratory a computer-controlled stimulation system termed active physical therapy (APT) to stress muscle in a manner similar to what would occur during voluntary exercise training.

B. Leg Exercise System

The basic device is a weight trainer for muscle exercise, similar in many respects to the weight lifting equipment available at a modern gymnasium. As indicated previously, weights in a weight tray are connected through a cable system to a harness that can be placed around the appropriate appendage, allowing muscles to operate. When a given muscle (such as the quadriceps muscle of the leg) contracts, the weights are lifted up and down. Operationally, a strap can be placed around the ankle of the subject and secured by Velcro fasteners. The leg is connected (through a steel cable connected to the strap) to the weight tray under the back of the chair. The subject generally cannot either feel the limb or see the weight that is being lifted. Therefore, a sensor along the length of the cable provides a computer with information about the position of the leg. The computer used in these studies was an Apple II microcomputer (Apple Computer Company, Berkeley, California). The computer–sensor interface was supplied by an A/D converter (Petrofsky et al., 1983b, 1984a).

The program to operate the chair involves a tracking algorithm that continually changes the stimulation voltage applied through the D/A converter and then senses the position of the leg to determine whether it is going up or down at the appropriate rate. The rate is determined by a ramp generator to provide a triangular wave with a rise time of 3 seconds and a fall time of 3 seconds. The output of this ramp generator is set to match the gain of the sensor for the leg, so that the computer program tries to match the target analog voltage to the actual input of the A/D converter. The computer program matches the synthetically generated ramp and tries to have the leg do the same thing. If the ramp voltage is going up faster than the leg is moving, stimulation voltage is proportionately reduced.

C. Protocol and Results

Training on the leg exerciser system resulted in a significant increase in quadriceps muscular strength (Petrofsky and Phillips, 1983). A group of four subjects were exercised (for 15 minutes a day, 3 days per week, for a month), lifting weights equivalent to 25% of their maximum voluntary strength. Each of the four subjects extended the foreleg (contracting the quadriceps) slowly at uniform velocity over a 3-second period, then lowered the foreleg downward (relaxing the quadriceps) over the same period. Six seconds was allowed between the exercise cycles. The results of these experiments showed an average increase for the group of 1.5 inches in leg size and an average increase for the group of almost 100% in strength (13–25 lb) over the one-month period.

D. Discussion

Although the primary goal of active physical therapy was the physical reconditioning of the SCI individual, it is also necessary to prepare the SCI subject to withstand the stresses associated with active ambulation. Once the subjects had physically reconditioned themselves, it was possible to use feedback control technology to coordinate movement during standing and walking (initially in a paraplegic subject).

Such developments as the leg exerciser system are a necessary antecedent to practical application of a portable walking system. The active physical therapy system described in this chapter can form the basis for an improvement of physical therapy for paraplegic and quadriplegic subjects insofar as it makes possible the reversal of some of the deconditioning effects associated with life in a wheelchair (Phillips and Petrofsky, 1983).

II. Cardiovascular Circulatory Dynamics with the Exercise Bicycle System*

A. Introduction

Cardiovascular and circulatory dynamics are quite different in chronic cervical spinal cord injury (quadriplegics) and in normal individuals, as described in Section III of Chapter 4. This is

*From D. Danopulos et al., Changes in cardiovascular circulatory dynamics after a twelve-week active bicycle rehabilitation in young tetraplegics. Reprinted with permission from *Journal of Neurological and Orthopedic Medicine and Surgery*, 7:179–184, 1986. Copyright 1986, American Academy of Neurological and Orthopedic Surgeons.

mainly due to the disruption of the sympathetic reflex loop at the cervical cord level.

As a consequence, homeostatic cardiovascular adjustment to head-up tilting is very poor, or absent, with no rise in plasma catecholamine levels, which are already extremely low at rest in quadriplegics (Guttmann et al., 1963; Bebarge et al., 1974; Mathias et al., 1975). Blood pressure is usually low and the resting heart rate is slow (Danopulos et al., 1985), probably due to predominant vagal tone. Some authors, however, reported no difference in heart rate from normals (Mathias and Frankel, 1983).

Quadriplegics also show low cardiac index (CI). However, left ventricular performance [measured by stroke volume (SV), ejection fraction (EF), systolic time intervals (STI), and ratio of preejection period to left ventricular ejection time (PEP/LVET)] was found to be normal (see Section III of Chapter 4).

Biochemical changes seen in quadriplegics can be summarized as those which are present due to prolonged immobilization (e.g., osteoporosis and atrophy of the nonused muscles). This is due to the complete lack of sympathetic activity following disruption of the sympathetic reflex arc at the cervical cord level.

This section will discuss the changes observed in the cardiovascular dynamics of quadriplegics before and after a 12-week period of functional electrical stimulation (FES) induced bicycle training. Obviously the benefits from a sustained and well-designed active rehabilitation program go far beyond those seen and observed in the cardiovascular system.

B. Methods

The subjects of this study were six sedentary young male quadriplegics with a mean age of 22 and with an average of 2.58 years since spinal cord injury. All subjects had closed cervical spinal cord transections at the levels of C4–C7. All lesions were clinically complete, with total loss of motor and sensory function below the lesion. The subjects volunteered to participate in the study, which was the first step of a program of research into the long-term rehabilitation of spinal cord injury.

This phase of rehabilitation consisted of 12 weeks, three times a week for intensive stationary bicycle training. Prior to entering the bicycle ergometry training program, the subjects had had extensive physical examinations and laboratory evaluations. They had 6 weeks of systematic FES-induced leg training with increasing weight lifting (to > 20 lb) to improve strength of the quadriceps and other leg muscles on an isokinetic leg trainer. They had been declared in sufficiently good condition to carry out the bike exercise as far as the bicycle training protocol required.

TABLE 7.1. Bicycle ergometer study: Resistive loads.[a]

	Week											
ID	1	2	3	4	5	6	7	8	9	10	11	12
1	0	0	0	1/8	1/4	1/4	1/4	1/4	1/4	1/4	1/4	1/4
2	0	0	1/8	1/8	1/4	1/4	1/4	1/4	1/4	1/4	1/4	1/4
3	0	0	1/8	1/8	1/8	1/4	1/4	1/4	1/4	1/4	1/4	1/4
4	0	1/8	1/4	3/8	3/8	3/8	3/8	3/8	3/8	3/8	3/8	1/2
5	0	0	1/8	1/4	1/4	1/4	1/4	1/4	3/8	3/8	3/8	3/8
6	0	1/8	1/4	3/8	3/8	3/8	3/8	3/8	3/8	3/8	3/8	1/2

[a]Values are in kiloponds (kp) frictional resistance; pedal frequency = 50/min; wheel/pedal ratio = 3/1; diameter of wheel = 20 in. (0.5 m); workload = kp × 3 × 50 × (0.5 × 3.1416) = kilopond-meter/min.

Subjects were prepared with appropriate placement of gluteal, hamstring, and quadriceps muscle electrodes for FES. This placement was done by trained members of the bicycle ergometer team, in accordance with a previously designated placement protocol.

Subjects were transferred (carried) to the bicycle ergometer by at least two team members. Leg and foot placement was checked and safety belts (waist and chest if needed) were applied and adjusted for the subjects' comfort. ECG electrodes were applied (for visual monitoring of heart rate and rhythm). Strips of ECG recording were made at approximately one-minute intervals.

Resting heart rate and arterial pressures were measured. Thresholds for electrical stimulation were determined. The subjects were then given 2 minutes of passive pedaling.

At the end of passive pedaling, arterial pressure and heart rate were again determined. If all physiological parameters were within acceptable limits, active pedaling with computer-controlled FES was initiated (Table 7.1). During the active pedaling, heart rate and arterial pressure were obtained at about 2-minute intervals. All the measurements were made by the same individual.

The subject was exercised for 15 minutes, or until he was fatigued to the point that he could not maintain a cycle speed of 10 km/h with maximal stimulation, or until one of the following physiological criteria was exceeded: (a) heart rate up to 85% of calculated maximal value or (b) arterial pressure reached 200 mmHg systolic, or 120 mmHg diastolic.

C. Hemodynamic Evaluation

Systolic time intervals (STI) were obtained from recording simultaneously electrocardiogram, phonocardiogram, and carotid pulse tracing on a multichannel oscilloscopic recorder (Electronics for Medicine "E for M" model VR-12) using photographic paper at a speed of 100 mm/s as previously reported by others (Weissler et al., 1968, 1969) and by us (Danopulos et al., 1985). Heart sounds were recorded by a Cambridge transducer microphone placed on the left anterior chest wall at the level of the second or third intercostal space, assuring clear delineation of the initial high frequency vibration of the first and second heart sounds. The carotid pulse was recorded at the point of maximum pulsation with an "E for M" pressure pulse transducer. The transducer was adjusted to obtain a clear delineation of the onset of the rapid rise and the point of the nadir of the incisura (dicrotic notch) of the pulse tracing. The ECG lead was selected to record a clear onset of the QRS complex.

To obtain the total electromechanical systole (QS_2), the interval between the onset of the QRS complex of the ECG to the first high frequency vibration of the aortic component of the second sound in the phonocardiogram was measured. Left ventricular ejection time (LVET) was measured as an interval from the beginning of the rapid upstroke of the carotid pulse tracing to the incisura. The preejection period (PEP) was calculated by subtracting LVET from QS_2. The PEP represents the time that is required from the left ventricle to raise the intraventricular pressure from the end-diastolic to the aortic level preceding systolic ejection. Stroke volume (SV), cardia output (CO), cardiac index (CI), and ejection fraction (EF) were calculated from echocardiographic measurements. Echocardiograms were recorded on "E for M" M-mode multichannel oscilloscopic recorder model VR-12 and on a two-dimensional recorder (Hewlett-Packard, model Ekoline 5000).

Measurements obtained included left ventricular end-diastolic dimension (LVEDd) measured from the endocardial surface of the left ventricular posterior wall to the interventricular septum. The ventricular septal thickness (LVSTh) and left ventricular free wall (posterior wall) thickness (LVPWTh) as well as (LVEDd) were all measured at the onset of the QRS complex after the participants were lying relaxed on the table. The left ventricular end-systolic dimension (LVESd) was measured as the smallest left ventricular dimension without reference to the electrocardiogram (Sahn et al., 1978; D'Cruz, 1983).

End-diastolic as well as end-systolic LV volumes were calculated as the cube of the respective dimensions. Stroke volume was obtained as the difference between the end diastolic and end systolic volumes. Cardiac output was calculated by multiplying the stroke volume with heart rate, and cardiac index as the product of the CO divided by the body surface area. Ejection fraction was calculated by dividing SV by LV end-diastolic volume. Left ventricular mass was calculated by using the formula:

$$[(\text{LVIDd} + 2\text{LVPW Th})^3 - \text{LVIDd}^3] \times 1.05$$

D. Experimental Protocol and Measurements

The whole procedure was carefully explained to each subject and an informed consent was obtained, approved by the Institutional Review Board of Wright State University.

Hemodynamic study was performed in two stages: the first before and the second within a week after the 12-week training program had been completed. The measurements were obtained in resting (supine) position, 15 minutes after the participants were lying relaxed on the table.

For the systolic time intervals, 10 consecutive cardiac cycles were measured and the data averaged. For echocardiography, three cycles were analyzed on a beat-to-beat basis and averaged for each participant. The results were compared and

TABLE 7.2. Physical characteristics of six subjects before and after 12 weeks' exercise.

				D		E		F		G	
ID	A	B	C	Before	After	Before	After	Before	After	Before	After
1	21	C5	5	1.80	1.84	61.9	64.1	106/74	110/66	44	53
2	20	C4,C5	2.5	1.84	1.84	65.9	65.0	103/69	105/67	42	40
3	18	C5,C6	1	1.89	1.90	68.0	70.9	109/75	118/78	53	53
4	21	C6,C7	1	1.52	1.55	44.0	46.4	85/57	94/60	65	69
5	26	C4,C5	4	1.98	1.99	70.0	70.0	100/64	110/65	48	55
6	23	C7	2	1.84	1.87	63.2	64.0	88/62	94/67	45	48
Mean:	22.0±3.0		2.58±1.48	1.81±0.14	1.83±0.14	62.2±8.6	63.4±8.1	98.5±9.0	105.2±8.8	49.5±7.8	53.0±8.7
								66.8±6.4	67.2±5.4		

Columns: A = age (years); B = level of spinal cord injury; C = time passed (years); D = body surface area (m²); E = weight (kg); F = blood pressure (mmHg); G = heart rate (beats/min)

Source: Wright State University School of Medicine, Cox Heart Institute, Cardiopulmonary Performance Laboratory.

the differences statistically analyzed by the Student *t*-test. Differences in *p* value below 0.05 were considered significant.

E. Results

Tables 7.2, 7.3, and 7.4 show all the data gathered from this study. The 12-week bicycle training did not induce any significant changes in left ventricular muscular mass (168 ±24 g before; 175 ±21.3 g after training). Left ventricular posterior wall thickness (LVPWTh) and ventricular septal thickness (VSTh) remained unchanged. Left ventricular end-diastolic dimensions changed very little, increasing from 46.5 ±3.25 mm before to 47.7 ±2.75 mm after; LVES dimensions changed from 27.7 ±4.19 mm before to 28.5 ±4.86 mm after. These were statistically not significant.

Systolic time intervals as preejection period (PEP), left ventricular ejection time (LVET), electromechanical systole (QS_s) and their indices, which were within normal limits before training, remained almost unchanged after exercise training (Table 7.3).

The subjects gained an average of 1.22 kg, with the average weight increasing from 62.2 ±8.6 kg to 63.4 ±8.1 kg at the end of the study. There was an average increase of the heart rate by 4 beats/min at the end of the exercise program, with the resting heart rate rising from 49.5 ±7.8 beats/min before to 53 ±8.7 beats/min after training. The posttraining heart rate was still considered bradycardic. The systolic blood pressure increased from 98.5 ±9 mmHg to 105.2 ±8.8 mmHg, while the diastolic remained almost unchanged with 66.8 ±6.4 mmHg before and 67.2 ±5.4 mmHg after.

Ejection fraction remained in the same high performance level, with 78.67 ±7.06% before and 73.8 ±5.5% after training. The stroke volume showed little change, increasing from 79.3 ±12.4 cm³ to 80.5 ±13.2 cm³ after training. But CI increased almost 8% after training, rising from 2.12 ±0.12 to 2.29 ±0.18 L/min/m². This is still considered low when compared with normal controls (Danopulos et al., 1985). The PEP/LVET ratio remained almost unchanged.

F. Discussion

The overall results of the 12-week FES-induced bicycle training in quadriplegics indicated no changes in left ventricular dimensions, left ventricular mass, and systolic time intervals. However, there was a definite increase in the resting systolic blood pressure, the heart rate, and the cardiac index. While the changes were relatively small, they were nevertheless statistically significant, and the trend seems to be in the opposite direction from the training effect in normal individuals. The significant weight gain correlates with the increased body muscle mass as the result of reversal of atrophy of the leg muscles, also indicated by

TABLE 7.3. Echocardiographic and systolic time interval characteristics of six subjects before and after 12 weeks' exercise.

	LVET		PEP		LVEDd		LFESd		LVMss		LVPWTh		VSTh	
ID	Before	After	Before	After	Before	After	Before	After	Before	After	Before	After	Before	After
1	360	310	100	100	46	47	25	25	148	154	8	8	8	8
2	377	375	100	105	48	49	28	25	185	192	9	9	8	8
3	320	354	110	115	49	50	34	36	192	198	9	9	9	9
4	328	314	98	96	40	42	22	22	138	149	9	9	9	9
5	324	410	100	99	50	50	32	32	198	198	9	9	9	9
6	358	467	102	96	46	48	25	31	148	159	8	8	10	10
Mean:	345±21.5	371.7±54.7	101.7±3.9	101.8±6.6	46.5±3.25	47.7±2.7	27.7±4.19	28.5±4.86	168±24.0	175±21.3	8.7±0.47	8.7±0.47	8.8±0.69	8.8±0.69

Columns: LVET = left ventricle ejection time (ms)
PEP = preejection period (ms)
LVEDd = left ventricle end-diastolic dimensions (mm)
LVESd = left ventricle end-systolic dimensions (mm)
LVMss = left ventricular mass (g)
LVPWTh = left ventricular posterior wall thickness (mm)
VSTh = ventricular septal thickness (mm)

Source: Wright State University School of Medicine, Cox Heart Institute, Cardiopulmonary Function Laboratory.

TABLE 7.4. Hemodynamic data of six subjects before and after 12 weeks' exercise.

	SV		CO		CI		EF		PEP/LVET	
ID	Before	After	Before	After	Before	After	Before	After	Before	After
1	82	82	3.60	4.34	2.00	2.35	84	79	0.294	0.322
2	89	95	3.73	3.70	2.03	2.02	81	80	0.265	0.280
3	77	79	4.08	4.19	2.15	2.20	65	63	0.343	0.324
4	54	54	3.51	3.72	2.30	2.48	84	73	0.298	0.305
5	92	92	4.42	5.06	2.23	2.54	74	74	0.322	0.305
6	82	81	3.70	4.05	2.00	2.15	84	74	0.279	0.268
Mean:	79.3±12.4	80.5±13.2	3.84±0.31	4.18±0.46	2.12±0.12	2.29±0.18	78.67±7.06	73.8±5.5	0.300 ±0.030	0.290 ±0.030

Columns: SV = stroke volume (cm^3) EF = ejection fraction (%)
CO = cardiac output (L/min) PEP/LVET = ratio %
CI = cardiac index ($L/mm/m^2$)

Source: Wright State University School of Medicine, Cox Heart Institute, Cardiopulmonary Performance Laboratory.

the girth increase of the thigh muscles. The increased cardiac index is probably a reflection of the increased blood flow to supply the leg muscles, reversed from atrophy to show a considerable increase in active muscle.

Increased resting heart rate and systolic blood pressure probably reflect some reversal of the lack of sympathetic drive and predominant vagal tone of the heart and circulatory system. The explanation for this is not clear and easy. In quadriplegics, energy expenditure and oxygen consumption rise more slowly than in normal individuals during exercise. This is because of the lack of sympathetic input. In normal individuals during dynamic exercise, plasma catecholamine levels increase as a reflection of increased sympathetic output. During dynamic exercise, norepinephrine and epinephrine increase in a ratio of 200–400%. Also, during orthostatis, normally there is a predominant increase of norepinephrine as part of the homeostatic regulation (Sever et al., 1980).

In quadriplegics with cervical spinal cord injury, the sympathetic reflex loop is disrupted. While peripheral autonomic nerve endings remain intact (Mathias et al., 1976), central sympathetic output is not transmitted. As a result, resting plasma norepinephrine levels are very low. They change very little, if any, during head-up tilt and exercise (Guttmann et al., 1963; Bebarge et al., 1974; Mathias et al., 1975). This is also true for epinephrine, which is secreted mainly in the adrenal medulla and is also activated by sympathetic drive.

Discharge of epinephrine from the adrenal medulla may be stimulated also by other factors, such as insulin-produced hypoglycemia (Vendsalu, 1960; Goldfien et al., 1961). However, epinephrine levels are still very low in quadriplegics. Thus, factors other than sympathetic drive contribute to the maintenance of resting blood pressure and possibly to the increase of systolic pressure subsequent to dynamic exercise training in quadriplegics.

As we have seen, a high stroke volume despite relatively low cardiac output maintains the systolic blood pressure within relatively physiological levels in quadriplegics. The high stroke volume is maintained by a high ejection fraction, which is higher than in normal controls (Danopulos et al., 1985), in the presence of bradycardia. However, quadriplegics have orthostatic blood pressure decreases because of the lack of sympathetic drive of the capacitance and resistance vessels, and, thus, lack of catecholamines during orthostatic changes and dynamic exercise, resulting sometimes in unsatisfactory homeostatic balance of the cardiovascular system.

Probably quadriplegics do as well as they do in the chronic stage as a result of adaptation of the renin–angiotensin system, as part of the adjustment of homeostatic mechanisms. Plasma renin activity was found to be high in tetraplegics, with further increase during head-up tilt and dynamic exercise. The product of renin, angiotension II, is a direct vasoconstrictor and also stimulates central sympathetic drive. Renin also activates aldosterone production, leading to sodium and water retention.

Renin release is probably activated by the decreased circulating perfusion pressure of the kidneys. During head-up tilt, pooling of blood in the lower body compartments will further decrease renal perfusion pressure and increase plasma renin release. The same probably is the mechanism of the renin release during passive and active exercise in quadriplegics.

In recent reports, it was shown that norepinephrine enhances tubular sodium and fluid reabsorption via the adrenergic receptors in the proximal tubules (Besarab et al., 1977; Bello-Reuss et al., 1980). It might be reasonable to conclude that impaired sympathetic function in quadriplegics leads to defective water and sodium retention, resulting in diminished circulating blood volume. Also, the decreased oxygen requirements of the nonfunctioning musculature of quadriplegics can account for decreased circulating blood volume.

G. Summary

Differences in exercise training results between normals and quadriplegics have been reported. These results may be summarized as follows.

In normal individuals, regular intensive exercise training leads to resting bradycardia, increase of left ventricular end-diastolic dimensions, and some left ventricular posterior wall thickness, as well as septal thickness, depending on the type and duration of the exercise training (Morganroth et al., 1975; Demaria et al., 1978; Ikaheimo et al., 1979; Nishimura et al., 1980; Snoeckx et al., 1983). Training effect tends to increase blood volume in normals very easily (Convertino et al., 1980; Nadel, 1985). This early adaptive reaction results in a better and quicker building up of performance by assuring an early increase in stroke volume. This is accomplished through an increased ventricular preload, better oxygen uptake, and better heat distribution during exercise due to increased blood volume (Fortney et al., 1983).

In spinal-cord-injured persons, there is a close relationship between the level of the lesion and capacity/endurance performance during exercise. This is reflected by the difference in maximum oxygen uptake between paraplegics and quadriplegics, the latter having the lowest maximum O_2 uptake (Gass and Camp, 1979; Rhodes et al., 1981). This is probably the result of the lack of sympathetic drive of a large part of the muscles in quadriplegics, while in paraplegics some of the muscles remain innervated. Despite this, quadriplegics still show exercise effect probably mainly by activation of the renin–angiotension system. How this is accomplished deserves further study.

III. Muscular, Respiratory, and Cardiovascular Responses with the Exercise Bicycle System*

A. Introduction

Active physical therapy (APT) is the use of paralyzed muscles as actuators so that the muscles themselves participate in the physical reconditioning of a previously deconditioned individual. This differs from conventional rehabilitation therapy, in which the paralyzed muscles are passive and manipulated by factors external to themselves. Closed-loop functional electrical stimulation (FES) is utilized to activate the paralyzed muscles so that they are made to perform useful work (Petrofsky and Phillips, 1984, 1985). An introduction to APT that reports preliminary results of physically reconditioning muscular strength and endurance is available (Petrofsky and Phillips, 1983).

The equipment utilized in active physical therapy includes an FES isokinetic leg trainer (Phillips and Petrofsky, 1983; Petrofsky et al., 1984a) and an FES exercise bicycle ergometer (Phillips and Petrofsky, 1983; Petrofsky et al., 1984b). Of particular interest is FES bicycle ergometry, since this more aerobic form of exercise has the potential to improve cardiopulmonary fitness.

Specifically, Phillips et al. (1984) have shown for four quadriplegics that $\dot{V}_{O_2}$ (at the end of 15 minutes of fatiguing FES bicycle exercise) averaged 0.69 L/min (see Figure 6.11), whereas after

---

*From C.A. Phillips et al., Muscular respiratory and cardiovascular responses of quadriplegics to an FES bicycle ergometer conditioning program. Reprinted with permission from *International Journal of Rehabilitation Research*, 12:147–157, 1989. Copyright 1989, Heidelberger Verlagsanstalt und Druckerei, Heidelberg, Germany.

15 minutes of isokinetic leg exercise, $\dot{V}_{O_2}$ averaged only 0.43 L/min (see Figure 5.8) for the four quadriplegics.

Arm crank ergometry does not appear to be an effective form of aerobic exercise for quadriplegics. Phillips et al. (1984) demonstrated that after 15 minutes of fatiguing voluntary arm crank ergometry, $\dot{V}_{O_2}$ averaged between 0.29 and 0.45 L/min for four quadriplegics (see Figure 5.8 and 6.11, respectively), about half the value obtained from FES bicycle ergometry.

Cardiovascular and pulmonary dynamics are quite different in chronic cervical spinal cord injury (quadriplegics) when compared to normal individuals. This is mainly due to (a) the disruption of the sympathetic reflex loop at the cervical cord level and (b) loss of some respiratory muscle function. Danopulos et al. (1985) evaluated the cardiovascular dynamics in nine young quadriplegics compared to five controls. These were non-exercise-trained subjects who were evaluated by echocardiography in the supine position and in the 40° head-up tilt position.

Systolic and diastolic blood pressures were lower in quadriplegics (106/69 mmHg) than in controls (127/78 mmHg). During head-up tilt, blood pressure decreased in quadriplegics to 82/54 mmHg while it remained at 123/85 mmHg for controls. Resting-supine heart rate was 53 $\pm$7/min in quadriplegics and 80 $\pm$9/min in controls. Heart rate increased in quadriplegics to 76 $\pm$14/min (50%) and in controls to 86 $\pm$10/min (7%) during head-up tilt.

Danopulos et al. (1985) also reported that resting cardiac index (CI) was low in quadriplegics, being 2.28 $\pm$0.2 L/min/m^2 and 3.00 $\pm$0.2 L/min/m^2 in controls. It decreased in quadriplegics during head-up tilt to 1.84 $\pm$0.3 L/min/m^2 and in controls to 2.85 $\pm$0.2 L/min/m^2. Resting-supine stroke volume was high in quadriplegics (80 $\pm$14 mL) and normal in controls (71 $\pm$1 mL).

Selected cardiorespiratory parameter changes that occur in quadriplegics during FES bicycle ergometer exercise (when compared to paraplegics and controls) have been reported by Petrofsky et al. (1985). Cardiac output increased in all three subject groups during the exercise. The increase in cardiac output was similar for the two groups of paralyzed subjects and was roughly proportional to workload. This increase in cardiac output was not statistically different from that of the control group during exercise at the same load. Therefore, for a given absolute level of exercise, there was no significant difference in cardiac output in all three subject groups ($p > 0.05$). Ventilation ($\dot{V}_E$) increased in proportion to the workload for all three subject groups. For a given workload, there was no statistical difference between the groups of subjects in terms of equivalent ventilatory rate ($\dot{V}_E$).

Another form of FES bicycle exercise has appeared in the literature, specifically the FES outdoor exercise bicycle (Petrofsky et al., 1983a). However, no reports have appeared with respect to the cardiopulmonary effects of such aerobic exercise. This could be due (in part) to the fact that the level of exercise would be difficult to quantitate or control as the bicycle is pedaled over varying grades and terrains.

There appears to be only one report of any actual training effects of FES bicycle ergometry on quadriplegic individuals (Danopulos et al., 1986). The results of this preliminary study have shown that quadriplegics can demonstrate some cardiovascular training effects.

In normal individuals, regular intensive exercise training leads to resting bradycardia, increase of left ventricular end-diastolic dimensions, and some left ventricular posterior wall thickness, as well as septal thickness. These changes depend on the type and duration of the exercise training (Morganroth et al., 1975; Demaria et al., 1978; Ikaheimo et al., 1979; Nishimura et al., 1980; Snoeckx et al., 1983). Training effect tends to increase blood volume in the normal individual very early (Fortney et al., 1983; Nadel, 1985). This early adaptive reaction results in a better and quicker improvement of performance by ensuring an early increase in stroke volume. An increased ventricular preload, better oxygen uptake, and better heat distribution during exercise are due to increased blood volume (Fortney et al., 1983).

In spinal-cord-injured persons, there is a close relationship between the level of the lesion and the exercise performance (capacity and endurance). This is reflected by the difference in maximum oxygen uptake (max $\dot{V}O_2$) between paraplegics and quadriplegics (the latter having the lowest max $\dot{V}O_2$ (Gass and Camp, 1979; Rhodes et al., 1981). This probably results from the lack of sympathetic drive of a large part of the muscles in quadriplegics, while in paraplegics some of the muscles remain

TABLE 7.5. Subject characteristics: FES bicycle ergometer conditioning program for quadriplegics.

Subject	Age (years)	Sex	Injury level	Time since injury (months)
A	18	M	C5,C6	16
B	21	M	C5,C6	62
C	23	M	C7	23
D	22	M	C6,C7	39
E	20	M	C4,C5	18
F	25	M	C6	41
G	26	M	C4,C5	37

innervated. Despite this, quadriplegics still show exercise effect.

Section II described only a few selected cardiovascular parameters; it did not consider a more comprehensive range of cardiovascular parameters (including cardiovascular reflexes). Nor did it evaluate training effects on either muscular or pulmonary parameters.

The primary objective of this section, therefore, is to systematically evaluate a comprehensive spectrum of muscular, respiratory, and cardiovascular parameters. The secondary objective is to identify those physiological parameters that could be used as indicators of the effect of exercising on the FES bicycle ergometer. To accomplish these objectives, this section will examine the changes in these parameters before and after an 8-week program of FES bicycle ergometry on seven quadriplegic subjects.

B. Methods

This experiment was performed on a group of seven quadriplegics (C4–C7), aged 18–26 years. All these subjects were males and had complete lesions (Frankel class A). Their time since injury varied from 16 to 62 months. Subject characteristics are summarized in Table 7.5.

All subjects were given a stringent health screening, including x-rays, CT scans (to detect excessive osteoporosis), neurological and muscular evaluation, clinical laboratory testing, a complete physical, special pulmonary and cardiovascular evaluation, and psychological evaluation (Phillips, 1987).

None of these subjects had received any active FES-induced exercise prior to these experiments. Since one objective of the experiments was to identify physiological parameters that could be used as indicators of effect of exercising on the FES bicycle ergometer, these subjects did not receive the standard FES leg trainer exercise that usually precedes the aerobic bicycle ergometer phase of the exercise program (Phillips, 1987).

Before the beginning of and following the end of the exercise program, all subjects were subjected to a series of measurements that were primary to the purpose of the experimentation. These measurements included:

1. Muscle size (as indicated by leg girth)
2. Pulmonary function measurements
   a. Vital capacity (VC)
   b. Forced vital capacity (FVC)
   c. Peak expiratory flow (PEF)
   d. Forced inspiratory capacity (FIC)
   e. Peak inspiratory flow (PIF)
   f. Expiratory reserve (ER)
   g. Inspiratory capacity (IC)
   h. Minute ventilation (MV)
   i. Tidal volume (TV)
   j. Respiratory rate (RR)
   k. Cardiac output (CO_2 Fick method)
   l. Forced expiratory volume at 1 second (FEV_1)
3. Hemodynamic parameters (at rest)
   a. Heart rate (RR)
   b. Systolic pressure (SP)
   c. Diastolic pressure (DP)
   d. Mean pressure (MP)
   e. Cardiac output (impedance cardiography) (CO_z)
   f. Peripheral resistance (PR)
4. Echocardiographic parameters (at rest)
   a. Heart rate (HR)
   b. P-R interval (PRI)
   c. Left ventricular ejection time (LVET)
   d. Preejection period (PEP)
   e. Stroke volume (SV)
   f. Cardiac output (CO_z)
   g. Cardiac index (CI)
5. Cardiovascular reflex data
   a. Cold pressor test
   b. Diving reflex test
   c. Tilt table test

The exercise program that was evaluated in this experiment amounted to 24 exercise sessions, performed three times per week, for an experiment duration of 8 weeks. Sessions were scheduled on

Monday, Wednesday, and Friday; thus, at least 48 hours elapsed between sessions. With this frequency of repetition of exercise, no accumulated fatigue was evidenced in any of the subjects. FES bicycle ergometry was performed on a system (Phillips and Petrofsky, 1983; Petrofsky et al., 1984b) designed and constructed at Wright State University.

Individual exercise sessions were performed with zero load on the bicycle ergometer. The pedaling rate was 50 pedal revolutions per minute. Since the subjects were totally untrained, many of them were incapable of continuing the exercise for the full 15 minutes at the beginning of the experiment. The pedaling was discontinued when the subject could no longer sustain the pedaling at 35 rpm. The length of time the subject could continue the pedaling was taken as a measure of the "endurance" of the individual.

Heart rate and arterial pressure were measured at approximately 2-minute intervals during the exercise sessions. The heart rate was found from an ECG strip taken at the appropriate time, and the arterial pressure was measured by the auscultatory method. Arterial pressure and heart rate limits at which the exercise was to be terminated were set (Phillips, 1987). Termination of exercise for this reason did not occur during this experiment.

Leg size was measured as the girth (circumference) of the leg at various distances along the leg, with reference to the upper edge of the patella. These distances were 10 and 20 cm above the patella (thigh measurements) and 20 cm below the patella (calf measurements). The calf measurements were included to determine whether stimulus "spillover" was causing significant exercising of the muscles of the lower leg.

Pulmonary function measurements were made with a Gould model 9000 Pulmonary Function Laboratory. All results were produced by the onboard computer of the equipment, and hard copy output was obtained for individual assessments.

Cardiovascular measurements made in this experiment were performed by the staff of the Department of Cardiology at the Cox Heart Institute, in Kettering, Ohio. Impedance cardiography was done with a Minnesota Impedance Cardiograph, which produced a hard copy output from the onboard computer. The Cox Heart Institute staff does not feel that absolute values for stroke volume and cardiac output obtained by this method are accurate but believes that they can be used for comparative purposes.

Echocardiography was performed with a Hewlett-Packard two-dimensional echocardiograph. The sonic probe was positioned with the instrument in scan mode, and recordings were made in the M mode. Computer programs for analysis of the results of echocardiography were still in preparation, so the results reported in this experiment were calculated manually.

Cold pressor tests, diving reflex tests, and tilt table tests were performed in the usual manner. Results of those tests were reported in terms of changes of heart rate, systolic pressure, and diastolic pressure. Changes reported in the tilt test were produced when the body was tilted from supine to a 60° (from horizontal) position.

All data were entered into a Columbia MPC1600 computer manually, using a data base manager (TIM IV). When all data had been entered, data lists were converted to ASCII format for use in statistical analysis. The statistical analyses were done in the same computer, using the statistical package MICROSTAT.

The analyses reported herein were performed by application of a paired Student's t-test. Thus, the subjects served as their own controls. The probability of the before and after groups belonging to the same population was computed, and probabilities of 0.05 or less were considered indicative of significant change.

C. Results

The results of these experiments are shown in Tables 7.6 through 7.10.

From Table 7.6, we see that the girth of the thighs (both left and right) was significantly increased by the exercise program. This increase in size was not reflected in the measurements of girth of the calf muscles.

Of the several parameters included in the respiratory function data (Table 7.7), only three were shown conclusively to be changed. The forced vital capacity (FVC), forced inspiratory capacity (FIC), and the FEV_1 were all increased posttraining over their pretraining values. The minute ventilation and the respiratory rate were also changed, possibly significantly. Variance of these two parameters is great, however, and the significance is less clear.

TABLE 7.6. Leg girth data for quadriplegics.

Site	Mean (cm) Before training	After training	SD	Significance
Left thigh, 10 cm	38.9	41.1	2.00	$p < 0.05$*
Left thigh, 20 cm	44.4	46.5	2.61	$p < 0.05$*
Right thigh, 10 cm	38.7	40.4	2.43	$p < 0.10$
Right thigh, 20 cm	44.3	46.2	2.47	$p < 0.05$*
Left calf	31.1	30.9	0.63	N.S.
Right calf	31.1	31.4	0.81	N.S.

*Statistically significant.

In Table 7.8, we see that standard cardiovascular vital signs show changes in some areas but not in others. The mean heart rate was slightly greater after the exercise program than before, and in this specific experiment the variance was low enough that the difference is significant.

Cardiac output (as measured by impedance cardiography) was significantly increased by the exercise training, and this difference is reflected in the change in cardiac index, as expected. By comparing the cardiac output data with those obtained by echocardiography (Table 7.9), it will be seen that impedance cardiography provides figures that

TABLE 7.7. Respiratory function data for quadriplegics.

Parameter	Before	Percent of norm	After	SD	Significance
VC	3.47 L	58.4	3.56 L	0.36	N.S.
FVC	3.23 L	54.6	3.42 L	0.15	$p < 0.05$*
FEV_1	2.77 L	59.1	3.07 L	0.24	$p < 0.05$*
PEF	7.15 L/min	71.1	7.73 L/min	1.56	N.S.
FIC	3.30 L	55.6	3.42 L	0.16	$p < 0.05$*
PIF	5.98 L/min		6.43 L/min	1.76	N.S.
ER	0.93 L		0.90 L	0.30	N.S.
IC	2.54 L		2.65 L	0.61	N.S.
MV	7.96 L/min		10.97 L/min	4.76	$p < 0.10$
TV	0.54 L		0.63 L	0.20	N.S.
RR	14.3 beats/min		16.6 beats/min	4.10	$p < 0.10$
CO (Fick)	4.28 L/min		3.76 L/min	1.41	N.S.

*Statistically significant.

TABLE 7.8. Hemodynamic data for quadriplegics.

Parameter	Before training	After training	SD	Significance
Heart rate	51.4 beats/min	54.5 beats/min	3.98	$p < 0.05$*
Systolic pressure	105 mmHg	104 mmHg	6.12	N.S.
Diastolic pressure	66 mmHg	68 mmHg	6.21	N.S.
Mean arterial pressure	79 mmHg	80 mmHg	5.67	N.S.
Cardiac output	4.14 L/min	4.47 L/min	0.28	$p < 0.05$*
Peripheral resistance[a]	19.5	18.4	1.53	$p < 0.10$

[a]Units are millimeters of mercury per liter per minute.
*Statistically significant.

are about 3% higher. The accuracy of cardiac output methods is rarely considered to be greater than about 5%, even with the best of methods.

Finally, there is a possibly significant decrease in the calculated peripheral resistance posttraining (Table 7.8). This would result from the significant increase in cardiac output posttraining, while the mean arterial blood pressure remains relatively constant pre- and posttraining.

The echocardiographic data, seen in Table 7.9, show that the P-R interval was shortened slightly but significantly. There is a statistically significant increase in heart rate. Otherwise, the major difference seen is the increase in cardiac output and, of course, cardiac index. Thus, the relative changes of cardiac output with the impedance method and the echocardiographic method are quite comparable.

The statistically significant difference in heart rate change in the cold pressor test is in a direction such that there is minimal change posttraining, compared to a significant increase pretraining. A similar difference in the amount of change of heart rate in the tilt test is seen, and these two parameters may be considered to be statistically significant (Table 7.10).

D. Discussion

This section has identified a number of parameters in standard physiological measurement that are changed significantly by a bicycle ergometry exer-

TABLE 7.9. ECG and echocardiographic data for quadriplegics.

Parameter	Before training	After training	SD	Significance
Heart rate	51.4 beats/min	54.5 beats/min	3.98	$p < 0.05$*
P-R interval	0.186 second	0.170 second	0.01	$p < 0.05$*
LVET	0.336 second	0.332 second	0.02	N.S.
PEP	0.104 second	0.102 second	0.005	N.S.
Stroke volume	76.7 mL	77.4 mL	2.63	N.S.
Cardiac output	3.88 L/min	4.18 L/min	0.29	$p < 0.05$*
Cardiac index	2.21 L/mm/m²	2.36 L/mm/m²	0.13	$p < 0.05$*

*Statistically significant.

TABLE 7.10. Cardiovascular reflex data for quadriplegics.

Parameter[a]	Before	After	SD	Significance
Cold pressor				
HR change	5.07	0.49	4.75	$p < 0.05$*
SP change	8.25	3.75	7.23	$p < 0.10$
DP change	10.25	8.25	6.59	N.S.
Diving reflex				
HR change	0.91	−1.44	7.79	N.S.
SP change	2.50	0.25	5.80	N.S.
DP change	1.75	−1.25	5.95	N.S.
Tilt table test				
HR change	6.87	−1.40	11.27	$p = 0.05$*
SP change	−4.25	−3.75	9.72	N.S.
DP change	−1.75	−3.75	7.48	N.S.

[a]Units: HR, beats/min; SP, DP, mmHg.
*Statistically significant.

cise program. It should be noted that the exercise program reported in this section was not a long program (only 24 sessions for each subject), and it was conducted at a very low intensity. Exercise sessions were performed with zero load on the ergometer. There is no assurance that the intensity of the exercise, its duration, or the pattern in which it was administered is optimal for physical conditioning of the deconditioned paralyzed individual. It is distinctly possible, for example, that changes in statistically significant physiological parameters might have been larger with higher intensity and longer duration exercise. Moreover, borderline statistical changes ($p \leq 0.10$) in certain physiological parameters might have become statistically significant ($p < 0.05$), with an optimal pattern of FES bicycle exercise.

Regarding the muscular response of the paralyzed extremity, the increase in thigh circumferential measurements was not reflected in the calf muscle measurements. This would indicate that spillover from the stimulated upper leg muscles to the nonstimulated lower leg muscles was minimal.

Three pulmonary parameters (FVC, FIC, and FEV_1) were all increased posttraining at a statistically significant level. It should be noted that these three parameters are descriptive primarily of the ability to move air rapidly. In pulmonary mechanics terminology, they reflect alterations to lung resistive properties (airway and chest wall) as contrasted with lung compliance properties. The direction of the changes would indicate a relative decrease in the pulmonary resistance values posttraining (resulting in higher peak flow rates). The mechanism of this change in the quadriplegic (i.e., functionally sympathectomized) individual is uncertain (see below).

A small but significant increase in the resting heart rate was observed posttraining. This trend is in the opposite direction from the training effect observed in normal individuals. Increased resting heart rate probably reflects some reversal of the lack of sympathetic drive and predominant vagal tone of the heart and circulatory system. The explanation for this is not certain but has been discussed in some detail in Section II. Such a reversal (even slight) would also cause slight bronchodilatation, which could lead to a reduction of pulmonary resistance.

If this hypothesis were correct, the reversal could be either humorally mediated or neurally mediated. To test this hypothesis thoroughly, quadriplegics would need to be tested both pretraining and posttraining for resting serum catecholamine levels and 24-hour urinary catecholamine excretion. They would also need to be tested pre- and posttraining for changes in sympathetic autonomic efferent neural activity using both cold pressor tests and dermatomal sweating response tests.

There was a definite increase in the cardiac output and associated cardiac index with FES bicycle exercise training of quadriplegics. It can be assumed that significant increase in upper leg girth (Table 7.6) correlates with an increase in the muscle mass. The increased cardiac index is probably a reflection of the increased blood flow to supply the leg muscles reversed from their previously atrophic state. However, both cross-sectional CT scans of the upper leg muscles pre- and posttraining and regional blood flow studies of the legs of quadriplegics (pre- and posttraining) will be required before this can be stated definitively.

The heart rate response to both the cold pressor test and tilt table test indicates a minimal response posttraining. The mechanism of this response is not clear. Danopulos et al. (1985) have described in detail the cardiovascular response of nine quadriplegics (nontrained) to a 40° head-up tilt table test. These subjects (as did our own subjects pretraining) all responded with a significant increase in heart rate. The minimal heart rate response that we have observed in the quadriplegics posttraining is impressive, however, when it is considered that these are changes in heart rate that occur when the cardiovascular system is stimulated.

In conclusion, this section has identified those parameters that are most likely to be worth measuring in subsequent studies of the effect of exercise on physical conditioning. The experimental results also show clearly that it is possible to obtain statistically significant changes with a relatively small number of subjects.

It is not out of order to conclude from the results of these experimental observations that a program of bicycle ergometry exercise enhances the physical condition of paralyzed persons, thus increasing their quality of life and expectations of further developments in their condition.

References

Bebarge, O., Christensen, N.J., Corbett, J.L., Eidelman, B.H., Frankel, H.L., and Mathias, C.J. Plasma catecholamines in tetraplegics. *Paraplegia*, 12:44–49, 1974.

Besarab, A., Silva, P., Landsberg, L., and Epstein, F.H. Effect of catecholamines on tubular function in the isolated perfused rat kidney. *Am. J. Physiol.* 233: F39–F45, 1977.

Bello-Reuss, E. Effect of catecholamines on fluid reabsorption by the isolated proximal convoluted tubule. *Am. J. Physiol.* 238:F347–F352, 1980.

Convertino, V.A., Greenleaf, J.E., and Bernauer, E.M. Role of thermal and exercise factors in the mechanism of hypervolemia. *J. Appl. Physiol.* 48:657–664, 1980.

Danopulos, D., Kezdi, P., Stanley, E.L., Petrofsky, J.S., Phillips, C.A., and Meyer, L.S. Cardiovascular circulatory dynamics in subjects with chronic cervical spinal cord injury in supine and head-up tilt position. *J. Neurol. Orthop. Med. Surg.* 6:265–270, 1985.

Danopulos, D., Kezdi, P., Stanley, E., Petrofsky, J., Stacy, R., Phillips, C., and Meyer, L. Changes in cardiovascular circulatory dynamics after a twelve week active bicycle rehabilitation in young tetraplegics. *J. Neurol. Orthop. Med. Surg.* 7:179–184, 1986.

D'Cruz, L.A. *Echocardiographic Diagnosis*. Macmillan, New York, 1983.

Demaria, A.N., Neumann, A., Garrett, L., Fowler, W., and Mason, D.T. Alterations in ventricular mass and performance induced by exercise training in man evaluated by echocardiography. *Circulation*, 57:237–244, 1978.

Fortney, S.M., Wenger, C.B., Bove, J.R., and Nadel, E.R. Effect of blood volume on forearm venous and cardiac stroke volume during exercise. *J. Appl. Physiol.* 55:884–890, 1983.

Gass, G.C., and Camp, E.M. Physiological characteristics of trained Australian paraplegic and tetraplegic subjects. *Med. Sci. Sports Exercise*, 11:256–259, 1979.

Goldfien, A., Moore, R., Zileli, S., Havens, L.L., Boling, L., and Thorn, G.W. Plasma epinephrine and norepinephrine levels during insulin-induced hypoglycemia in man. *J. Clin. Endocr. Metab.* 21:296, 1961.

Guttmann, L. Munro, A.F., Robinson, R., and Walsh, J.J. Effect of tilting on the cardiovascular responses and plasma catecholamine levels in spinal man. *Paraplegia*, 1:4–18, 1963.

Ikaheimo, M.J., Palatsi, I.J., and Takkunen, J.T. Noninvasive evaluation of the athletic heart: Sprinters versus endurance runners. *Am. J. Cardiol.* 44:24–30, 1979.

Mathias, C.J., and Frankel, H.C. Autonomic failure in tetraplegia. In *Autonomic Failure*, Vol. I, R. Bannister, Ed. Oxford University Press, Oxford, pp. 453–488, 1983.

Mathias, C.J., Christensen, N.J., Corbett, J.L., Frankel, H.L., Goodwin, T.J., and Peart, W.S. Plasma catecholamines, plasma renin activity and plasma aldosterone in tetraplegic man, horizontal and tilted. *Clin. Sci. Mol. Med.* 49:291–299, 1975.

Mathias, C.J., Christensen, N.J., Corbett, J.L., Frankel, H.L., and Spalding, J.M.K. Plasma catecholamines during paroxysmal neurogenic hypertension in quadriplegic man. *Circ. Res.* 39:204–208, 1976.

Morganroth, J., Maron, B.J., Henry, W.L., and Epstein, S.E. Comparative left ventricular dimensions in trained athletes. *Ann. Intern. Med.* 82:521–524, 1975.

Muller, E.A. Das Arbeitsmaximum bei statischer Haltearbeit. *Arbeitsphysiologie*, 5:605, 1932.

Nadel, E.R. Physiological adaptations in aerobic training. *Am. Sci.* 73:334-343, 1985.

Nishimura, T., Yoshihisa, Y., and Chuichi, K. Echocardiographic evaluation of long-term effects of exercise on left ventricular hypertrophy and function in professional bicyclists. *Circulation*, 61:832-840, 1980.

Peckham, P.H. Control of contraction strength of electrically stimulated muscle by pulse width and frequency modulation. *Eng. Med. Biol.* 18:116, 1976.

Petrofsky, J.S., and Phillips, C.A. Constant velocity contractions in skeletal muscle of the cat. *Med. Biol. Eng. Comput.* 17:583-592, 1979.

Petrofsky, J.S., and Phillips, C.A. The influence of recruitment order and fiber composition on the force-velocity relationship of skeletal muscle in cat. *Med. Biol. Eng. Comput.* 18:381-390, 1980.

Petrofsky, J.S., and Phillips, C.A. Influence of temperature, initial length and recruitment order on the velocity of contraction of skeletal muscle in the cat. *J. Biomech.* 14:297-306, 1981.

Petrofsky, J.S., and Phillips, C.A. Active physical therapy—A modern approach for rehabilitation of the disabled. *J. Neurol. Orthop. Surg.* 4:165-173, 1983.

Petrofsky, J.S., and Phillips, C.A. The use of functional electrical stimulation for rehabilitation of spinal cord injured patients. *CNS Trauma*, 1:57-72, 1984.

Petrofsky, J.S., and Phillips, C.A. Closed-loop control of movement of skeletal muscle. *CRC Crit. Rev. Biomed. Eng.* 13:35-96, 1985.

Petrofsky, J.S., Rinehart, J.S., and Lind, A.R. Isometric strength and endurance of slow and fast muscles in the cat. *Fed. Proc.* 35:291, 1976.

Petrofsky, J.S., Heaton, H.H., and Phillips, C.A. Outdoor bicycle for exercise in paraplegics and quadriplegics. *J. Biomed. Eng.* 5:292-296, 1983a.

Petrofsky, J.S., Heaton, H.H., Phillips, C.A., and Glaser, R.M. Modification of the Apple computer as an exercise device for paralyzed muscle. *Coll. Microcomput.* 1:97-104, 1983b.

Petrofsky, J.S., Heaton, H.H., and Phillips, C.A. Leg exerciser for training of paralyzed muscle by closed-loop control. *Med. Biol. Eng. Comput.* 22:298-303, 1984a.

Petrofsky, J.S., Phillips, C.A., and Heaton, H.H. Bicycle ergometer for paralyzed muscle. *J. Clin. Eng.* 9:13-19, 1984b.

Petrofsky, J.S., Phillips, C.A., and Hendershot, D. Cardiorespiratory stresses which occur during dynamic exercise in paraplegics and quadriplegics. *J. Neurol. Orthop. Surg.* 6:252-258, 1985.

Phillips, C.A. The medical criteria for active physical therapy: Physician guidelines for patient participation in a program of functional electrical exercise. *Am. J. Phys. Med.* 66:269-286, 1987.

Phillips, C.A., and Petrofsky, J.S. Velocity of contraction of skeletal muscle as a function of activation and fiber composition: A mathematical model. *J. Biomech.* 13:549-558, 1980.

Phillips, C.A., and Petrofsky, J.S. Computer controlled movement of paralyzed muscle: The medical perspective. *Proceedings of the First Vienna International Workshop on Functional Electrostimulation*, paper 5.2, Oct. 19-22, 1983.

Phillips, C.A., Petrofsky, J.S., Hendershot, D.M., and Stafford, D. Functional electrical exercise: A comprehensive approach for physical conditioning of the spinal cord injured individual. *Orthopedics*, 7:1112-1123, 1984.

Rhodes, E.C., McKenzie, D.C., Coutts, K.D., and Rogers, A.R. A field test for the prediction of aerobic capacity in male paraplegics and quadriplegics. *Can. J. Appl. Sport Sci.* 6:182-186, 1981.

Sahn, D.J., Demaria, A., Kisslo, J., and Weyman, A. Recommendations regarding quantitation in M-mode echocardiography: Results of a survey of echocardiographic measurements. *Circulation*, 58:1072-1083, 1978.

Sever, P.S., Gordon, D., and Vecht, R.J. Catecholamines, left ventricular function and isometric exercise. In *Hypertension: Mechanisms and Management*, A. Distler and T.L. Phillips, Eds. Springer, Heidelberg, pp. 55-63, 1980.

Snoeckx, L.H.E.H., Abeling, H.F.M., Lambregts, J.A.C., Schmitz, J.J.F., Verstappen, F.T.J., and Rene-Man, R.S. Cardiac dimensions in athletes in relation to variations in their training program. *Eur. J. Appl. Physiol.* 52:20-28, 1983.

Solomonow, M., Foster, J., Eldred, E., and Lyman, J. Proportional control of paralyzed muscle, *Fed. Proc.* 37:215, 1978.

Thoma, H., Holle, J., Moritz, E., and Stohr, H. Walking after paraplegia: A principle concept. *Proceedings of the Sixth International Symposium on External Control of Human Extremities*, Dubrovnik, pp. 71-80, 1978.

Vendsalu, A. Studies on adrenalin and noradrenalin in human plasma. *Acta Physiol. Scand.* (Suppl. 173) 49:1-77, 1960.

Vrbova, G. The effect of motorneurone activity on the speed of contraction of striate muscle. *J. Physiol.* 169:513-526, 1963.

Weissler, A.M., Harris W.S., and Schoenfeld, C.D. Systolic time intervals in heart failure in man. *Circulation*, 37:149-159, 1968.

Weissler, A.M., Harris, W.S., and Schoenfeld, C.D. Bedside techniques for the evaluation of ventricular function in man. *Am. J. Cardiol.* 23:577-583, 1969.

Part 3
Ambulatory Exercise Rehabilitation

8 Ambulatory Exercise Rehabilitation: Introduction and Overview

I. Synopsis of Ambulatory Exercise Systems, Procedures, and Effects

A. Background

Interest in electrical stimulation for ambulation of spinal cord injury (SCI) individuals began with the report of "electrical splinting" of a paraplegic patient by Brindley et al. (1978) more than a decade ago. Since then, however, a number of problems associated with developing a practical electrical stimulation system for SCI have been encountered (Stallard et al., 1989). The most practical approach to date appears to combine electrical stimulation of paralyzed muscles with lower extremity orthotic bracing, as first proposed by investigators from two universities (Petrofsky et al., 1985, 1986). However, only recently has this approach been reduced to physician-prescribable, commercially available components.

B. Basic Device

The ambulation system consists of four interacting components: surface electrodes (e.g., Medtronic models 3793 and 3795, from the Neurological Division, Minneapolis) are applied to selected lower extremity muscle groups; commercially manufactured, small and portable electrical muscle stimulators (EMS: e.g., from NTRON)* are worn on a belt and connected to the electrodes; EMS override switches (e.g., NTRON Part 473 D) are attached to a standard four-point walker and connected to the EMS units (to allow the patient manual control of the electrical stimulation); balance assistance is provided (via the walker); finally, the physician-prescribed and orthotist-fitted reciprocating gait orthosis from Durr-Fillauer Medical, Inc.,† is applied to the paralyzed lower extremities.

This medically prescribed system has been reported to produce stand-up and sit-down (from a wheelchair or other chair) as well as forward and backward ambulation in a paraplegic (Phillips, 1990) and a quadriplegic (Phillips, 1989) patient.

C. Indications and Contraindications for Usage

The system is limited to patients with paralysis from spinal cord injury who have fairly complete lesions (very little motor and sensory function below the level of the lesion). The upper injury level appears to be C6/C7 and the lower level is at T11/T12. The medical criteria by which patients are selected (or rejected) for initial prescription of the system, as well as the physician guidelines for periodic monitoring of patient progress (or lack thereof), have been reported (Phillips, 1987).

**NTRON Electronics EMS-8100: Patient Instruction Booklet* is available from NTRON, Inc., 104 Industrial Boulevard, Sugarland, TX 77478.

†The Orthopedic Division of Durr-Fillauer (P.O. Box 1678, Chattanooga, TN 37401) publishes *LSU Reciprocating Gait Orthosis: A Pictorial Description and Applications Manual.*

D. Patient Preconditioning

The selected patients must be preconditioned with electrical stimulation (ES) exercise prior to ES ambulation (Phillips, 1987). Such exercise consists of ES isokinetic leg training and ES bicycle ergometry as first described in 1983 (Petrofsky and Phillips, 1983; Phillips and Petrofsky, 1983). The systems, the Regys I and Ergus I Electrical Stimulation Bicycle Ergometers,* which are now commercially available, have been reported to produce acute exercise effects on the SCI patients (Phillips et al., 1984), as well as chronic conditioning effects (Danopulos et al., 1986; Pollock et al., 1986; Phillips et al., 1989).

E. Therapeutic Results

It is now generally acknowledged that there are significant potential health benefits when a paraplegic is able to stand and walk rather than remain sedentary in a wheelchair (Rowley and Edwards, 1987). Cardiopulmonary efficiency has been reported to improve significantly in both a paraplegic and a quadriplegic patient who engaged in daily ES ambulation over a period of months (Hendershot et al., 1987). Furthermore, cardiopulmonary (oxygen) demand when ambulating with ES and a lower extremity (LE) orthosis is lower than when ambulating with a LE orthosis alone (Hirokawa et al., 1990; Phillips and Hendershot, 1991). Finally, SCI patients who regularly stand and walk with ES and an LE orthosis have more positive psychological attitudes and personality characteristics than a similar group of nonwalkers (Bell et al., 1987). Many possible long-term therapeutic effects are not known, however.

F. Risks

A study by one group of investigators found that ES for exercise of paralyzed muscle in SCI patients is safe (Ragnarsson et al., 1986). However, other investigators report that some bone fractures have occurred in their series of patients (Phillips and Petrofsky, 1985). A major predisposing factor appears to be disuse osteoporosis.

Some investigators report that using ES to improve muscle function in SCI patients can cause occasional skin burns (Balmaseda et al., 1987). These authors conclude that this can be avoided, however, with careful technique. Other possible long-term risks are not known.

G. Conclusions

Physician-prescribable electrical stimulation (with commercially available components) for ambulation of SCI patients is now a clinical treatment reality. For medically qualified and properly preconditioned SCI patients, there appear to be selected (cardiopulmonary and psychological) therapeutic benefits. Other potential long-term physiological and psychological therapeutic benefits are not known. Neither are the potential quality of life or vocational rehabilitation benefits known. However, there are also some risks. In addition, there may be long-term risks, not known at this time.

Electrical stimulation for ambulation of SCI patients is a new clinical treatment modality. Only with the acquisition of extensive patient information, collected from numerous medical practitioners, can we ultimately assess both the risk–benefit ratio and the cost–benefit ratio.

II. Overview of Ambulatory Exercise Rehabilitation

By utilizing commercially available electrical muscle stimulators (EMS) for functional electrical stimulation (FES) and interfacing them with a reciprocating gait orthosis (RGO), a new system has been developed; it is described herein as an EMS-RGO. Various advantages of the system include (a) commercially available subsystems from various manufacturers and (b) subsystems recommended for applications such as gait training. The simplified system currently employed features two EMS units worn on a belt, controlled by remote switches and interfacing to electrodes placed over the hamstring and gluteal muscle groups of each leg. The two EMS units (for stimulation of the hip extensors) function primarily during ambulation. Each EMS unit is powered by a 9 V alkaline transistor battery that provides approxi-

*Therapeutic Technologies, Inc., 5709 Johns Road, Tampa, FL 33634.

mately 2 miles (> 10,000 feet) of walking before replacement is necessary. The system operation is reported here for two paraplegic and two quadriplegic individuals.

A. Introduction

A major thrust of our laboratory during the past few years has been to make FES walking technology available to the medical community for use by the general spinal-cord-injured population. The limitations of our past approach (Petrofsky et al., 1985, 1986) have been (a) the separate portable power pack (consisting of four Kodavision 8 V batteries and eight lead-acetate 2 V batteries) that weighs approximately 8 pounds and must be carried on the back; (b) the need to modify a conventional walker with special push-button switches for system control; and (c) the need for specialized and custom-built electronic package and power pack.

This section describes a new system that overcomes these limitations. By utilizing commercially available electrical muscle stimulators as the FES component (in combination with the RGO), the new system termed EMS-RGO has been developed. The advantages of such a system are (a) the power pack is reduced to a single 9 V alkaline battery contained within each EMS; (b) remote control switches (available as standard accessories with each EMS) can be attached directly to the walker without special modification; and (c) a total of two EMS units, each weighing less than 6 ounces, with battery, provide complete level-surface walking functions. Each component of the EMS-RGO system is commercially available from various manufacturers, recommended for applications such as "gait training," and a physician-prescribable technology (i.e., available as a prescription item from a licensed physician).

B. Materials

The EMS-RGO system consists of the following components.

1. Reciprocating Gait Orthosis (RGO)

A commercially manufactured RGO (Durr-Fillauer, Inc.) is applied to the lower extremities of the paraplegic or quadriplegic. This is a bilateral hip-knee–ankle–foot orthosis (HKAFO), which utilizes two interconnecting (reciprocating) cables at the hips. With weight shifted to the ipsilateral leg, stimulation is applied to the ipsilateral hip extensors (gluteal and hamstring muscles). Through the action of the cables, this results in hip flexion (leg forward) of the contralateral (non-weight-bearing) leg (Figure 8.1).

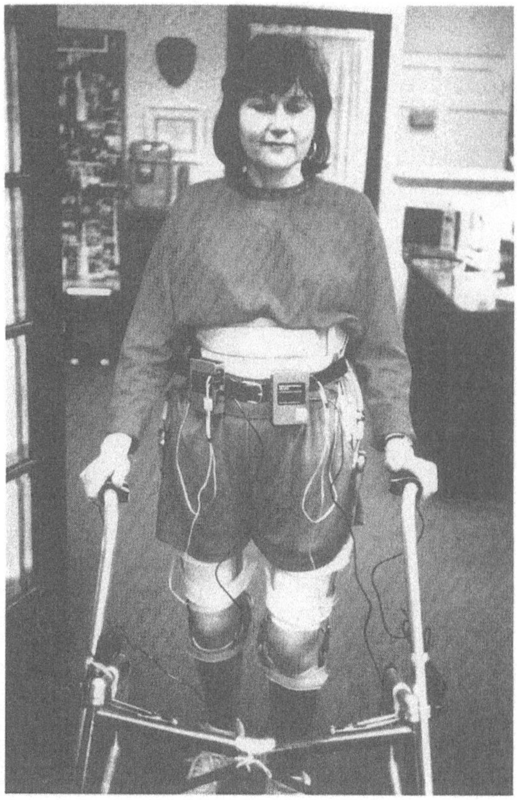

FIGURE 8.1. The EMS-RGO with the RGO worn outside the clothing of a paraplegic patient.

2. Electrical Muscle Stimulator

The EMS unit employed in this study was the EMS-8100 muscle stimulator (NTRON). Two such units were employed, hereafter designated as A and B. They were worn in the front of a belt (encircling the waist) as shown in Figure 8.2. The initial operating parameters were adjusted for one set of values and the operating parameters were subsequently adjusted during the course of the study for optimal individual performance.

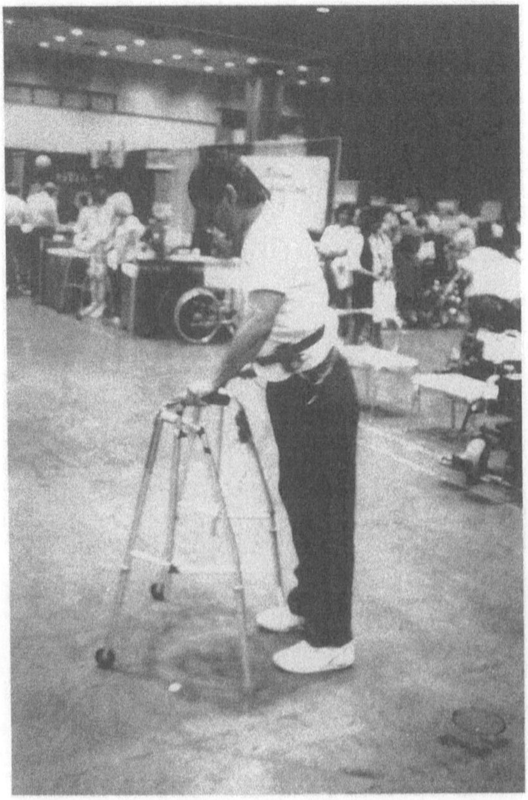

FIGURE 8.2. EMS units worn on the front of the belt are adjusted by a quadriplegic patient (RGO is under the clothing).

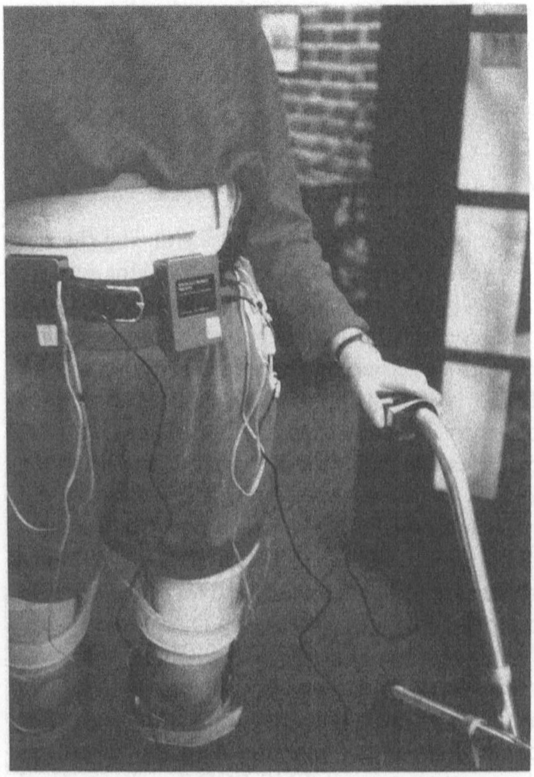

FIGURE 8.3. Configuration of the remote control switches located near the handle of a conventional walker (as used by paraplegic patient).

3. Switches

A commercially available remote control switch (NTRON part 473D) was part of the system. This slide-type switch, normally "off," will slide to and hold in the "on" state and activate the EMS unit. One remote control switch was connected to each of the EMS units (A and B) for a total of two remote control switches (remote switches A and B). Remote switch A was mounted near the right handle of a conventional walker, and remote switch B was mounted near the left handle. These switches were mounted such that they would be thumb activated (Figure 8.3). The palm and remaining fingers rested on and gripped the handles.

4. Batteries

A 9 V alkaline transistor battery was used with each EMS unit designed to operate with a 9 V battery; 7.2 V rechargeable Ni-Cad batteries were tried but did not provide sufficient voltage amplitude to meet the current demands of the application. Battery capability was tested in a continuing series of experiments in which the battery voltage was measured at the terminals before and after each experimental session. The activities of that session were also recorded. When the battery could no longer provide sufficient voltage output to meet the current demands of the applications, the results were calculated.

5. Electrodes

These studies utilized four 2 in. × 4 in. carbon rubber electrodes (Medtronic) and four 1½ in. × 1½ in. carbon rubber electrodes from the same manufacturer. One carbon rubber electrode of each size is placed over each gluteal muscle group and each

TABLE 8.1. Characteristics of the SCI subjects in the EMS-RGO study.

Subject	Age (years)	Sex	Level of injury	Duration (years)	Functional class (Frankel)
G.B.	28	M	C7	6	C
D.M.	25	M	C7	5	C
S.R.	30	M	T5	7	B
S.T.	41	F	T5	5	A

hamstring muscle group (for a total of eight electrodes, four of each type). Liqui-Cor electrode gel is applied first, and both adhesive patches and adhesive tape are utilized to secure the electrodes. EMS unit A is connected to the right hip extensors and EMS unit B is connected to the left hip extensors.

C. Methods

Four spinal-cord-injured persons participated in this study. Their ages, sex, level of injury, time since injury and Frankel functional class are given in Table 8.1. The EMS-RGO system interacts with the patients to perform specific tasks including forward and backward walking, and turning. These procedures are described elsewhere (Phillips, 1989, 1990).

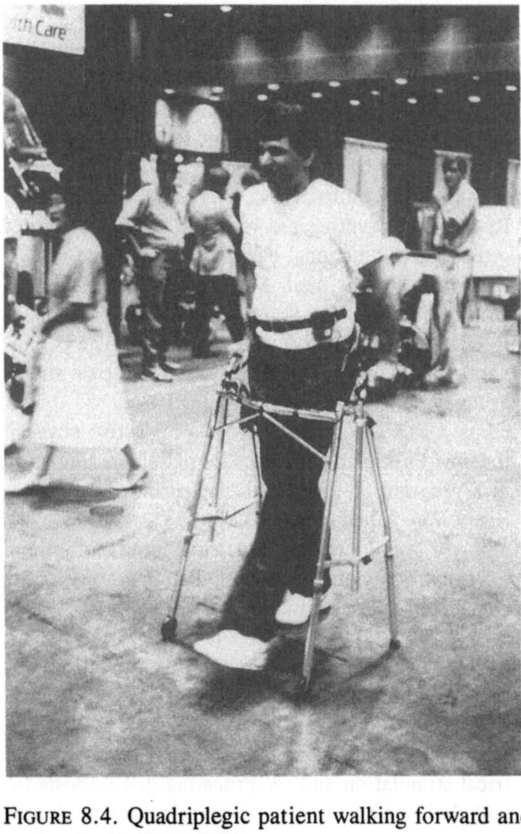

FIGURE 8.4. Quadriplegic patient walking forward and outfitted with EMS-RGO.

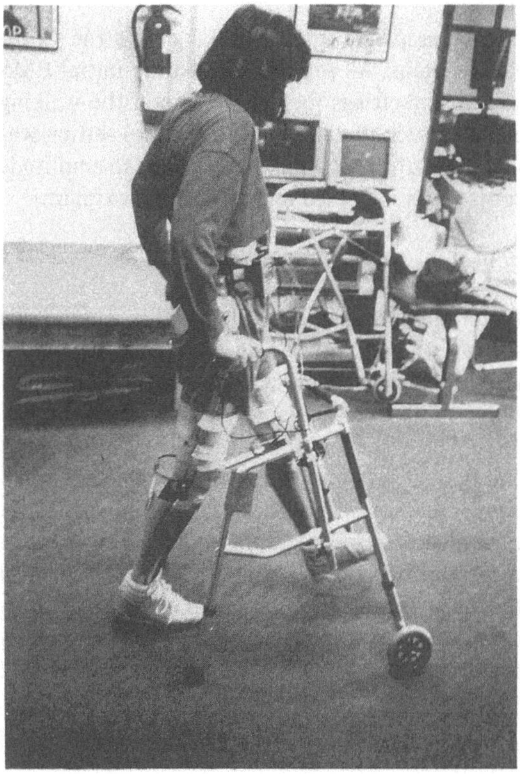

FIGURE 8.5. Paraplegic patient walking forward while utilizing the EMS-RGO.

D. Results

The results are divided into walking performance when utilizing the EMS-RGO system and battery performance.

1. *Walking Performance*

The two paraplegic and two quadriplegic individuals routinely covered between 1.0 and 1.6 kilometers per experimental session at velocities ranging from 1.2 to 2.0 km/h. The quadriplegic patient shown ambulating with the EMS-RGO in Figure 8.4 is wearing the RGO under his clothing. In Figure 8.5, the RGO is on the outside of the clothing of the paraplegic subject walking with the EMS-RGO system. All subjects altered the initial EMS parameter settings (before the end of the walking episode) such that the frequency (rate) setting was increased to 100 Hz (maximum) and the amplitude setting was increased to 9.0 V (the maximum).

2. *Battery Performance*

Battery performance was calculated as amount of a specific activity per volt, since with alkaline batteries the output voltage at the terminals is proportional to the current delivery capability of the battery. With respect to the batteries of EMS units A and B, there was approximately 10,400 feet of walking per one volt drop at the battery terminals. These figures also approximate the useful lifespan of the respective batteries, since after about a 1 V drop of the output terminal voltage, there is no longer sufficient current delivery capability for satisfactory performance of the EMS-RGO system.

E. Discussion

As with any treatment modality, the patient must medically qualify for the technology being prescribed. Thus there are medical criteria that must be satisfied for prescription of active physical therapy including walking exercise with the EMS-RGO. The initial evaluation of the patient candidate includes history, physical examination, lab tests, report summary, and an FES prescription (level of clearance), presented in more detail in Chapters 11 and 12.

In the Wright State program, the FES leg trainer is currently considered to be the entry level FES exercise modality. Progression to the FES bicycle ergometer requires adequate flexion and medial–lateral stability at the knee, as well as minimal dorsiflexion at the ankle (see Chapter 11, Table 11.1). Progression to the EMS-RGO and walking exercise then requires muscular, cardiovascular, and respiratory system function sufficient for 30 minutes of continuous FES bicycling at minimal (0 kp·m) load. Other medial criteria for walking exercise with the EMS-RGO are summarized in Chapter 11 (Table 11.1).

A functional compromise has been made in simplifying the EMS-RGO system from four to six EMS units down to two EMS units. Specifically, the stand-up and sit-down capabilities are no longer available as described in Chapter 9.

References

Balmaseda, M.T., Jr., Fatehi, M.T., Koozekanani, S.H., and Sheppard, J.S. Burns in functional electric stimulation: Two case reports. *Arch. Phys. Med. Rehabil.* 68:452–453, 1987.

Bell, R., Kopf, E.H., and Toadvine, J. A study of personality changes in quadriplegics engaged in a functional electrical stimulation (FES) program. *J. Neurol. Orthop. Med. Surg.* 8:353, 1987.

Brindley, G.S., Polkey, C.E., and Rushton, D.N. Electrical splinting of the knee in paraplegia. *Paraplegia,* 16:428, 1978.

Danopulos, D., Kezdi, P., Stanley, E., et al. Changes in cardiovascular circulatory dynamics after a twelve week active bicycle rehabilitation in young tetraplegics. *J. Neurol. Orthop. Med. Surg.* 7:179–184, 1986.

Hendershot, D.M., Moore, M.L., and Phillips, C.A. Cardiopulmonary conditioning when walking with and without FES in the paralyzed. *Fed. Proc.* 46:680, 1987.

Hirokawa, S., Grimm, Le, T., et al. Energy consumption in paraplegic ambulation using the reciprocating gait orthosis and electrical stimulation of the muscles. *Arch. Phys. Med. Rehabil.* 71:687–694, 1990.

Petrofsky, J.S., and Phillips, C.A. Active physical therapy—A modern approach for rehabilitation of the disabled. *J. Neurol. Orthop. Surg.* 4:165–173, 1983.

Petrofsky, J.S., Phillips, C.A., Larson, P., and Douglas, R. Computer synthesized walking: An application of orthosis and functional electrical stimulation (FES). *J. Neurol. Orthop. Med. Surg.* 6:219–230, 1985.

Petrofsky, J.S., Phillips, C.A., Douglas, R., and Larson, P. Computer-controlled walking system: The combination of an orthosis with functional electrical stimulation. *J. Clin. Eng.*, 11:121–133, 1986.

Phillips, C.A. The medical criteria for active physical therapy: Physician guidelines for patient participation in a program of functional electrical rehabilitation. *Am. J. Phys. Med.* 66:269–286, 1987.

Phillips, C.A. Functional electrical stimulation and lower extremity bracing for ambulation exercise of the spinal cord injured individual: A medically prescribed system. *Phys. Ther.* 69:842–849, 1989.

Phillips, C.A. Physician directed ambulation for rehabilitation of spinal cord injury. *Surg. Rounds Orthop.* 4:41–47, 1990.

Phillips, C.A., and Hendershot, D.M. Functional electrical stimulation and reciprocating gait orthosis for ambulation exercise in a tetraplegic patient: A case study. *Paraplegia*, in press, 1991.

Phillips, C.A., and Petrofsky, J.S. Computer controlled movement of paralyzed muscle: The medical perspective. *Proceedings of the First Vienna International Workshop on Functional Electrostimulation*, paper 5.2, 1983.

Phillips, C.A., and Petrofsky, J.S. A review of fracture cases in spinal cord injured individuals participating in closed-loop functional electric stimulation experiments. *J. Neurol. Orthop. Med. Surg.* 6:247–251, 1985.

Phillips, C.A., Petrofsky, J.S., Hendershot, D.M., and Stafford, D. Functional electrical exercise: A comprehensive approach for physical conditioning of the spinal cord injured individual. *Orthopedics*, 7:1112–1123, 1984.

Phillips, C.A., Danopulos, D., Kezdi, P., and Hendershot, D. Muscular, respiratory and cardiovascular responses of quadriplegic persons to an FES bicycle ergometer conditioning program. *Int. J. Rehab. Res.* 12:147–157, 1989.

Pollack, S., Haas, F., and Ragnarsson, K.T. Endurance training by functional electrical stimulation of muscles paralyzed by spinal cord injury. *Arch. Phys. Med. Rehabil.* 67:658, 1986.

Ragnarsson, K.T., Pollack, S., O'Daniel, W., Edgar, R., Petrofsky, J., and Nash, M. Clinical evaluation of computerized functional electrical stimulation (FES) in spinal cord injury: A pilot multicenter study. *Arch. Phys. Med. Rehabil.* 67:628, 1986.

Rowley, D.I., and Edwards, J. Editorial: Helping the paraplegic to walk. *J. Bone Jt. Surg.* 69-B:173–174, 1987.

Stallard, J., Major, R.E., and Patrick J.H. A review of the fundamental design problems of providing ambulation for paraplegic patients. *Paraplegia*, 27:70–75, 1989.

9 Ambulatory Exercise Systems

I. Paraplegic Walking System*

A. Introduction

Since the Brindley et al. (1978) report, considerable effort has been focused on the application of neural prostheses for standing and walking in the spinal-cord-injured individual. Such neural prostheses require an intact neuromuscular system below the level of the spinal cord lesion so that the patient can respond to electrical stimulation. However, central command has been lost, and the great majority of patients remain permanently paralyzed below the level of injury.

Some problems have limited the use of FES technology for ambulation of the spinal-cord-injured individual (Gruner, 1986). One serious problem with FES technology is that the subject is not protected from falling. Postural instability may result from stimulation of only a few muscle groups (out of the total population of paralyzed muscles). Furthermore, electronic components may occasionally fail, resulting in partial or total loss of the tonic electrical activity (which maintains the upright posture).

A reciprocating gait orthosis has been employed to resolve these problems. The RGO is a long-leg brace with hip and knee joints, pelvic band, and thoracic supports. It is composed of plastics and lightweight aluminum (Douglas et al., 1983; McCall et al., 1983; Yngve et al., 1984). The pelvic band and thoracic support offer better postural stability for mid- and high level paraplegics than is the case with other braces.

The cabling system is a significant feature of the RGO. Two cables are attached to the anterior and posterior aspects of the hip joints. A reciprocal action at the hips is produced by shifting upper body weight (via posterior movement of the shoulders). This results in forward thrust of the pelvis (hip extension) followed by a transfer of weight onto the arms (and onto the walker). There is a slight rotation of the axis of the body. In order to walk with FES and the RGO, electrical stimulation of contralateral thigh muscles produces contralateral hip extension (of the anteriorly placed leg) so that the posterially placed ipsilateral leg is moved forward (by action of the interconnecting cables). The operation and application of this FES-RGO system is described elsewhere (Petrofsky et al., 1985b, 1986). The FES-RGO (computer-directed FES combined with an RGO) walking system is used in our laboratory as the more energy efficient system (Hendershot et al., 1987) compared to free-standing FES walking or braces-alone walking (Edwards et al., 1986).

A major thrust of our laboratory during the past year has been to make this FES-RGO technology available to the medical community and to facilitate subsequent use by the general spinal-cord-injured population. The disadvantages of our previous system (Petrofsky et al., 1985b, 1986) have been:

*From C.A. Phillips, Electrical muscle stimulators in combination with a reciprocating gait orthosis for ambulation of paraplegics. © 1989 Reprinted from *Journal of Biomedical Engineering*, 11:338–344, 1989; by permission of the publishers, Butterworth–Heinemann Ltd., Westbury House, Bury Street, Guilford, Surrey GU2 5BH, United Kingdom.

1. A heavy and separate portable power pack (consisting of four Kodavision 8 V batteries and eight lead-acetate 2 V batteries). This power pack weighs approximately 8 pounds and is carried on the back.
2. Modification of a conventional walker with special push-botton switches for system control.
3. A heavier, bulkier FES electronic package when the sit-down/stand-up capability was combined with the basic walking system.
4. Specialized and custom-built electronic package and power pack.

By utilizing commercially available electrical muscle stimulators (EMS) as the FES component (in combination with the RGO), a new system has been developed and is termed an EMS-RGO. Advantages of such a system are:

1. The power pack is now reduced to a single 9 V alkaline battery contained within each EMS.
2. Remote control switches (available as standard accessories with each EMS) can be attached directly to the walker without special modification.
3. A total of four EMS units (each weighing less than 6 ounces, with battery) provide complete stand-up/sit-down and walking functions.
4. Each component of the EMS-RGO system is (a) commercially available from various manufacturers, (b) recommended for applications such as "gait training," and (c) technology that is available as a prescription item from a licensed physician.

The purpose of this section is to describe the new system which overcomes the disadvantages of our previous system with the advantages just enumerated. Specifically, the successful application of this system to a paraplegic subject is reported and battery capability is evaluated.

B. Materials

A spinal-cord-injured patient participated in these studies. S.R. is a 30-year-old, male, T5 level paraplegic who was injured 7 years ago. The patient sustained a motor-complete, but sensory-incomplete injury and is considered as Frankel class B.

TABLE 9.1. Technical specifications of the EMS-8100.[a]

Parameter	Description
Power	0–80 mA continuously adjustable (1000 Ω load), each channel
Rate	20–100 pulses per second (Hz)
Width	60–800 μs, continuously adjustable
Rise	0–10 seconds
Fall	0–5 seconds
Duty cycle	On time, 0–30 seconds; off time, 0–30 seconds; continuously adjustable
Synchronous/reciprocal modes	Channels operate in a simultaneous or alternating fashion, per set duty cycle
Power source	9 V alkaline battery (supplied); 7.2 V rechargeable battery (optional)
Waveform	Symmetrical rectangular pulsatile
Manual control option	Remote jack allows for manual control by hand or foot switches (optional)
Size	2⅜ × 3⅜ × 1¼ in. (6 × 9.5 × 3.2 cm)
Weight	5¾ oz (160 g) with battery
Warranty	5 years parts, 1 year labor (excludes case, cables, battery, and electrodes)

[a]All parameters subject to variation due to normal production tolerances; specifications subject to change without notice.

1. *Electrical Muscle Stimulator*

The unit employed in this study was the EMS-8100 muscle stimulator (NTRON). The technical specifications are given in Table 9.1. Four such units were employed, hereafter designated as A, B, C, and D. They were worn in the front of a belt (encircling the waist) with units A and C to the right of the buckle and units B and D to the left (Figure 9.1).

The initial operating parameters were adjusted for one set of values for units A and B and another set of values for units C and D (see Table 9.2). These operating parameters were subsequently adjusted during the course of the study for optimal individual performance.

2. *Switches*

A commercially available remote control switch (NTRON part 473D) was part of the system. This is a slide-type switch, normally "off," which will slide to and hold in the "on" state. Pressing down on the switch slide, while in the "on" position, activates a spring return to return the slide to the "off" position. The switch connects to a length of cable that is terminated by a subminiature phone plug. This plug inserts into a manual override jack on the side

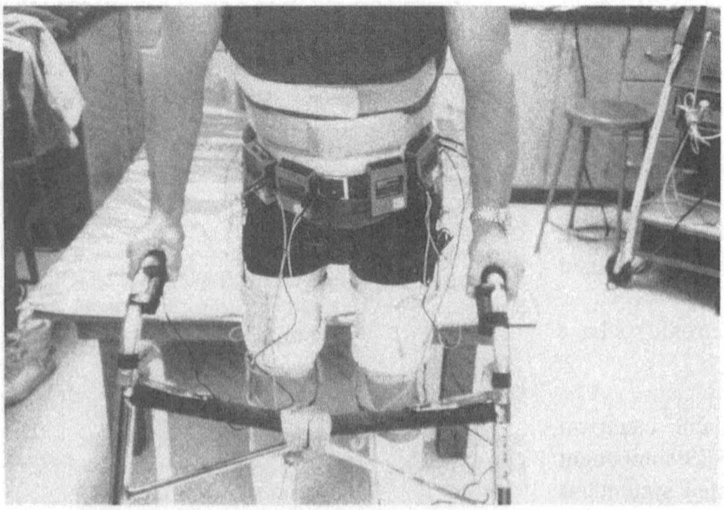

FIGURE 9.1. Four of the EMS units arrayed on the belt of a paraplegic patient.

of the EMS unit. When the switch is "off," the contacts on the phone plug are open and no stimulation current is delivered through the EMS unit's channels A and B to the electrodes. When the switch is "on," the contacts on the phone plug are shorted and stimulation current is delivered via both channels of the EMS device to the electrodes.

One remote control switch was connected to each of the EMS units (A, B, C, and D) for a total of four remote control switches. These will be referred to as remote switches A, B, C, and D, respectively. Remote switches A and B were mounted on the superior aspect of the right and left handles, respectively, of a conventional walker. They were activated by the patient's thumbs. Remote switches C and D were mounted on the lateral (and outside) aspect of the right and left handles, respectively . They were activated by the patient's index fingers. The palm and remaining three fingers rested on and gripped the handles. An example of the mounting configuration is shown in Figure 9.2.

TABLE 9.2. Initial operating parameters for the paraplegic walking system.

Parameter	Units A/B	Units C/D
Rise time	2.0 seconds	0 (instantaneous)
Rate (frequency)	100 Hz	50 Hz
On time	30 seconds	30 seconds
Off time	0	0
Fall time	5 seconds	0 (instantaneous)
Pulse width	800 μs	800 μs
Amplitude (thumbwheel reading)	7.0	7.0

3. Batteries

Nine-volt alkaline transistor batteries were used, since each EMS unit was designed to operate with 9 V batteries; 7.2 V rechargeable Ni-Cad batteries were tried, but did not provide sufficient voltage amplitude to meet the particular current demands of the application.

Battery capability was tested in a continuing series of experiments in which the battery voltage was measured at the terminals before and after each experimental session. The activities of that session were also recorded. When the battery could no longer provide sufficient voltage output to meet the current demands of the application, the results were calculated. The total activity performed was divided by the total battery voltage drop as an indication of battery capability. This type of analysis is reasonably valid, since for 9 V alkaline batteries the fall in voltage at the terminals is proportional to the decrease in current supplying capability (Enercell, 1985).

4. Electrodes

These studies utilized ten 2 in. × 4 in. carbon rubber electrodes (Medtronic model 3793, Neurological Division, Minneapolis) and four 1½ in. × 1½ in. carbon rubber electrodes (Medtronic model 3795).

A transcutaneous transducer garment, TTG, has also been successfully tested and is discussed in Section III of this chapter.

As shown in Figure 9.3, three 2 in. × 4 in. carbon rubber electrodes are placed over each quadriceps muscle group (for a total of six electrodes). Liqui-Cor electrode gel is first applied, and both adhesive patches and adhesive tape are utilized to secure the electrodes.

As shown in Figure 9.4, one 2 in. × 4 in. carbon rubber electrode and one 1½ in. × 1½ in. carbon rubber electrode are placed over each hamstring muscle group and each gluteal muscle group (not shown) for a total of eight electrodes, four of each type. Electrode gel and adhesive are used as before.

Referring to Figure 9.5, EMS unit A is connected to the right quadriceps, EMS unit B is connected to the left quadriceps, EMS unit C is connected to the right hip extensors, and EMS unit D is connected to the left hip extensors.

5. Lower Extremity Orthosis *(see page 146)*

C. Methods of Operation

The EMS-RGO system interacts with the subject to perform a variety of tasks including standing up, walking, turning, and sitting down.

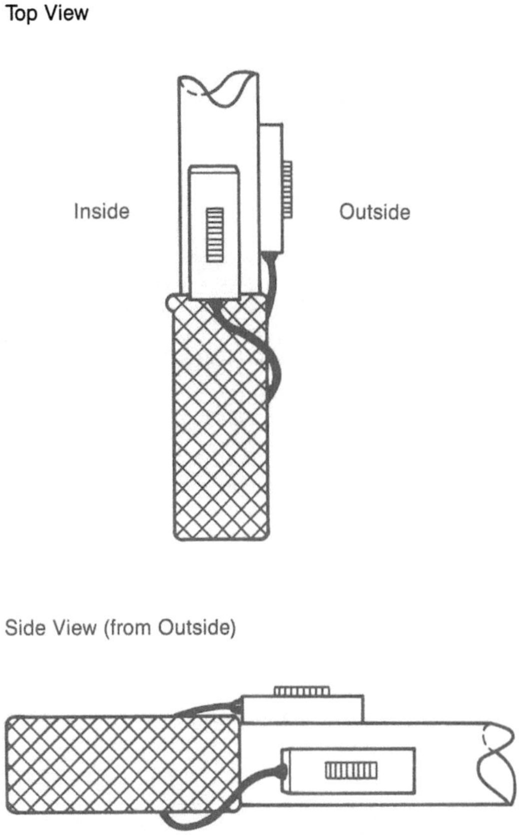

FIGURE 9.2. Configuration of the remote control switches located near the handles of a conventional walker (as used by the paraplegic patient).

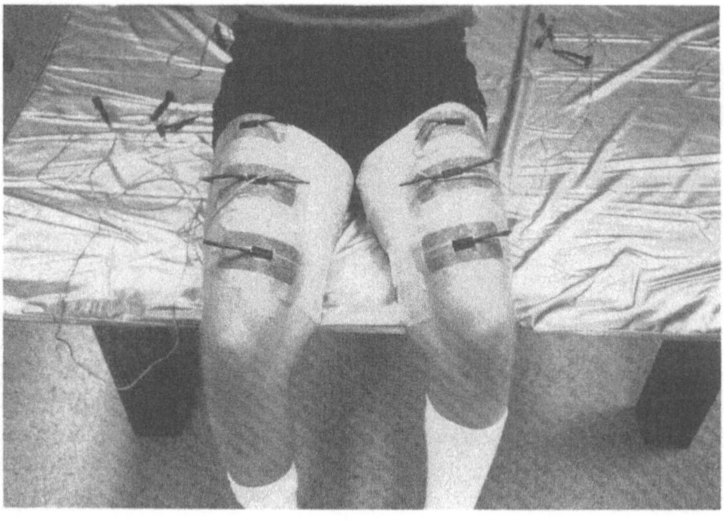

FIGURE 9.3. Electrode configuration over the anterior aspect of the legs (quadriceps muscles).

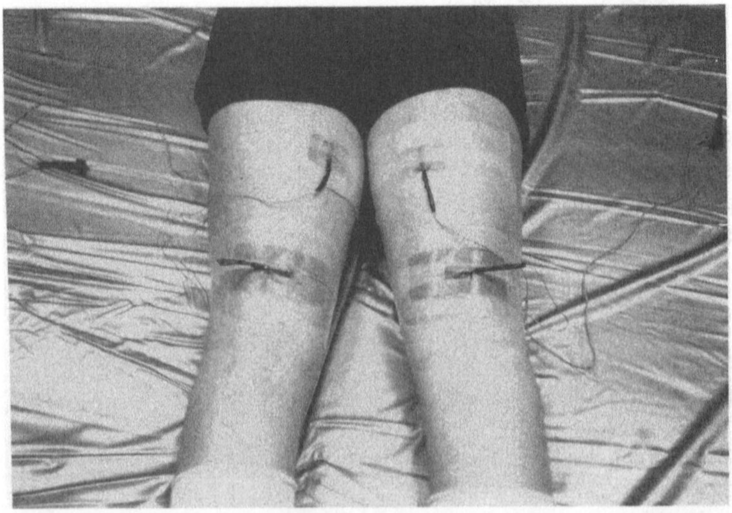

FIGURE 9.4. Electrode configuration over the hip extensor muscles: hamstring muscle group. Gluteal muscle group not shown.

1. Stand-up procedure (Figure 9.6)
   a. Patient with forelegs bent slightly backward at knees. Patient leaning forward at hips (over knees).
   b. Seat elevated so that hip is flexed less than 90°.
   c. Activate EMS units A and B in the constant-on mode.
   d. Patient rises as though in a "forward/upward dive" configuration (on balls of feet).
   e. When erect, patient inclines backward (on heels) to lock knees.
   f. Activate EMS units C and D in continuous mode for hip lock–hold only until locked.
   g. Deactivate EMS units C and D.
   h. Deactivate EMS units A and B–allow time to "ramp down."
   i. Proceed to "walk" mode.
2. Sit-Down Procedure
   a. Patient standing upright about 4–6 in. forward of sitting platform.
   b. Unlock hip locks with each hand separately (other hand on walker).
   c. Activate EMS units C and D, continuous or intermittent mode, as necessary to maintain upright balance.
   d. Activate EMS units A and B in continuous mode–allow time to "ramp up."
   e. Release knee locks.
   f. Deactivate EMS units A and B.
   g. Proceed to "sit down" as stimulation "ramps down."
3. Forward walking procedure (Figure 9.7). Let us begin with the right foot somewhat forward of the left foot and weight equally distributed on the feet. We now wish to take a left step *forward*.
   a. Weight is shifted so that all weight bearing is on the right foot.

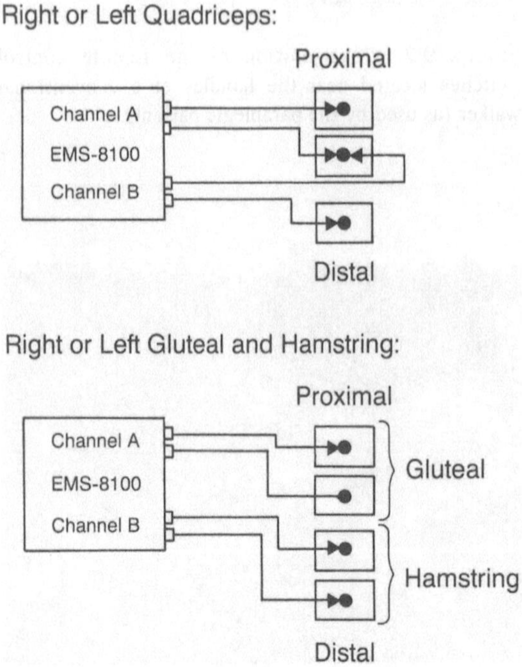

FIGURE 9.5. Interfacing of the EMS units with the electrode configuration over the various muscle groups of a paraplegic patient.

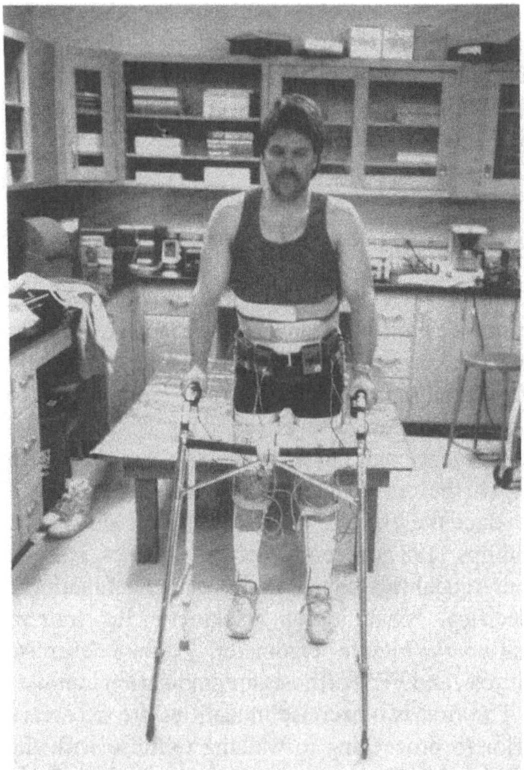

FIGURE 9.6. Paraplegic patient standing erect and outfitted with the EMS-RGO.

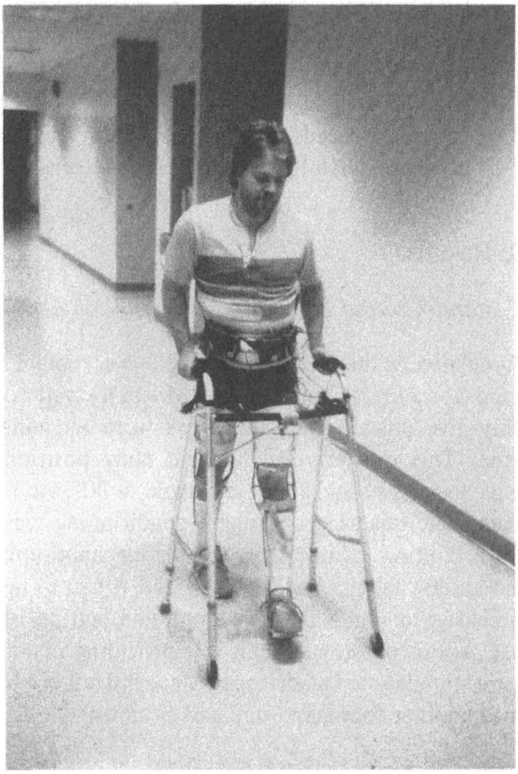

FIGURE 9.7. Paraplegic patient walking while utilizing the EMS-RGO.

b. EMS unit C is activated, resulting in right hip extension.
c. Through the reciprocal connection of the cross-connected hip cables, there is contralateral (left hip) flexion.
d. The left leg swings forward, and simultaneously EMS unit C is deactivated.
e. Weight is redistributed equally on both feet, with the left foot now somewhat forward of the right foot.

In order to continue forward walking, we can now repeat the procedure by shifting weight to the left foot and activating EMS unit D as appropriate. By alternating the shifting of weight and activation of the appropriate unit, the person walks forward.

4. Backward walking procedure. Returning to the right foot somewhat forward of the left foot and the weight equally distributed on the feet, we now wish to take a right step *backward*.
   a. Weight is shifted so that all weight bearing is on the left foot.
   b. EMS unit C is activated, resulting in right hip extension.
   c. The right leg now swings backward, and simultaneously EMS unit C is deactivated.
   d. Weight is redistributed equally on the feet, with the right foot now somewhat behind the left foot.

In order to continue backward walking, we can now repeat the procedure by shifting weight to the right foot and activating EMS unit D as appropriate. By alternating the shifting of weight and activation of the appropriate unit, the person walks backward.

5. Turning procedure. The same turning procedure is employed whether the person is making a 90° or a 180° turn.
   a. No FES is necessary.
   b. Pattern follows a "military maneuver" style.

c. RIGHTWARD TURN is on *heel* of right foot and *ball* of left foot.
d. LEFTWARD TURN is on *heel* of left foot and *ball* of right foot.
e. Pattern proceeds as sections of an arc, in between which the walker is repositioned.

D. Results

1. Stand-Up and Sit-Down

Cumulatively, the patient performed 45 acceptable stand-up's and sit-down's in the EMS-RGO system. Only five unsuccessful stand-up's were encountered. This was primarily due to body position (e.g., feet not behind knees, hip angle > 90°, etc.). Inadequate quadriceps strength conditioning was not a limiting factor. After about four stand-up's and sit-down's, the EMS units A and B had to be increased to the maximum amplitude setting of 9.0. At this point, each unit was providing maximum stimulators. Quadriceps fatigue did not occur after another four stand-up's and sit-down's.

2. Walking

The patient routinely averaged 0.6 kilometer per experimental session at velocities ranging from 0.8 to 1.6 km/h. Experimental time constraint, rather than muscle fatigue, was usually the limiting factor. The patient did increase the initial EMS parameter settings (before the end of the walking episode) such that the frequency (rate) setting was increased to 70 or 85 Hz (maximum is 100 Hz) and the amplitude setting was increased to 9.0 (the maximum).

3. Battery Performance

Battery performance was calculated as amount of a specific activity per volt, since with alkaline batteries the output voltage at the terminals is proportional to the current delivery capability of the battery.

With respect to the batteries of EMS units A and B, there were approximately 36 stand-up/sit-down's per volt drop at the battery terminals. With respect to the batteries of EMS units C and D, there was approximately 3.1 kilometers of walking per volt drop at the battery terminals.

These figures also approximate the useful lifespan of the respective batteries, since after about a 1 V drop of the output terminal voltage, there is no longer sufficient current delivery capability for satisfactory performance of the EMS-RGO system.

E. Discussion

The EMS-RGO described above provides walking exercise as one phase in a comprehensive program of active physical therapy. In APT, the electrically stimulated muscles act upon their environment to produce the exercise effect. This is in contrast to conventional physical therapy (no FES), in which the environment acts upon the muscle to produce the exercise effect (Phillips et al., 1984). Phillips (1987) recently defined APT as a set of four modalities capable of producing functional electrical rehabilitation: isokinetic leg trainer, stationary bicycle ergometer, outdoor exercise bicycle, and FES–orthosis for ambulation exercise.

The first two exercise modalities are necessary prior to proceeding to walking exercise with the EMS-RGO. The isokinetic leg training stimulates the quadriceps group of each leg. Various weights are connected to the foreleg via an ankle harness, and repetitive isovelocity contractions are produced at the knee. The isokinetic leg trainer has previously been described in detail (Petrofsky et al., 1984a). Numerous groups have confirmed that this type of exercise increases the strength of the quadriceps muscles (Petrofsky and Phillips, 1983b; Collins et al., 1984; Robinson et al., 1986).

The FES bicycle ergometer utilizes six channels of stimulation, one each to the quadriceps, hamstring, and gluteal groups of each leg. The paralyzed legs are stimulated to pedal at 50 rpm; progressive loads are adjusted via a friction belt attached to a flywheel. The FES bicycle ergometer has been described in detail elsewhere (Petrofsky et al., 1984c).

Investigators have reported physical conditioning (training) effects including increased pedaling endurance (Petrofsky et al., 1985a), improved resting blood pressure levels (Phillips et al., 1984), increases in resting heart rate and cardiac output (Danopulos et al., 1986), and finally increases in maximal workloads and minute ventilation (Moore et al., 1987).

This preconditioning is necessary to physically train the patient to a sufficient level to proceed to ambulation exercise. However, the patient must medically qualify for the technology being prescribed. Specifically, there are medical criteria to be met before prescription of active physical therapy, including walking exercise with the EMS-RGO. The initial evaluation of the patient candidate includes history, physical examination, lab tests, report summary, and an FES prescription (level of clearance) as previously described by Phillips (1987).

The useful lifespan of the EMS-RGO on one set of 9.0 V alkaline batteries is about 36 stand-up/sit-down's and about 3 kilometers of walking. New battery technology, such as lithium batteries, can achieve much higher current densities for comparable package sizes. These batteries are now regularly utilized in commercially available photographic equipment. They commonly have 1200 milliampere-hour current ratings. The battery connectors, however, are often specific to the camera equipment for which they are intended. These would have to be modified to interface with the EMS units. Some lithium battery package sizes are larger than the conventional 9 V alkaline cell, but others are of comparable size. Lead acetate batteries and gel cells (as used in our previous system) are definitely too heavy and bulky to satisfy everyday user requirements.

Patient compliance may be a problem in a system where so many electrodes (total of 14) need to be applied over the surface of the body. Not only is the application of those electrodes a somewhat tedious and cumbersome task, but good positioning over appropriate motor units is important for optimal performance of the system. New stimulation technology (TTG*) allows stimulation delivery at multiple sites by means of a customized garment (similar to leotards). The actual electrical conducting surfaces are interwoven into the fabric of the garment. The patient now simply dons the garment and could then interface with the EMS units via a single universal connector. This new system is the subject of a separate report (Phillips et al., 1991).

*From Bio-Stimu Trend Corporation, 14851 NW 27th Avenue, Opa Locka, FL 33054.

II. Quadriplegic Walking System*

A. Introduction

Different investigators have utilized functional electrical stimulation for standing and walking through a variety of lower extremity neural (FES) prostheses (e.g., Brindley et al., 1978; Kralj et al., 1980; Vodovnik et al., 1981; Marsolais and Kobetic, 1983; Holle et al., 1984; Petrofsky et al., 1984b). In 1982 computer-controlled standing and walking with FES was introduced by the Wright State University group. The investigators utilized closed-loop control (Petrofsky and Phillips, 1983a; Petrofsky et al., 1984b) and their approach has been reviewed (Petrofsky and Phillips, 1985). Closed-loop control requires that data from sensors modify the output of a controller to help coordinate walking. Sensors are placed on the hips, knees, and/or ankles of the paralyzed individuals to provide positional data for the computer controller (Petrofsky and Phillips, 1983a, 1985). Closed-loop control was developed using animal experiments (Petrofsky and Phillips, 1979; Phillips and Petrofsky, 1980, 1981), although it was originally proposed on humans (Vodovnik et al., 1967; Stanic and Trnkoczy, 1974).

More recently, a portable FES walking system has been developed and utilized in combination with a lower extremity orthosis (Petrofsky et al., 1985b, 1986). With such a combination system, selected patients have been able to regularly walk distances of one mile (Hendershot et al., 1987).

A major thrust of our laboratory during the past year has been to make this technology available to the medical community for use by the general spinal-cord-injured population. The limitations of our past approach (Petrofsky et al., 1985b, 1986) have been:

1. A heavy and separate portable power pack (consisting of four Kodavision 8 V batteries and eight lead-acetate 2 V batteries). This power pack weighs approximately 8 pounds and is carried on the back.

*From C.A. Phillips, Functional electrical stimulation and lower extremity bracing for ambulation exercise of the spinal cord injured individual: A medically prescribed system. Reprinted from *Physical Therapy*, 69:842–849, 1989, with the permission of the American Physical Therapy Association.

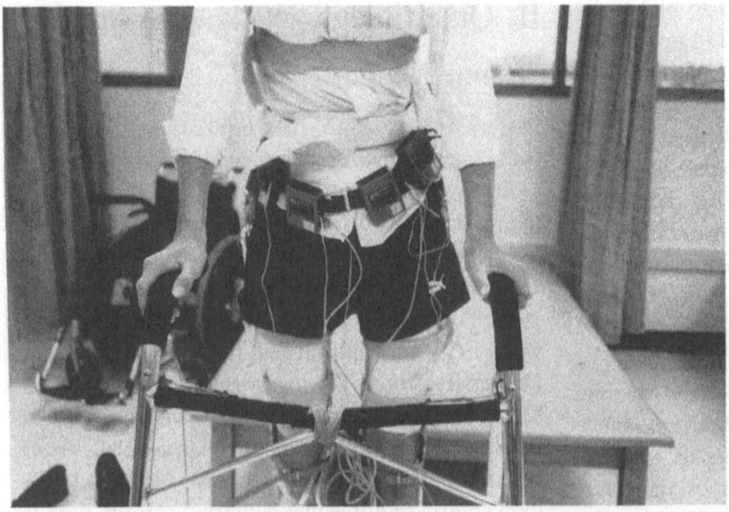

FIGURE 9.8. Four electrical stimulator units worn on the front of the belt of a quadriplegic patient.

2. A heavier, bulkier FES electronic package when the sit-down/stand-up capability was combined with the basic walking system.
3. Specialized and custom-built electronic package and power pack.

This section describes a new system that overcomes these limitations. By utilizing commercially available electrical muscle stimulators in combination with the gait orthosis, a new system has been developed. The advantages of the system are:

1. The power pack is now reduced to a single 9 V alkaline battery contained within each electrical stimulator.
2. A total of six electrical stimulators (each weighing less than 6 ounces, with battery) provide complete stand-up/sit-down and walking functions.
3. Each component of the walking system is
   (a) commercially available from various manufacturers
   (b) recommended for applications such as "gait training," and
   (c) medically prescribed equipment, hence available as a prescription item from a physician.

The purpose of this section is to determine whether commercially available, physician-prescribed electrical stimulators, when appropriately interfaced and combined with orthotic bracing, will allow a quadriplegic patient to stand up, sit down, and functionally walk on level ground.

B. Materials and Methods

A spinal-cord-injured individual participated in this study. G.B. is a 28-year-old, male, C7 level quadriplegic who was injured 6 years ago. He sustained an incomplete injury with some sensation and slight (but functionally useless) movement below the level of injury. A somewhat modified system has been described for a paraplegic patient in Section I of this chapter.

1. Electrical Muscle Stimulator

The unit employed in this study was the EMS-8100 muscle stimulator (NTRON). The technical specifications were given in Table 9.1. Six such units were employed, hereafter designated as A, A', B, B', C, and D. Units A, A', B, and B' were worn in the front on a belt (see Figure 9.8) and units C and D were worn on the back of the belt.

The initial operating parameters were adjusted for one set of values for units A, A' and B, B' and another set of values for units C and D (see Table 9.3). These operating parameters were subsequently adjusted during the course of the study for optimal individual performance.

2. Switches

The patient had some movement of the thumb, index, and middle finger of the left hand, but no thumb or finger movement of the right hand. In this case, one slide switch and two push-button

TABLE 9.3. Initial operating parameters for quadriplegic walking systems.

Parameter	Units A/A′/B/B′	Units C/D
Rise time	1.0–2.0 seconds	0 (instantaneous)
Rate (frequency)	100 Hz	70–85 Hz
On time	30 seconds	30 seconds
Off time	0	0
Fall time	5 seconds	0 (instantaneous)
Pulse width	800 μs	800 μs
Amplitude (thumbwheel reading)	7.0–8.0	7.0–8.0

switches were mounted near the left handle of a conventional walker. Two push-button switches were mounted in-line and adjacent to each other on the medial side aspect near the walker handle. These were connected by ribbon cable to a subminiature cable connector (male), which interconnected with a subminiature cable connector (female). The latter connector separated as two cables, each terminating with a subminiature phone plug that then connected to the manual override jacks of electrical stimulator units C and D.

One remote slide switch was mounted near the left handle and oriented 45° obliquely between the medial side aspect and the superior aspect (see Figure 9.9). The remote slide switch cable was divided into four cables connected in parallel. Each cable was then terminated with a subminiature phone plug, which connected to the manual override jacks of electrical stimulator units A, A′, and B, B′.

In operation, the anterior and posterior push-button switches were actuated with the index and middle fingers, while the slide switch was actuated with the thumb. The left palm and remaining fingers assisted with positioning and gripping of the walker handle.

3. Batteries

Nine-volt alkaline transistor batteries were used.

4. Electrodes

These studies utilized twelve 2 in. × 4 in. carbon rubber electrodes (Medtronic model 3793), and four 1½ in. × 1½ in. carbon rubber electrodes (Medtronic model 3795).

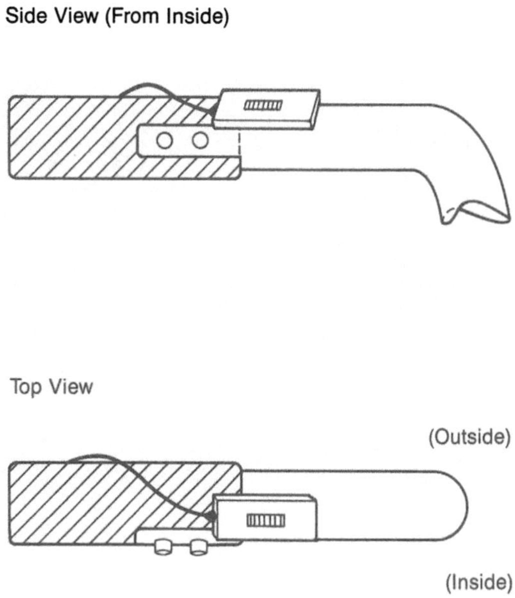

FIGURE 9.9. Configuration of the remote control switches located near the handle of a conventional walker (as used by quadriplegic patient).

As shown in Figure 9.10, four 2 in. × 4 in. carbon rubber electrodes are placed over each quadriceps muscle group (for a total of eight electrodes). Liqui-Cor electrode gel is first applied, and both adhesive patches and adhesive tape are utilized to secure the electrodes.

As shown in Figure 9.11, one 2 in. × 4 in. carbon rubber electrode and one 1½ in. × 1½ in. carbon rubber electrode are placed over each gluteal

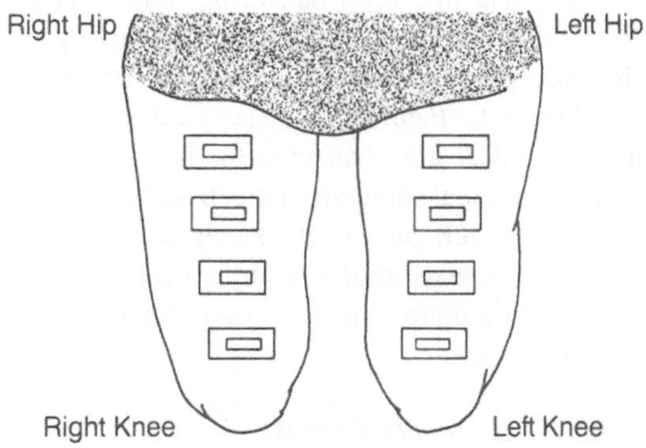

FIGURE 9.10. Electrode configuration over the anterior aspect of upper half of the legs (quadriceps muscles).

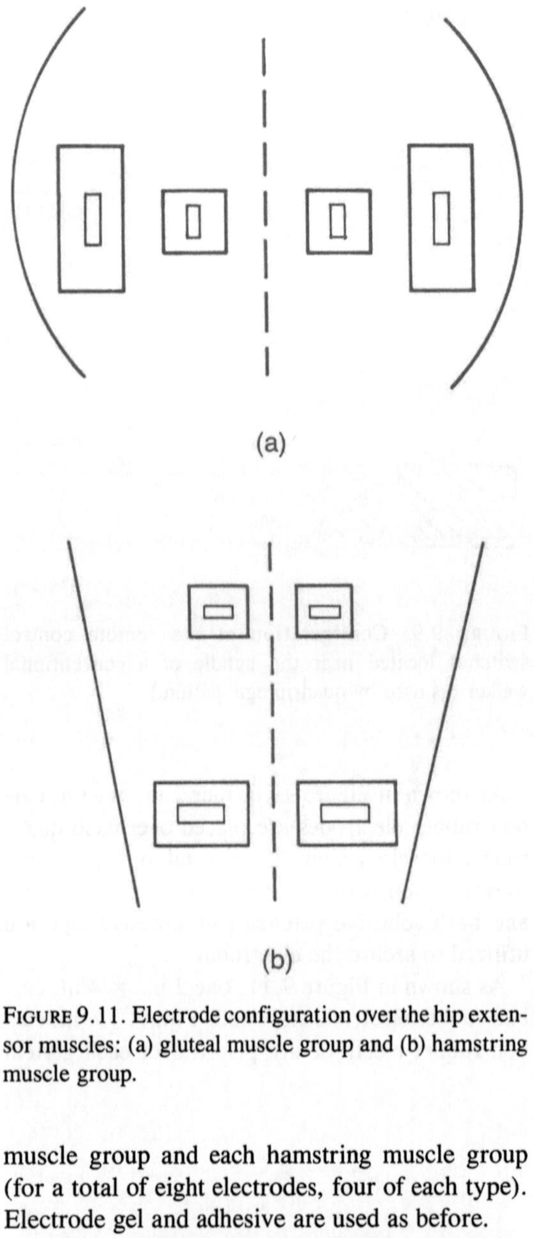

FIGURE 9.11. Electrode configuration over the hip extensor muscles: (a) gluteal muscle group and (b) hamstring muscle group.

muscle group and each hamstring muscle group (for a total of eight electrodes, four of each type). Electrode gel and adhesive are used as before.

Referring to Figure 9.12, electrical stimulator units A and A′ are connected to the right quadriceps. Electrical stimulator units B and B′ are connected to the left quadriceps. Electrical stimulator unit C is connected to the right hip extensors, and electrical stimulator unit D is connected to the left hip extensors.

5. *Lower Extremity Orthosis*

The particular lower extremity orthosis used in this project is a hip–knee–ankle–foot orthosis

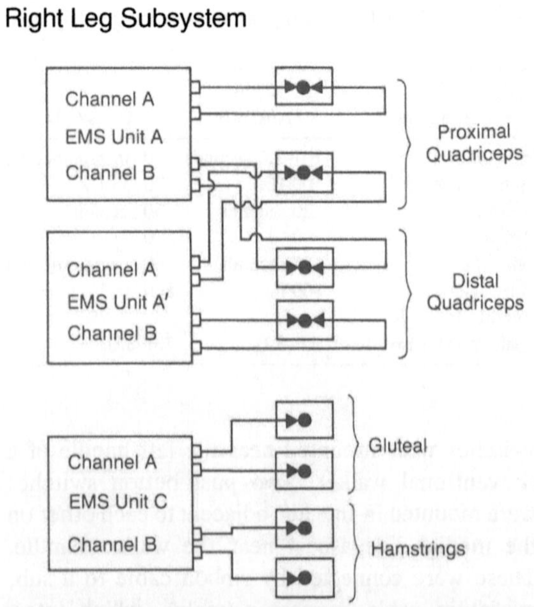

FIGURE 9.12. Interfacing of the electrical stimulator units with the electrode configuration over the various muscle groups.

(HKAFO). There is both a right and left orthosis (LSU-RGO): see footnote, p. 129). Coupling occurs between the left and right orthoses in such a manner that hip extension in either one tends to force the other hip joint into flexion. This provides coordinated motion between the legs and makes possible a "reciprocating" gait.

A Bowden cable transmits the force from one orthosis to the other. Disconnects at each hip joint permit simultaneous flexion for sitting. A custom-molded plastic pelvic girdle is used.

The posterior thigh shells are of polypropylene molded over plaster models of the patient, as are the ankle–foot sections. Reinforcements of carbon fiber inserts about the malleoli sections permit the use of thin polypropylene. Aluminum offset knee joints are used. Each of the KAFOs is provided with a drop lock and retention ball spring on the lateral side.

6. *System Costs*

Table 9.4a reports the initial cost (for a 5-year period) for an FES-RGO program consisting of five walking sessions per week. Table 9.4b reports the recurring costs on an annual basis for an FES-

TABLE 9.4. Cost per year for 5 walking sessions per week.

Quantity	Item	Cost
	a. Initial cost	
6	NTRON EMS-8100 units at $796/unit	$ 4780
1	RGO (manufactured and orthotist fitted)	3900
	Total initial cost	$ 8680
	b. Recurring cost (annual)	
	PT charges, at $110/session for 250 sessions	$27,500
	Electrodes, etc., for $6.50/session for 250 sessions	1,625
	Total recurring cost for 1 year	$29,125

RGO walking program composed of five sessions per week. Finally, Table 9.5 reports the cost for each individual walking session that utilizes the FES-RGO system.

Overall, these tables summarize the outpatient physical therapy costs associated with the prescription of the electrical stimulation system for ambulation of the spinal cord injury (quadriplegic) patient. Comparable figures have been published for the outpatient PT costs associated with the prescription of a similar system for paraplegic patients (Phillips, 1989b).

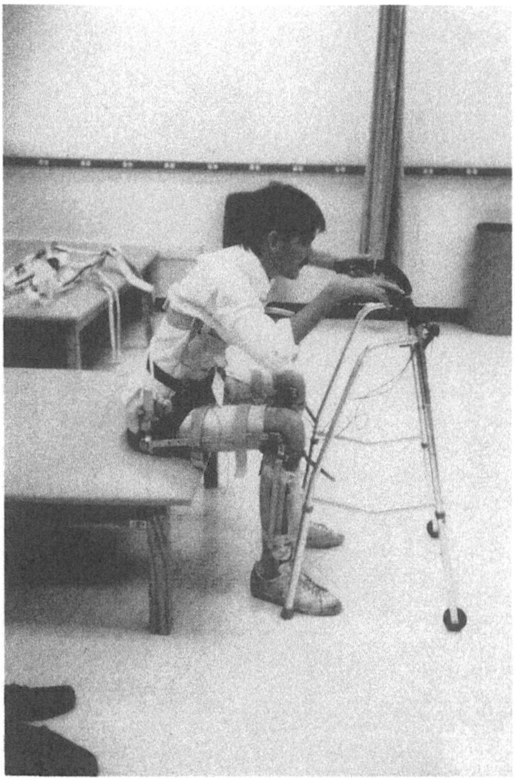

FIGURE 9.13. Quadriplegic patient in sitting position just prior to standing.

7. *Procedural Note*

The walking system interacts with the subject to allow the subject to perform a variety of tasks including standing up, walking, turning, and sitting down. These procedures are reviewed in Section I, except that when unit A is mentioned the reader should also include unit A′, and when unit B is mentioned, also include unit B′.

C. Results

The quadriplegic individual performed more than 20 acceptable repetitions of stand-up and sit-down exercises in the walking system (see Figures 9.13 and 9.14). About five unsuccessful stand-up's were encountered, primarily due to body position (e.g., feet not behind knees, hip angle > 90°). Inadequate quadriceps strength conditioning was a limiting factor for this quadriplegic patient. After two or three stand-up's and sit-down's, the electrical stimulator units A, A′ and B, B′ had to be increased to the maximum amplitude setting of 9.0. At this point, all four units were providing maximal stimulation. Quadriceps fatigue then occurred after another two or three stand-up's and sit-down's.

The patient routinely averaged about 0.8 kilometer per experimental session at walking velocities ranging from 1.2 to 2.0 km/h (see Figure 9.15). Muscle fatigue was usually the limiting factor (after 30 minutes to 1 hour) for this quadriplegic individual. The individual altered the initial electrical stimulation parameter settings (before the end of the walking episode) such that the frequency (rate) setting was increased to 100 Hz (maximum) and the amplitude setting was increased to 9.0 (the maximum).

TABLE 9.5. Cost for each additional walking session.

PT charges per session	$110.00
Electrodes, etc., per session	6.50
Total cost per session	$116.50

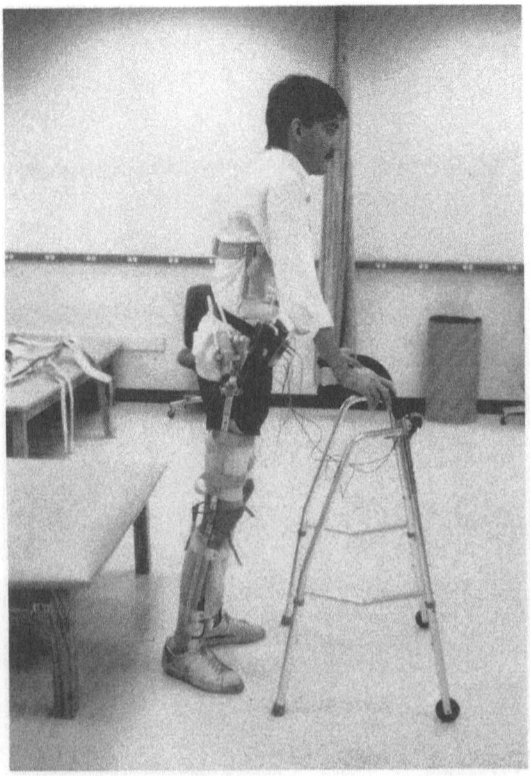

FIGURE 9.14. Quadriplegic patient standing erect and outfitted with walking system.

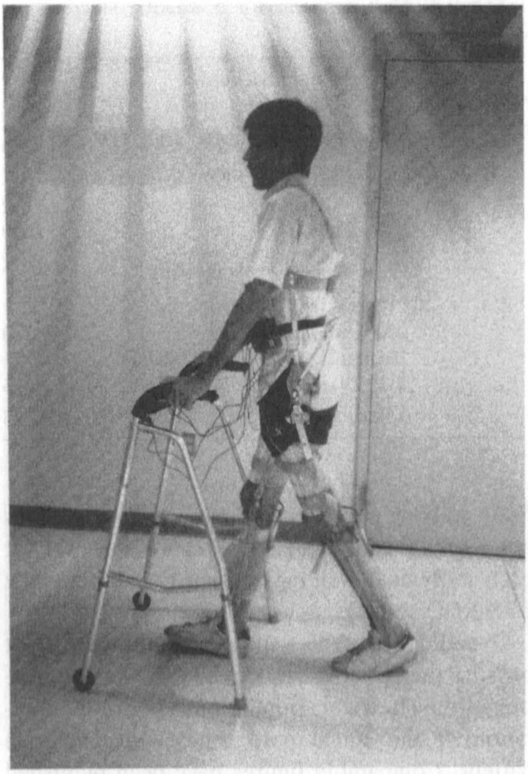

FIGURE 9.15. Quadriplegic patient walking with the walking system.

In addition to the actual walking time, time was needed for the preparation of the patient. At the beginning of each walking session, 30 to 45 minutes was required to attach the electrodes, apply the orthosis, and attach the stimulation units. At the conclusion of each walking session, another 30 minutes was required for removal of the gait orthoses, electrodes, and stimulation units.

D. Discussion

A majority of earlier FES walking systems have only utilized electrical stimulation by itself (Brindley et al., 1978; Kralj et al., 1980; Vodovnik et al., 1981; Marsolais and Kobetic, 1983; Holle et al., 1984; Petrofsky et al., 1984b). Petrofsky and Phillips et al. were the first to combine electrical stimulation with reciprocating gait orthotic devices (Petrofsky et al., 1985b, 1986). This section has reported that commercially available electrical muscle stimulators can now be utilized as the FES component (in combination with the gait orthoses) to allow walking in the paralyzed individual.

This section has also reported that commercially available, physician-prescribed electrical stimulators, when appropriately interfaced and combined with orthotic bracing, do allow a quadriplegic patient to perform stand-up, sit-down, and functional (level-ground) walking. Specifically, the quadriplegic individual successfully performed more than 20 stand-up's and sit-down's with the system. Furthermore, the patient routinely averaged 0.8 kilometer of walking at each experimental session at velocities ranging from 1.2 to 2.0 km/h.

Muscle fatigue rather than battery power was the limiting factor in all the experimental trials. As fatigue approached, stimulation parameters were increased to their maximal settings.

In any one experimental session, the patient was able to perform only about four to six stand-up's and sit-down's. About 0.8 kilometer was traversed per experimental session. However, Phillips (1989a) has shown for a quadriplegic individual that the system is capable of about 40 stand-up's and sit-down's and about 4 kilometers of total distance covered before the 9 V alkaline batteries need to be replaced.

The advantages of this approach (compared to conventional braces alone) are that (a) good upper extremity strength is not necessary to produce the reciprocating gait pattern; (b) stand-up and sit-down procedures can be performed in an easy and coordinated manner by the subject alone; and (c) improved cardiopulmonary conditioning is achieved, as reflected by reduced energy expenditure (Phillips, 1989b).

The disadvantages of this approach (compared to conventional braces alone) are that (a) more time and effort are required to apply electrodes and connect the stimulation units; (b) the subject's walker must be modified to accommodate the remote control switches connected to the stimulation units; and (c) additional cost is required to purchase the electronic equipment.

As with any treatment modality, the patient must medically qualify for the equipment being prescribed. Specifically, there are medical criteria that must be satisfied for prescription of walking exercise with the EMS-RGO. The initial evaluation of the patient includes history, physical examination, laboratory tests (e.g., ECG, pulmonary function testing, serum chemistry, and urinalysis), report summary, and an FES prescription (level of clearance) as previously described by Phillips (1987).

In the Wright State program, the isokinetic leg exerciser is currently considered to be the entry-level FES exercise modality. Progression to the FES bicycle ergometer requires adequate flexion and medial–lateral stability at the knee, as well as minimal dorsiflexion at the ankle [Table 1 of Phillips (1987)]. Progression to the walking system and walking exercise then requires muscular, cardiovascular, and respiratory system function sufficient for 30 minutes of continuous FES bicycling at minimal (zero kp•m) load. Other medical criteria for walking exercise with the walking system are summarized in Table 1 of Phillips (1987).

The writing of the walking system prescription proceeds after the fulfillment of two prerequisites: completion of the preliminary conditioning phases of a program, and satisfaction of the medical criteria for walking exercise with the EMS-RGO. The prescription is composed of two parts.

The first part is prescription of the reciprocating gait orthosis and preliminary gait training. As indicated above, a pictorial description and applications manual (LSU-RGO) is available from the manufacturer. The prescription is to be filled by a certified orthotist, knowledgeable and competent in the fitting of this type of orthosis. Finally, arrangements must be made with a physical therapist for gait training using the orthosis.

The second part is prescription of the electrical muscle stimulators. Technical information on the electrical stimulators is available as a patient instruction booklet from NTRON, the manufacturer (see footnote, p. 129). Next, the physician prescribes the electrical stimulators; this is usually a "hands-on" demonstration, since the manufacturer provides the electrical stimulator units to the physician on a consignment basis. The information provided here should provide sufficient detail (see Section II.B) for the successful interfacing of electrical stimulator units, electrodes (or TTG), and the gait orthosis. Finally, suitable arrangements should be made for gait training in the walking system.

III. A Systems Approach to Physician-Prescribed FES Ambulation

An electrical stimulation system for ambulation of spinal-cord-injured patients has been described (Sections I and II) and consists of physician-prescribable, commercially available stimulation units and a gait orthosis. This section describes the electrode delivery system, a transcutaneous transducer garment, which allows a variety of electrode configurations and stimulation patterns so that system optimization can be achieved.

A. Introduction

Interest in electrical stimulation for ambulation of spinal-cord-injury (SCI) individuals occurred with Brindley's report (1978) of "electrical splinting" of a paraplegic patient. However, a number of problems associated with developing a practical electrical stimulation system for SCI have been defined (Stallard et al., 1989). The most practical approach to date appears to combine electrical stimulation of

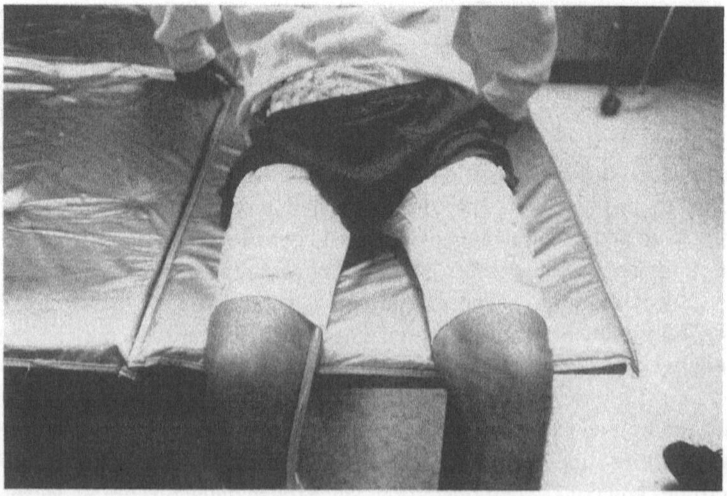

FIGURE 9.16. Patient seated and wearing the TTG (front view).

paralyzed muscles with lower extremity bracing, as first proposed by investigators from two universities (Petrofsky et al., 1986). Recently this approach has been reduced to physician-prescribable, commercially available components (Phillips, 1989b).

This medically prescribed system has been reported to produce stand-up and sit-down (from a wheelchair or other chair) as well as forward and backward ambulation. This has been described for a paraplegic subject (Section I) who sustained a motor-complete (but sensory-incomplete) lesion and was Frankel class B. This system has also been described for a quadriplegic subject (Section II) who sustained a motor-incomplete (but functionally useless) lesion and was Frankel class C.

One problem with such systems lies in the complexity of the electrode placement. To make the system more efficient, cosmetic, and easier to don and doff, electrically conducive clothing with embedded electrodes was reported (Petrofsky et al., 1986). These transcutaneous transducer garments were developed by Bio-Stimu Trend Corporation and offer the advantage of ease of use, no exposed wires, and machine washability (Figure 9.16). They have snap connectors that interface to the EMS for control of movement of major muscle groups of the leg necessary for walking (Granek and Granek, 1989).

As experience has been obtained with the medically prescribed FES ambulation system (Phillips, 1989b), it has also become apparent that various electrode configurations and alternative stimulation patterns are necessary for optimal patient performance when using the system. Illustrative examples follow.

B. Methods and Procedures

The system previously reported for a paraplegic subject (Phillips, 1990) consists of four EMS units, each with two channels (A and B), in accordance

TABLE 9.6. The 4 unit, Walk 1 mode system.

EMS unit	Channel	Electrode
1	A	Right quadriceps
1	B	Right quadriceps
2	A	Left quadriceps
2	B	Left quadriceps
3	A	Right gluteal
3	B	Right hamstring
4	A	Left gluteal
4	B	Left hamstring

TABLE 9.7. The 4 unit, Walk 2 mode system.

EMS unit	Channel	Electrode
1	A	Right quadriceps
1	B	Right quadriceps
2	A	Left quadriceps
2	B	Left quadriceps
3	A	Right quadriceps
3	B	Left gluteal/hamstring
4	A	Left quadriceps
4	B	Right gluteal/hamstring

TABLE 9.8. The 6 unit, Walk 1 mode system.

EMS unit	Channel	Electrode
1	A	Right quadriceps
1	B	Right quadriceps
2	A	Right quadriceps
2	B	Right quadriceps
3	A	Left quadriceps
3	B	Left quadriceps
4	A	Left quadriceps
4	B	Left quadriceps
5	A	Left gluteal
5	B	Left hamstring
6	A	Right gluteal
6	B	Right hamstring

TABLE 9.9. The 6 unit, Walk 2 mode system.

EMS unit	Channel	Electrode
1	A	Right quadriceps
1	B	Right quadriceps
2	A	Left quadriceps
2	B	Left quadriceps
3	A	Right quadriceps
3	B	Right quadriceps
4	A	Left quadriceps
4	B	Left quadriceps
5	A	Left gluteal
5	B	Left hamstring
6	A	Right gluteal
6	B	Right hamstring

with Table 9.6. This is the Walk 1 mode, in which an ipsilateral leg forward is obtained by stimulating the gluteal and hamstring muscle groups of the contralateral (stationary) leg. For more details, refer to Section I.

A system currently being evaluated for an SCI (paraplegic) subject also consists of four EMS units, each with two channels (A and B) as shown in Table 9.7. This is the Walk 2 mode, in which an ipsilateral leg forward is obtained by *simultaneously* stimulating the quadriceps muscle group of the ipsilateral (moving) leg *and* the gluteal and hamstring muscle groups of the contralateral (stationary) leg.

The system previously described for a quadriplegic subject (Phillips, 1989c) consists of six EMS units, each with two channels (A and B), in accordance with Table 9.8. This is the Walk 1 mode as described above.

The system currently being evaluated on an SCI (quadriplegic) subject also consists of six EMS units, each with two channels (A and B) as shown in Table 9.9. This is the Walk 2 mode as described above.

C. Results

It was observed that the subject, a C7 level quadriplegic, who was motor incomplete, but functionally useless (Frankel class C), was able to pre-gel his TTG without any assistance (Figure 9.17). Furthermore, the subject was able to don and doff

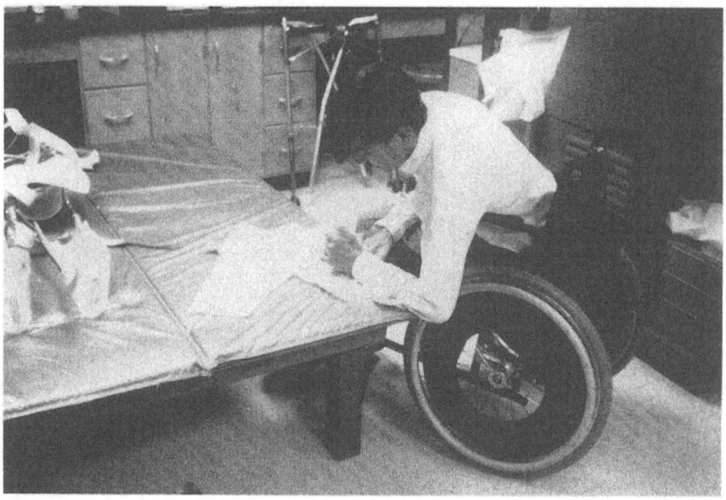

FIGURE 9.17. Quadriplegic patient pre-gelling the TTG electrode garment prior to donning it.

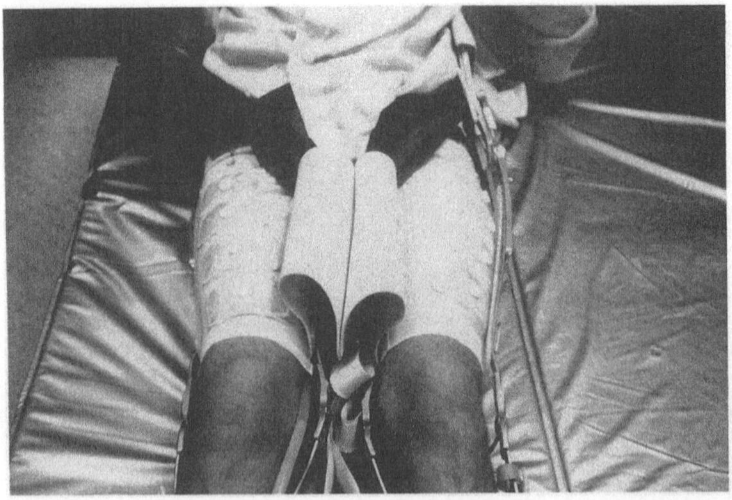

FIGURE 9.18. Quadriplegic patient after donning TTG and positioned in the RGO.

his TTG again without assistance (Figure 9.18). This means that functionally, the subject was able to prepare and position all 16 electrodes (and remove them after the testing session). This task would have been virtually impossible with 16 individual carbon rubber electrodes, gel, patches, and paper tape.

The interconnection between the EMS units and the TTG for the illustrative examples (Tables 9.6 through 9.9) are shown in Figures 9.19 through 9.24. Figure 9.19 indicates the interconnection of EMS units 1 and 2 on the anterior (quadriceps) aspect of the TTG for the configurations given in Table 9.6. Each EMS unit channel (A and B) has a separate anode (+) and cathode (○). Figures 9.20 and 9.21 indicate the interconnection of EMS units 1 through 4 on the anterior (quadriceps) aspect and EMS units 3 and 4 on the posterior (glutal/hamstring) aspect, respectively, of the TTG as shown in Table 9.7. Figures 9.22 and 9.23 indicate the interconnection of EMS units 1 through 4 on the anterior (quadriceps) aspect of the TTG for the

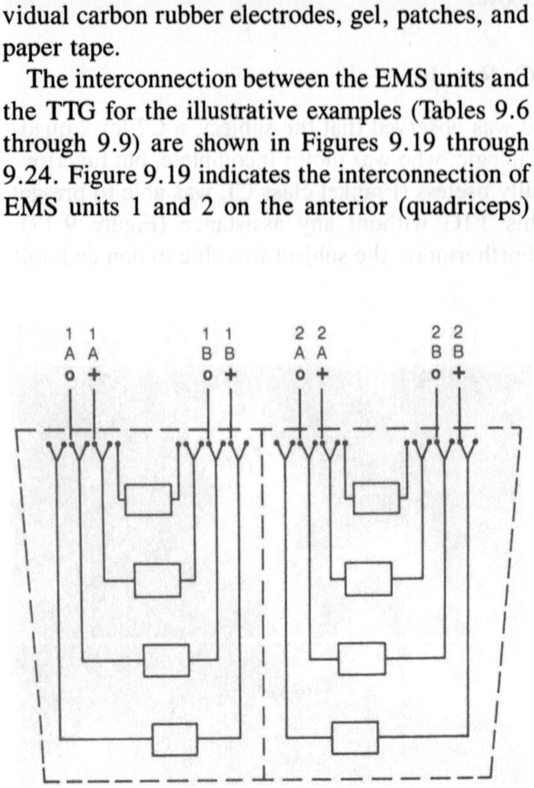

FIGURE 9.19. EMS unit connections to the TTG garment: four units, Walk 1 mode, quadriceps muscles.

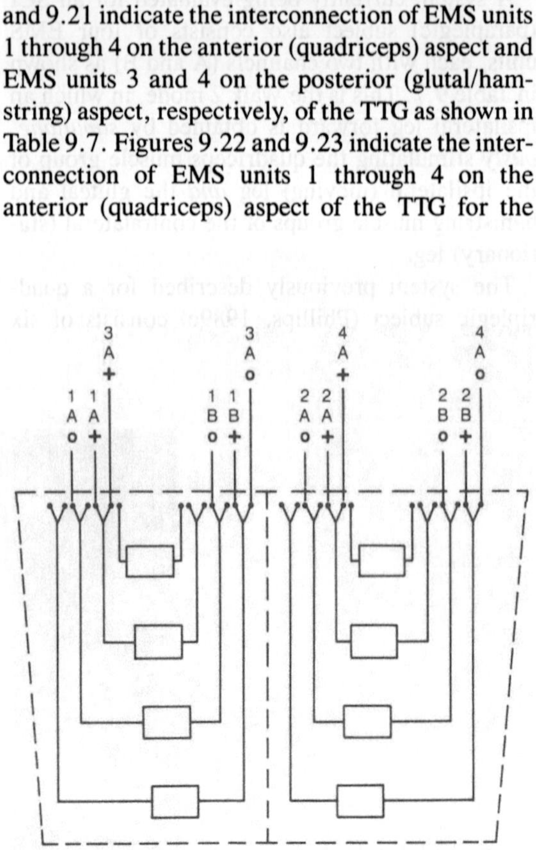

FIGURE 9.20. EMS unit connections to the TTG garment: four units, Walk 2 mode, quadriceps muscles.

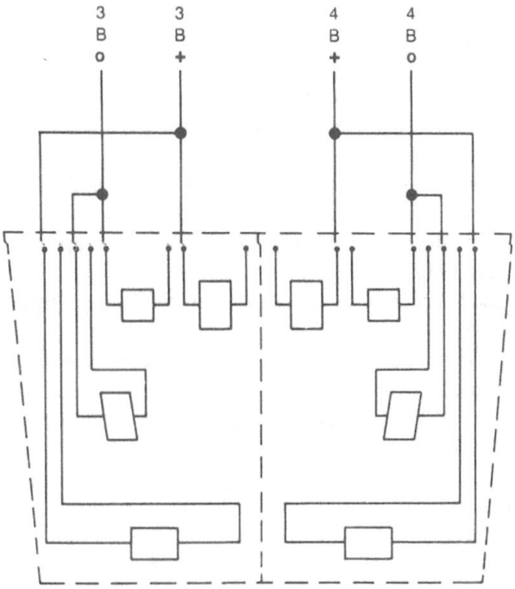

FIGURE 9.21. EMS unit connections to the TTG garment: four units, Walk 2 mode, gluteal/hamstring muscles.

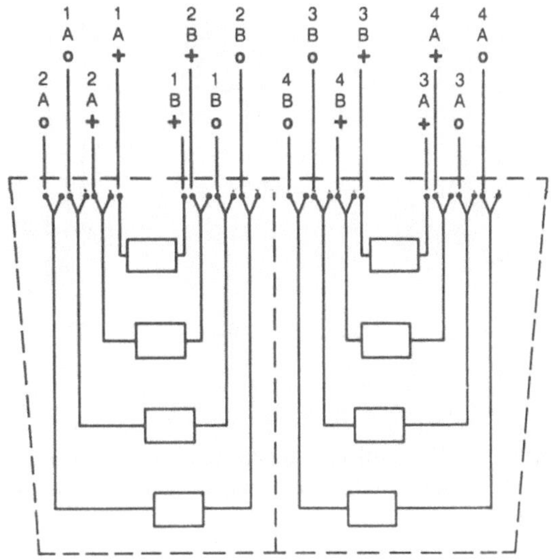

FIGURE 9.22. EMS unit connections to the TTG garment: six units, Walk 1 mode, quadriceps muscles.

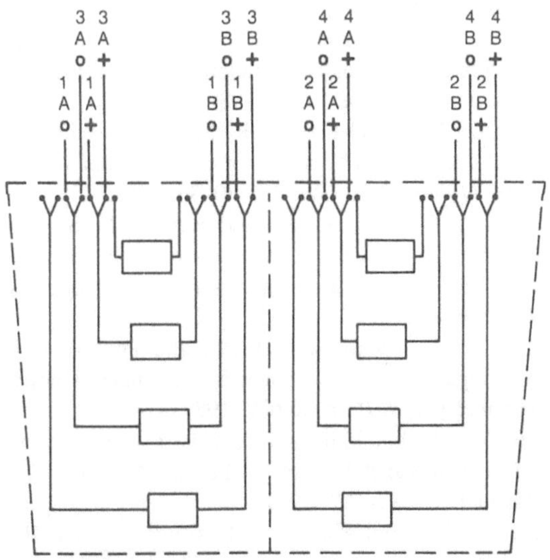

FIGURE 9.23. EMS unit connections to the TTG garment: six units, Walk 2 mode, quadriceps muscles.

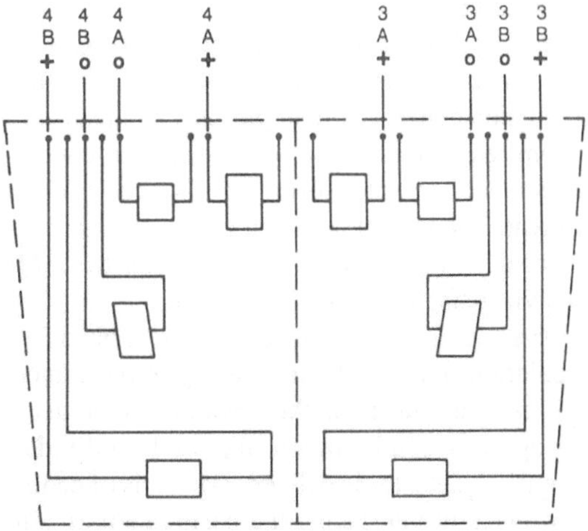

FIGURE 9.24. EMS unit connections to the TTG garment: four units, Walk 1 mode, gluteal/hamstring muscles.

configurations shown in Tables 9.8 and 9.9, respectively.

Figure 9.24 indicates the interconnections of EMS units 3 and 4 on the posterior (gluteal/hamstring) aspect of the TTG for the configuration given in Table 9.6. EMS units 5 and 6 interconnect on the posterior (gluteal/hamstring) aspect of the TTG (Figure 9.25) in a similar manner for the configurations given in Tables 9.8 and 9.9.

D. Discussion

Patient compliance may be a problem in a system where many individual electrodes need to be

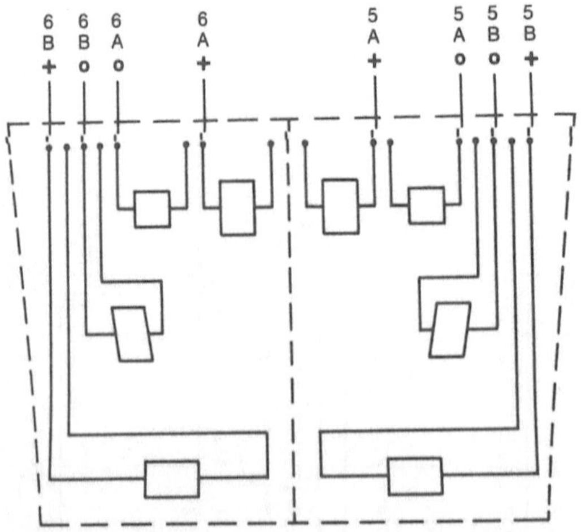

FIGURE 9.25. EMS unit connections to the TTG garment: six units, Walk 1 and 2 modes, gluteal/hamstring muscles.

applied to the surface of the body. The application of these electrodes is a somewhat tedious and cumbersome task. Also, good positioning over appropriate motor units is important for optimal performance of the system.

New stimulation delivery technology, such as the transcutaneous transducer garment (TTG), allows stimulation delivery at multiple sites by means of a customized garment. The actual electrical conduction surfaces are interwoven into the fabric of this garment. The patient now simply dons the garment and could (in the future) interface with the EMS units via a single universal connector.

A systems approach to medically prescribed FES ambulation requires the interfacing of basic modules, each of which has the flexibility to adjust to the optimal configuration for each individual patient. The electrode delivery system is of central importance in interfacing the basic modules. The TTG appears to allow such system optimization.

References

Brindley, G.S., Polkey, G.E., and Rushton, D.N. Electrical splinting of the knee in paraplegia. *Paraplegia*, 16:428–435, 1978.

Collins, S.R., Glaser, R.M., and Feinberg, N.D. Progressive training of paralyzed leg muscles with electrical stimulation. *Fed. Proc.*, Abstract 4262, 1984.

Danopulos, D., Kezdi, P., Stanley, E.L., et al. Changes in cardiovascular dynamics after a twelve week active bicycle rehabilitation in young tetraplegics. *J. Neurol. Ortho. Surg.* 7:179–184, 1986.

Douglas, R., Larson, P., D'Ambrosia, R., and McCall, R. LSU reciprocating gait orthosis. *Orthopedics*, 6: 834–839, 1983.

Edwards, B.G., Lew, R.D., and Marsolais, E.B. Relative energy costs of long-leg brace and FNS ambulation. *Proceedings of the Ninth RESNA Conference*, pp. 322–324, 1986.

Enercell Battery Guidebook. Master Publishing, Dallas, 1985.

Granek, H., and Granek, M. Transcutaneous transducer garments: An advancement in transcutaneous delivery and end user compliance. *Automedica* (London) 11: 19–23, 1989.

Gruner, J. Considerations in designing acceptable neuromuscular stimulation systems for restoring function in paralyzed limbs. *CNS Trauma J.* 3:37–47, 1986.

Hendershot, D.M., Moore, M.L., and Phillips, C.A. Cardiopulmonary conditioning when walking with and without FES in the paralyzed. *Fed. Proc.* 46:680, 1987.

Holle, J., Frey, M., Gruber, H., Kern, H., Stohr, H., and Thoma, H. Functional electrical stimulation of paraplegics. *Orthopedics*, 7:1145–1155, 1984.

Kralj, A., Bajd, T., and Turk, R. Electrical stimulation providing functional use of paraplegic patient muscles. *Med. Prog. Technol.* 7:3–9, 1980.

Marsolais, E.B., and Kobetic, R. Functional walking in paralyzed patients by means of electrical stimulation. *Clin. Orthop. Relat. Res.* 175:30–36, 1983.

McCall, R.E., Douglas, R., and Rightor, N. Surgical treatment in patients with myelodysplasia before using the LSU reciprocation gait system. *Orthopedics*, 6:843–848, 1983.

Moore, M.L., Hendershot, D.M., and Phillips, C.A. Cardiopulmonary conditioning by FES-induced bicycle ergometry. *Fed. Proc.* 46:679, 1987.

Petrofsky, J.S., and Phillips, C.A. Constant velocity contractions in skeletal muscle of the cat. *Med. Biol. Eng. Comput.* 17:583–592, 1979.

Petrofsky, J.S., and Phillips, C.A. Computer-controlled walking in the neurological paralyzed individual. *J. Neurol. Orthop. Surg.* 4:153–164, 1983a.

Petrofsky, J.S., and Phillips, C.A. Active physical therapy—A modern approach for rehabilitation of the disabled. *J. Neurol. Orthop. Surg.* 4:165–173, 1983b.

Petrofsky, J.S., and Phillips, C.A. Closed-loop control of movement of skeletal muscle. *CRC Crit. Rev. Biomed. Eng.* 6:35–94, 1985.

Petrofsky, J.S., Heaton, H.H., and Phillips, C.A. Leg exerciser for training of paralyzed muscle by closed-loop control. *Med. Biol. Eng. Comput.* 22:298–303, 1984a.

Petrofsky, J.S., Phillips, C.A., and Heaton, H.H. Feedback control system for walking in man. *Comput. Biol. Med.* 14:135–149, 1984b.

Petrofsky, J.S., Phillips, C.A., Heaton, H.H., and Glaser, R.M. Bicycle ergometer for paralyzed muscle. *J. Clin. Eng.* 9:13–19, 1984c.

Petrofsky, J.S., Phillips, C.A., and Hendershot, D. The cardiorespiratory stresses which occur during dynamic exercise in paraplegics and quadriplegics. *J. Neurol. Orthop. Surg.* 6:252–258, 1985a.

Petrofsky, J.S., Phillips, C.A., Larson, P., and Douglas, R. Computer-synthesized walking—An application of orthosis and functional electrical stimulation (FES). *J. Neurol. Othop. Med. Surg.* 6:219–230, 1985b.

Petrofsky, J.S., Phillips, C.A., Douglas, R., and Larson, P. A computer-controlled walking system: The combination of an orthosis with functional electrical stimulation. *J. Clin. Eng.* 11:121–133, 1986.

Phillips, C.A. The medical criteria for active physical therapy: Physician guidelines for patient participation in a program of functional electrical rehabilitation. *Am. J. Phys. Med.* 66:269–286, 1987.

Phillips, C.A. Electrical muscle stimulators in combination with a reciprocating gait orthosis. *Automedica* (London) 11:247-261, 1989a.

Phillips, C.A. Electrical stimulation for ambulation of selected paraplegics and quadriplegics. *J. Neurol. Ortho. Med. Surg.* 10:109–110, 1989b.

Phillips, C.A. Functional electrical stimulation and lower extremity bracing for ambulation exercise of the spinal cord injured individual: A medically prescribed system. *Phys. Ther.* 69:842–849, 1989c.

Phillips, C.A. Physician directed ambulation for rehabilitation of spinal cord injury. *Surg. Rounds Orthop.* 4:41–47, 1990.

Phillips, C.A., Hendershot, D.M., and Elzayat, S.G. A systems approach to medically prescribed FES ambulation after spinal cord injury. *Paraplegia*, in press, 1991.

Phillips, C.A., and Petrofsky, J.S. Velocity of contraction of skeletal muscle as a function of electrical activation and fiber composition. *J. Biomech.* 13:549–558, 1980.

Phillips, C.A., and Petrofsky, J.S. The passive element force–velocity relationship: Influence on maximal contractile element velocity. *J. Biomech.* 14:399–403, 1981.

Phillips, C.A., Petrofsky, J.S., Hendershot, D.M., and Stafford, D. Functional electrical exercise: A comprehensive approach for physical conditioning of the spinal cord injured individual. *Orthopedics*, 7:1112–1123, 1984.

Robinson, C.J., Bolam, J.M., Chinoy, M., et al. Response to surface electrical stimulation of the quadriceps in individuals with spinal cord injury. *Proc. RESNA*, 6:282–284, 1986.

Stallard, J., Major, R.E., and Patrick, J.H. A review of the fundamental design problems of providing ambulation for paraplegic patients. *Paraplegia*, 27:70–75, 1989.

Stanic, U., and Trnkoczy, A. Closed-loop positioning of hemiplegic patient's joint by means of functional electrical stimulation. *IEEE Trans. Biomed. Eng.* BME-21:365–370, 1974.

Vodovnik, L., Crochetiere, W.J., and Reswick, J.B. Control of a skeletal joint by electrical stimulation of antagonists. *Med. Biol. Eng.* 5:97–109, 1967.

Vodovnik, L., Bajd, T., Kralj, A., Gracanin, F., and Strojnik, P. Functional electrical stimulation for control of locomotor systems. *CRC Crit. Rev. Bioeng.* 6:63–131, 1981.

Yngve, D.A., Douglas, R., and Roberts, J.M. The reciprocation gait orthosis in myelomeningocele. *J. Pediatr. Orthop.* 4:304–310, 1984.

10
Ambulatory Exercise Procedures and Effects

I. Cardiopulmonary Effects During Quadriplegic Ambulation Exercise*

A. Introduction

Various problems have limited the use of functional electrical stimulation (FES) technology for ambulation of the spinal-cord-injured individual (Stallard et al., 1989). For example, the high energy cost associated with movement induced by FES results in rapid muscular fatigue (Marsolais and Edwards, 1988).

FES is optimally applied during the "swing phase" of gait so that the result is forward movement of the individual. However, since FES is also applied during the "stance phase" of gait, the upright posture is maintained. In this phase, there is no movement, but the tonic (isometric) contraction of the paralyzed muscle contributes significantly to fatigue.

Another serious problem associated with FES ambulation is that the subject is not protected from falling. Postural instability can result from stimulation of only a few muscle groups (out of the total population of paralyzed muscles). Furthermore, electronic components can occasionally fail, resulting in partial or total loss of the tonic electrical activity (which maintains the patient upright).

The reciprocating gait orthosis has been employed to resolve these problems. The RGO is a long-leg brace with hip and knee joints, a pelvic band, and thoracic support. It is composed of plastics and lightweight aluminum (Douglas et al., 1983; McCall et al., 1983; Yngve et al., 1984). The pelvic band and thoracic support offer better postural stability for mid- and high level paraplegics than is the case with other braces.

A cabling system is one significant feature of this orthosis. Two cables are attached to the anterior and posterior aspects of the hip joints. A reciprocal action at the hips is produced by shifting upper body weight (via posterior movement of the shoulders). This results in forward thrust of the pelvis (hip extension) followed by a transfer of weight onto the arms (and onto the walker). There is a slight rotation of the axis of the body. To achieve ambulation, FES of contralateral thigh muscles produces contralateral hip extension (of the anteriorly placed leg) so that the posteriorly placed ipsilateral leg is moved forward (by action of the interconnecting cables). The operation and application of FES with this gait orthosis system is described elsewhere (Petrofsky et al., 1985, 1986; Phillips, 1989a,b). The walking system (FES combined with a gait orthosis) is used in our laboratory because it is a more energy-efficient system (Hendershot et al., 1987; Hendershot and Phillips, 1988) than free-standing FES or braces-alone walking (Marsolais and Edwards, 1988).

This section will report the extent to which the cardiopulmonary stresses (as reflected by heart rate, blood pressure, oxygen consumption, and task cost) are significantly reduced for a quadriplegic individual when walking with electrical stimulation and a gait orthosis (as compared to walking with a gait orthosis alone).

*With D. Hendershot of Wright State University.

B. Methods

A spinal-cord-injured individual participated in this study. G.B. was a 28-year-old, male, C7 level quadriplegic who was injured 6 years before the study. He sustained an incomplete injury with some sensation and slight (but functionally useless) movement below the level of injury. G.B. had been walking in an FES-RGO system described previously (Petrofsky et al., 1985, 1986). A modified version of this system has been described and utilized by this quadriplegic subject (Phillips, 1989c) and also by a paraplegic subject (Phillips, 1990).

1. Progressive Velocity Walking

The subject participated in an 8-month training protocol in which he used electrical stimulation and an orthosis to walk on 5 days of every week. The subject would perform forward walking for a 65-minute interval divided into:

an initial 30-minute period
a 5-minute rest period
a final 30-minute period

Walking began at a baseline rate of 0.4 km/h and progressed in increments of 0.4 km/h (i.e., 0.4, 0.8, 1.2, 1.6, and 2.0 km/h). The walking involved progressive velocities. If the subject completed the initial 30-minute period at a given rate, the subject proceeded to walk at that rate *plus 0.4 km/h* during the final 30-minute period. If the subject did not complete the initial 30-minute period at the given rate, he was asked to walk at that rate *minus 0.4 km/h* during the final 30-minute period.

Depending on whether the subject did nor did not complete the final 30-minute period, the next session's initial velocity was calculated as the final 30-minute period's velocity *plus or minus* (respectively) 0.4 km/h. In this manner, the subject underwent progressive velocity walking.

The lap times were measured by a technician with a stopwatch. Since the lap distance was known, the lap velocity was manually calculated by dividing the former into the latter. This procedure has recently been automated by having the subjects themselves utilize a portable sensory feedback system, which provides the necessary information for subject control of gait at a constant velocity (Phillips et al., 1990).

2. Progressive Velocity Stress Testing

At least once a month the subject participated in a modified stress test. Initially, he walked at a velocity of 0.4 km/h for 4 minutes. For 30 seconds thereafter (and while the subject was still walking), expired air was collected and the cardiopulmonary system analyzed as described below. If the subject successfully completed the 4-minute walking period, there was a 5-minute rest period (to avoid cumulative fatigue effects). This was followed by another 4-minute walking period with the velocity incremented by 0.4 km/h. After each 4-minute walking period expired, air was collected and analyzed. There was then another 5-minute rest period. This protocol continued until the subject attained a velocity at which he could not complete the 4-minute walking period.

Two progressive velocity stress tests were done, on separate days. One test employed functional electrical stimulation and a gait orthosis, and the other test was with the gait orthosis alone (hereafter referred to as With FES and Without FES, respectively).

3. Cardiovascular Measurements

At the conclusion of the 30-second expired gas collection interval, the subject stood erect, but was no longer walking. The systolic and diastolic blood pressures were then determined by manual sphygmomanometry and auscultation of the right arm. Concurrently, the heart rate was determined by manual palpation of the radial pulse of the left arm.

4. Pulmonary Measurements and Task Cost

The ventilation equivalent ($\dot{V}_E$) and oxygen consumption ($\dot{V}_{O_2}$) were determined as follows. The 30 seconds of expired air (CO_2 and unused O_2) was sampled using blood gas syringes and expelled directly into the Beckman Medical Gas analyzer (model 7000 IV), where O_2 and O_2 percentages by volume were calculated for the number of breaths that were expired per 30-second unit of time ($\dot{V}_E$ in liters per half-minute). The correction factor (for sea-level barometric pressure and body temperature) was determined by a standard chart using barometric pressure (mmHg) and temperature (°C). The corrected $\dot{V}_E$ was obtained by multiplying the correction factor by 1.44 and by twice the

TABLE 10.1. Serial determination of heart rate (beats/min) at various walking velocities when using RGO alone (no FES).

Test date (1986–1987)	Resting	Velocity [km/h (mph)] 0.4 (¼)	0.8 (½)	1.2 (¾)	1.6 (1)	2.0 (1 ¼)
9/11	87	84	90	108	116	135
9/25	56	56	56	56	72	120
9/30	68	84	84	68	88	108
10/24	72	84	88	92	96	100
10/30	76	76	76	76	84	120
12/15	64	64	68	80	84	136
12/17	60	72	68	76	76	112
1/14	80	88	72	72	96	112
1/16	56	62	68	68	72	76
2/2	80	96	96	92	112	132
2/3	72	76	76	84	88	88
3/9	60	88	84	88	112	132
3/16	60	60	60	90	114	133
3/23	88	84	84	76	84	112
3/24	60	88	92	80	96	104
Mean:	69.3	77.5	77.4	80.4	92.7	114.7
SD:	11.0	12.2	12.0	12.6	15.1	17.9

$\dot{V}_E$ (so that the two 30-second time periods would equal one minute). The $\dot{V}_{O_2}$ was obtained by taking the corrected $\dot{V}_E$ times the percent true O_2 and dividing by 100. Oxygen uptake rate $\dot{V}_{O_2}$ was calculated as follows:

$$\dot{V}_{O_2} = \frac{(\text{cor } \dot{V}_E \text{ (\% true } O_2))}{100} \qquad (10.1)$$

and

$$\text{cor } \dot{V}_E = (2)(\dot{V}_E)\text{ (cor factor) } (1.44) \qquad (10.2)$$

where $\dot{V}_E$ is in liters per half-minute.

Task cost (Pierrynowski et al., 1981) was calculated as the ratio of useful work output (in kilograms times meters) per milliliter of oxygen consumed.

The overall equation is:

$$\frac{\text{kg} \cdot \text{m}}{\text{mL } O_2} = \frac{\text{wt(kg)} \times \text{vel(km/h} \times 0.0167\text{h/min)} \times 1000\text{(m/km)}}{\dot{V}_{O_2} \text{ (mL/min)}} \qquad (10.3)$$

Note that $\dot{V}_{O_2}$ was calculated from Equation (10.1) and was determined with FES and without FES, as described previously. The weight (in kilograms) represents the external weight of the system being utilized and the weight of the subject. This consisted of the gait orthosis and the walker (total of 5.45 kg) and the weight of the patient (66.55 kg) for a total weight of 72.0 kg.

For this particular individual and system, the more simplified equation becomes:

$$\frac{\text{kg} \cdot \text{m}}{\text{mL } O_2} = \frac{\text{vel (km/h)} \times 1200 \left(\dfrac{\text{kg} \cdot \text{m} \cdot \text{h}}{\text{km} \cdot \text{min}}\right)}{\dot{V}_{O_2} \text{ (mL } O_2\text{/min)}} \qquad (10.4)$$

C. Results

Table 10.1 indicates the measured heart rate as a function of walking velocity using just the gait orthosis and walker (Without FES mode). Table 10.2 indicates the measured heart rate as a function of walking velocity using the system (With FES mode). All data are also shown as obtained chronologically during the 8-month protocol, using the progressive velocity stress test protocol.

Table 10.3 shows the measured systolic and diastolic blood pressures as these relate to walking velocity using the RGO alone. The blood pressure data similarly are shown, but using the system (With FES) in Table 10.4. Note that systolic blood

TABLE 10.2. Serial determination of heart rate (beats/min) at various walking velocities when using the FES-RGO.

Test date (1986–1987)	Resting	Velocity [km/h (mph)] 0.4 (¼)	0.8 (½)	1.2 (¾)	1.6 (1)	2.0 (1 ¼)
9/5	80	80	86	90	96	126
9/8	92	92	92	112	100	124
9/24	90	96	72	84	128	132
9/29	92	96	96	96	112	124
10/15	76	84	104	126	138	140
10/17	84	56	80	60	96	104
12/18	64	68	96	100	100	112
1/9	56	60	56	64	65	84
1/12	64	72	80	100	90	88
1/28	52	60	64	64	60	68
2/4	56	72	64	76	68	76
3/17	62	60	80	76	88	88
3/19	76	72	88	92	96	100
Mean:	72.6	74.5	81.4	87.7	95.2	105.1
SD:	14.4	14.1	14.3	19.6	22.7	23.1
Significance:	N.S.	N.S.	N.S.	N.S.	N.S.	N.S.

pressure is significantly reduced when walking with FES and the orthosis (compared to orthosis alone) at the higher velocities of 1.2 ($p < 0.025$), 1.6 ($p < 0.005$), and 2.0 km/h ($p < 0.025$).

Table 10.5 relates the calculated oxygen consumption as a function of walking velocity using the gait orthosis and walker (Without FES). The oxygen consumption results are shown in Table 10.6 for the patient using the FES system with the orthosis. At the highest velocity of walking (2.0 km/h), the oxygen consumption is significantly reduced when walking with ES and the RGO compared to RGO walking alone ($p < 0.025$).

Table 10.7 reports the calculated task cost as a function of walking velocity using just the gait orthosis and walker, but no FES. Table 10.8 further

TABLE 10.3. Serial determination of blood pressure (mmHg) at various walking velocities when using RGO alone (no FES).

Test date (1986–1987)	Resting	Velocity [km/h (mph)] 0.4 (¼)	0.8 (½)	1.2 (¾)	1.6 (1)	2.0 (1 ¼)
9/11	88/56	96/56	114/62	128/74	130/82	122/78
9/25	100/68	110/70	120/72	130/80	120/82	120/80
9/30	102/70	88/72	90/76	110/80	102/66	84/54
10/29	90/65	98/65	95/60	98/78	100/72	100/70
10/30	94/70	100/72	96/70	108/78	106/86	114/82
12/15	95/65	100/65	116/80	118/85	122/98	138/100
12/17	96/54	102/68	104/74	110/74	102/82	124/86
1/14	104/72	106/76	118/80	130/82	126/82	138/82
1/16	110/90	108/87	105/84	109/86	110/82	115/90
2/2	116/86	110/74	114/80	118/78	118/84	108/82
2/3	90/68	108/64	108/64	122/74	112/66	106/70
3/9	90/62	110/70	110/68	130/70	122/72	120/70
3/16	103/70	108/76	122/70	128/68	132/80	130/80
3/23	98/72	90/72	86/74	90/72	118/82	122/90
3/24	98/58	94/70	102/72	102/78	112/80	110/80
Mean:	98.3/68.4	101.9/70.5	106.7/72.4	115.4/77.1	115.5/79.7	116.7/79.6
SD:	7.9/9.8	7.5/6.9	11.1/7.0	12.8/5.2	10.2/8.1	14.2/10.8

TABLE 10.4. Serial determination of blood pressure (mmHg) at various walking velocities when using FES-RGO.

Test date (1986–1987)	Resting	Velocity [km/h (mph)] 0.4 (¼)	0.8 (½)	1.2 (¾)	1.6 (1)	2.0 (1 ¼)
9/5	100/70	100/70	102/74	100/72	94/72	98/72
9/8	84/68	84/66	92/72	84/74	86/74	84/72
9/24	98/70	90/68	102/85	100/70	108/75	110/80
9/29	106/72	102/70	98/68	108/72	102/66	110/74
10/15	102/70	106/78	104/68	128/76	104/74	90/58
10/17	90/65	85/60	95/65	95/60	95/80	90/65
12/18	98/64	90/65	100/70	102/70	98/64	90/62
1/9	98/78	104/85	105/85	107/88	109/86	118/84
1/12	100/76	100/75	100/78	105/75	108/80	112/82
1/28	116/80	116/81	118/80	118/82	127/86	126/90
2/4	104/72	110/78	116/78	118/80	118/80	120/85
3/17	90/60	92/60	100/58	100/70	102/80	108/80
3/19	94/50	90/72	88/68	96/64	80/62	70/52
Mean:	98.5/68.8	97.6/71.4	101.5/73.0	104.7/73.3	102.4/75.3	102./73.5
SD:	8.1/7.9	9.9/7.8	8.3/8.0	11.5/7.3	12.5/7.8	16.4/11.5
Significance:	N.S./N.S.	N.S./N.S.	N.S./N.S.	$p<0.025$/N.S.	$p<0.005$/N.S.	$p<0.025$/N.S.

reports the calculated task cost as a function of walking velocity using the FES system. Statistical analysis of these results indicates that the task cost of walking with FES and the RGO is significantly improved at the highest walking velocity (2.0 km/h) compared to walking with just the RGO ($p < 0.025$).

D. Discussion

This section has provided specific information concerning the cardiopulmonary responses to ambulation exercise with the FES-RGO. It has been reported that the cardiopulmonary stresses (as indicated by blood pressure and oxygen con-

TABLE 10.5. Serial determination of oxygen consumption, $\dot{V}_{O_2}$ (L/min), at various walking velocities when using RGO alone (no FES).

Test date (1986–1987)	Resting	Velocity [km/h (mph)] 0.4 (¼)	0.8 (½)	1.2 (¾)	1.6 (1)	2.0 (1 ¼)
9/11	152	350	544	791	849	1129
9/25	105	234	519	627	823	888
9/30	123	376	623	640	968	892
10/29	130	238	243	520	437	978
10/30	48	406	339	765	797	1163
12/15	151	418	677	900	596	1060
12/17	139	556	617	640	1104	867
1/14	71	508	732	899	764	1834
1/16	117	500	759	1163	1053	1517
2/2	240	427	720	693	829	1098
2/3	157	323	666	765	878	1510
3/9	89	451	803	657	1105	1275
3/16	160	589	508	705	631	1111
3/23	127	506	687	874	850	1030
3/24	126	432	774	846	830	1043
Mean:	129	421	614	774	834	1160
SD:	44.3	105	160	164	185	271

TABLE 10.6. Serial determination of oxygen consumption, $\dot{V}_{O_2}$ (L/min), at various walking velocities when using FES-RGO.

Test date (1986–1987)	Resting	Velocity [km/h (mph)] 0.4 (¼)	0.8 (½)	1.2 (¾)	1.6 (1)	2.0 (1 ¼)
9/5	120	337	767	695	907	756
9/8	80	500	766	833	813	1117
9/24	140	507	705	659	978	1214
9/29	152	416	783	623	678	939
10/15	141	431	668	539	663	963
10/17	49	261	516	393	526	918
12/18	206	530	576	913	831	938
1/9	96	413	461	797	722	976
1/12	113	560	717	640	747	1204
1/28	64	461	766	787	742	813
2/4	160	329	634	724	821	1058
3/17	181	580	788	717	620	1052
3/19	173	546	511	841	891	920
Mean:	129	452	666	705	765	990
SD:	47.8	98.5	116	139	126	136
Significance:	N.S.	N.S.	N.S.	N.S.	N.S.	$p < 0.025$

sumption) are significantly less for this quadriplegic individual when walking with functional electrical stimulation and a gait orthosis than when walking with a gait orthosis alone.

Although measured heart rate did not change, systolic blood pressure was significantly reduced when walking with FES and orthosis (compared to orthosis alone) at the higher walking velocities of 1.2, 1.6, and 2.0 km/h. Furthermore, at the highest walking velocity (2.0 km/h), the oxygen consumption was significantly reduced when walking with FES and the RGO as contrasted with RGO walking alone.

TABLE 10.7. Serial determination of task cost $\left(\frac{\text{kg} \cdot \text{m}}{\text{mL } O_2}\right)$ at various walking velocities when using RGO alone (no FES).

Test date (1986–1987)	Velocity [km/h (mph)] 0.4 (¼)	0.8 (½)	1.2 (¾)	1.6 (1)	2.0 (1 ¼)
9/11	1.37	1.76	1.82	2.26	2.13
9/25	2.05	1.85	2.30	2.33	2.70
9/30	1.28	1.54	2.25	1.98	2.69
10/29	2.02	3.95	2.77	4.39	2.46
10/30	1.18	2.83	1.88	2.41	2.06
12/15	1.15	1.42	1.60	3.22	2.26
12/17	0.86	1.56	2.25	1.74	2.77
1/14	0.94	1.31	1.60	2.51	1.31
1/16	0.96	1.26	1.24	1.82	1.59
2/2	1.12	1.33	2.08	2.32	2.19
2/3	1.49	1.44	1.88	2.19	1.59
3/9	1.06	1.20	2.19	1.74	1.88
3/16	0.81	1.89	2.04	3.04	2.16
3/23	0.95	1.40	1.65	2.26	2.33
3/24	1.11	1.24	1.70	2.31	2.30
Mean:	1.22	1.73	1.95	2.43	2.16
SD:	0.38	0.74	0.38	0.68	0.43

TABLE 10.8. Serial determination of task cost $\left(\frac{kg \cdot m}{mL\ O_2}\right)$ at various walking velocities when using FES-RGO.

	Velocity [km/h (mph)]				
Test date (1986–1987)	0.4 (¼)	0.8 (½)	1.2 (¾)	1.6 (1)	2.0 (1 ¼)
9/5	1.42	1.25	2.07	2.12	3.17
9/8	0.96	1.25	1.73	2.36	2.15
9/24	0.95	1.36	2.19	1.96	1.98
9/29	1.15	1.23	2.31	2.83	2.56
10/15	1.11	1.44	2.67	2.90	2.49
10/17	1.84	1.86	3.66	3.65	2.61
12/18	0.91	1.67	1.58	2.31	2.56
1/9	1.16	2.08	1.81	2.66	2.46
1/12	0.86	1.39	2.25	2.57	1.99
1/28	1.04	1.25	1.83	2.59	2.95
2/4	1.46	1.51	1.99	2.34	2.27
3/17	0.83	1.22	2.01	3.10	2.28
3/19	0.88	1.88	1.71	2.15	2.61
Mean:	1.12	1.49	2.14	2.58	2.47
SD:	0.29	0.29	0.54	0.46	0.34
Significance:	N.S.	N.S.	N.S.	N.S.	$p < 0.025$

Task cost can be viewed as an index of the cardiopulmonary performance. It is the ratio of useful work output (in kilograms times meters) for each milliliter of oxygen consumed. The index, developed by Pierrynowski et al. (1981), may be considered to be somewhat similar to "miles per gallon" as an index of automobile performance: the higher the index, the better the performance. The results of this study indicate that the task cost of walking with FES and the RGO is significantly improved at the highest walking velocity (2.0 km/h) in contrast to walking with just the RGO.

The end result is that cardiopulmonary stresses are significantly less for this quadriplegic individual when walking with functional electrical stimulation and a gait orthosis as opposed to walking with a gait orthosis alone. This finding should translate into improved endurance during each walking session (and result in longer walking times). Consequently, the aerobic exercise effect (e.g., muscular endurance training and improved cardiopulmonary reserve) should be enhanced when a quadriplegic walks with FES and an RGO instead of an RGO alone. This certainly is a testable hypothesis and will be the subject of a future study.

Finally, the lowered cardiopulmonary stresses might allow some quadriplegics, who would be unable to walk with the RGO alone, to walk with electrical stimulation and the RGO. Results from one such individual have been recently reported by Solomonow et al. (1988). As functional electrical stimulation with orthotic bracing becomes more universally applied, additional case studies should particularly examine this specific issue.

II. Psychological Effects with Quadriplegic Ambulation Exercise*

Functional electrical stimulation improves muscle bulk, strength, range of motion, motor skill, and body appearance. This section will report on changes of personality and self-image of the patients undergoing FES.

A. The Study Population

A volunteer study group of 14 spinal-cord-injury quadriplegics (13 male, 1 female), ages 20–46, was selected randomly from a sample population of 50 SCI patients. These patients were involved in the

*From R. Bell et al., A study of personality changes in quadriplegics engaged in a functional electrostimulation (FES) program. American Academy of Neurological and Orthopedic Surgeons. Copyright 1987, reprint with permission from *Journal of Neurological and Orthopedic Medicine and Surgery*, 8:353, 1987.

functional electrostimulation therapy program at the Help Them Walk Again Foundation (HTWA) in Las Vegas, Nevada. This therapeutic approach through FES was developed by Petrofsky, Phillips, et al. (1985). All 14 patients in this study group had sustained spinal cord injuries within the range of C4–C7 as a result of accidents. Some of the accidents involved automobiles (9), motorcycles (2), diving (1), and fall (1).

B. The Study

The study group began the FES program in January 1983 and was engaged in the beginning of the orderly progress and expansion of that type of therapy at the time of the administration of a standard test. This test is called a 16 Personality Factor Questionnaire (16PFQ) developed by the Institute for Personality and Ability Testing (IPAT, 1986). The test measures 16 primary personality factors such as "enthusiasm," "self-discipline," and "drive." Combining the results of the primary factors produces a composite personality portrait. The pretest of the 16PFQ was performed in March 1984. The posttest of the 16PFQ was administered to the study group during early 1987. The study test group scores were evaluated over a period of years. They were also compared to a control group of non-FES patients undergoing other types of standard physical therapy (Runyon, 1977).

The posttest findings reflected substantially higher scores in the composite personality factor categories, significant at $\alpha = 0.05$, $t = 3.47$, $p > |t| = 0.0042$ in the validity scale; pretest x = 5.36, posttest x = 6.93. One factor listed as *G* or "conformity" approached statistical significance at $\alpha = 0.05$, with behavioral orientation toward "expediency" at $t = 2.12$, $p\ |t| = 0.0537$, pretest x = 5.43, posttest x = 5.00.

The change that occurred in the personality factors comprising the total behavioral orientation was measured in the study group during the period of study while new techniques of FES therapy were being phased into the therapy program. During the FES therapy period, the differentials were computed from mean values recorded by the group over several years. The findings demonstrated tendencies toward improvement of the primary personality factors: "self-assurance," "self-sufficiency," "self-discipline," "concrete thinking," "forthrightness," "enthusiasm," "drive" (tension), "boldness," "trust," "warmth," "tendermindedness" (sensitivity), "submissiveness," and "imagination." The composite personality factors augmented were "extravertism" and "leadership." "Faking good" was augmented only to a small degree. ("Faking good" means in this context that the patients believe that they have improved more than their personality tests and neurological exams demonstrate.) Factors diminished were "tough poise," "manipulation," "anxiety," "confabulation," "isolationism," and "neuroticism."

Those members of the study sample who had developed ambulation skills (walking) to a varied degree tended to present the foregoing behavioral orientations more than a control group of nonwalkers. The only primary factors that did not change between the study group and the control group were "intelligence" and "radicalism." Conservatism/liberalism also reflected little or no change in orientation during the study in either group. The "walkers" appeared to present responses that were more group oriented and relaxed.

In addition, an author-constructed opinion inventory questionnaire (OIQ) was administered to the study sample on or about the time of the administration of the 16PFQ posttest. The data forthcoming from OIQ suggest that the majority of the study sample perceived the HTWA FES program as the primary contributory factor producing an improvement in the behavioral profile.

In conclusion, as a functional electrical stimulation system of physical therapy results in improvement in muscle strength, muscle motion, posture, and physical appearance, there are concomitant improvements in personality, self-reliance, and self-sufficiency.

III. Training Effects with the Sensory Feedback System*

A. Introduction

To maintain a state of balance or to initiate a walking sequence, various types of feedback are employed to tell a person when the desired position has been attained. This feedback is needed continuously since, even when motionless, the person needs to make slight adjustments in muscle activity

*With R. Koubek of Wright State University.

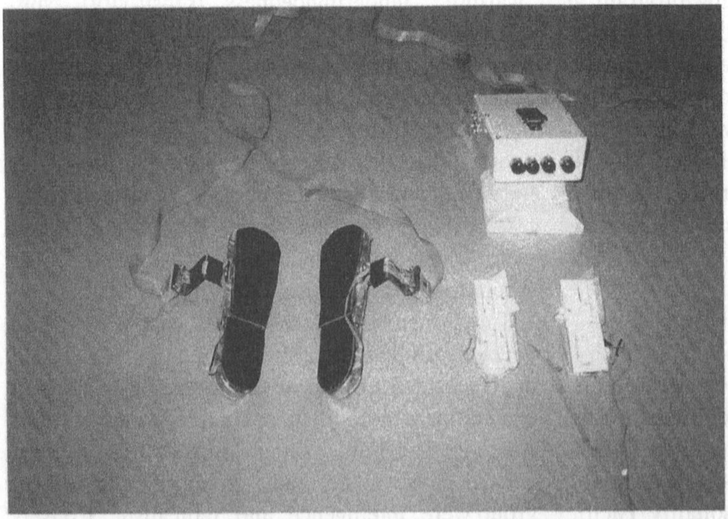

FIGURE 10.1. The complete sensory (tactile) feedback system (see text).

to avoid falling (Thomas and Whiney, 1959). Since the feedback information is used by the central nervous system to initiate and maintain movement (with skeletal muscle as the actuator), this network can be viewed as a closed-loop feedback control system (Petrofsky and Phillips, 1985).

The feedback for the human system occurs both from impulses generated reflexly by internal receptors in the muscles and joints, and by more externally apparent "sensory" feedback consisting of vision, touch, and audition. In this type of feedback the eyes, ears, and skin act as transducers, which convert outside information into impulses recognized by the central nervous system (Vander et al., 1975).

In a paraplegic individual, some of this information flow has been interrupted, since this person no longer has the link along the spinal cord to the lower alpha motor neurons. The skeletal muscle itself remains healthy but begins to atrophy with disuse (Coxe and Grubb, 1974). Various functional electrical stimulation orthoses, including those developed by the Wright State University group, have succeeded in providing a means for the paralyzed individual to generate movement and thus prevent further deterioration of the limbs (Phillips et al., 1984, 1989).

For a standing and walking orthosis to be continuous and automatic, a closed-loop sensory feedback system is desirable (Phillips, 1988). This system provides vibrotactile feedback in addition to the individual's own visual sensation.

Such a sensory (tactile) feedback system has been described (Phillips and Petrofsky, 1985a). The system consists of the following elements: foot load transducers, sensor signal conditioning, carrier oscillator, balanced modulator, interface signal conditioning, and a vibrocutaneous interface. The same study (Phillips and Petrofsky, 1985a) has determined the following: (a) stimulus threshold with respect to frequency, (b) spatial separation with respect to frequency ratios, and (c) spatial separation and frequency ratios with respect to tactile location.

The prototype sensory feedback system originally described by Phillips and Petrofsky (1985a) was subsequently modified for another study (Phillips and Petrofsky, 1986). The modified system (Figure 10.1) consisted of the following components: foot load transducers, signal processing unit, and the vibrocutaneous interface.

Results were reported for a T4 level (complete) paraplegic subject (Phillips and Petrofsky, 1985b). While upright and blindfolded (Figure 10.2), subjects experienced an initial period averaging 30 seconds, during which time they continuously adjusted their posture (and inclination at the hip varied to $\pm 5°$ peaks). This was followed by an average 5-minute period of prolonged standing, during which time hip inclination varied up to $\pm 1°$ peaks. The sensory feedback system was then deactivated while the subject was still blindfolded.

Within a period averaging 5 seconds, the subjects lost their balance (at which time hip inclination exceeded $\pm 10°$).

Visual feedback has been utilized for the maintenance of upright posture, balance, and walking in all lower extremity neural (FES) prostheses to date (e.g., Brindley et al., 1979; Kralj et al., 1980; Vodovnik et al., 1981; Marsolais and Kobetic, 1983; Holle et al., 1984; Petrofsky et al., 1984). The maintenance of upright posture and balance (in the presence of visual feedback) for the paraplegic subject who is using a neural (FES) lower extremity prosthesis has been mathematically modeled by Kralj (1986), Jaeger (1986), and Buhrman and Phillips (1989). These investigators have characterized the upright posture and balance situation in terms of the familiar inverted-pendulum model.

The most interesting aspect of the Phillips and Petrofsky (1985b, 1986) study was that upright body posture and balance can be maintained in a high level paraplegic without the use of visual feedback. This was considered an initial test of the usefulness of the sensory feedback system.

It subsequently became apparent, however, that when trying to walk, the subject became confused and possibly overloaded with the tactile information. The vibrocutaneous interface (which provided a proportional signal as the pressure on the foot surfaces varied) provided stimulation to four locations on the chest. This system apparently provided more information than the subject could interpret upon first exposure.

Dual processing code theory (Wickens, 1984) suggests that repeated training of consistent task components can become automated, thus requiring little or no cognitive resources. Because walking is made up of primarily consistent task components, it is expected that repeated training will result in improved performance and decreased cognitive load.

This section tests the hypothesis that sufficient information can be transferred [using the sensory (tactile) feedback system of Phillips and Petrofsky (1986)] to effectively determine foot position and anticipate the upcoming step. The effect of training is further examined in the experimental design by evaluating the cognitive workload relating to the interpretation of the feedback by the subject. The results reported in this section provide additional information for the development of FES orthoses to be used by spinal-cord-injured patient.

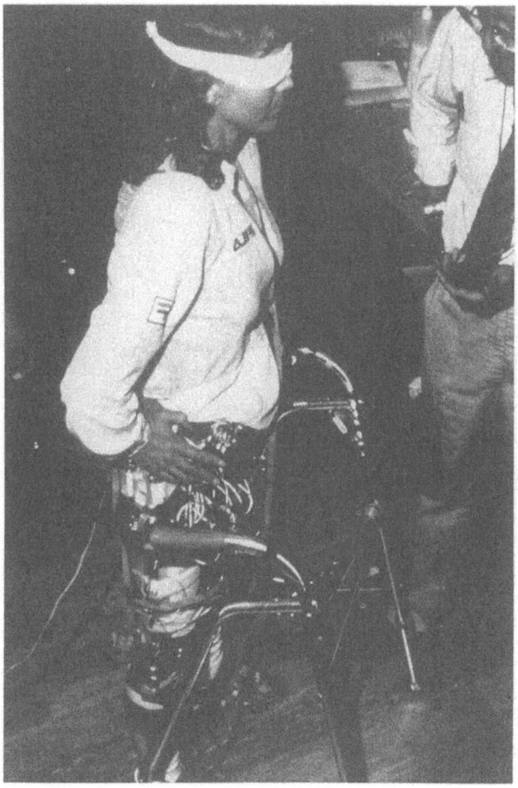

FIGURE 10.2. Paraplegic patient initially standing with the FES–orthosis and the sensory feedback system.

B. Methods

1. Design

The experiment was designed to permit an evaluation of the sensory (tactile) feedback system without the use of paraplegic subjects (see the Section III. 13.3, below). For the experiment, a nondisabled person was outfitted with the foot load transducers (FLTs). This person, hereafter referred to as the Walker, walked at a steady pace down a length of a hallway. A second nondisabled person was outfitted with the vibrocutaneous interface (VCI). This person, hereafter referred to as the Mimic, attempted to interpret the feedback provided by the FLTs through the VCI. The feedback was a mild, concentrated vibration, which varied with the foot pressure applied. The task of the Mimic was to walk in step with the Walker. The length of the stride was not important, but the synchronization of the footfalls was. The measure of

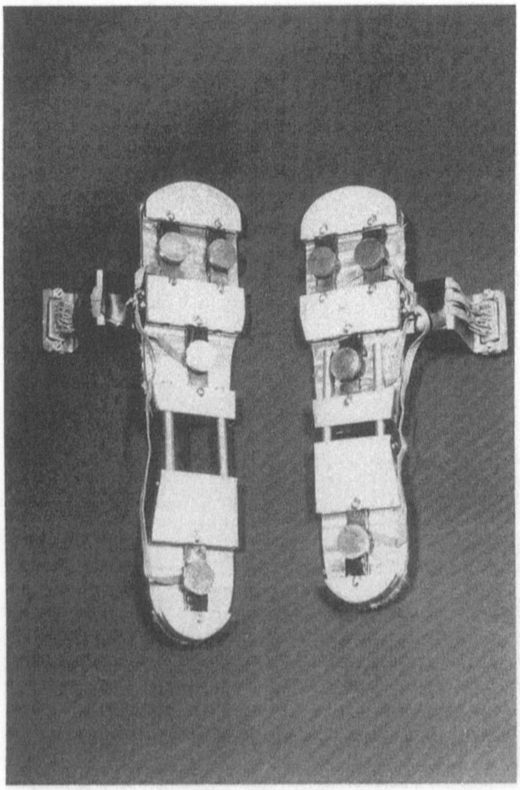

FIGURE 10.3. Right and left foot load plates, each with four foot load transducers (strain gages).

walking performance was the time delay between like footfalls of the Walker and the Mimic.

The experiment utilized a 2 × 3 repeated measure design. The Mimic's performance was evaluated (with and without an added cognitive load) at no training and at two additional levels of training. The independent variables in the experiment were cognitive load and training. The cognitive load was applied by having the Mimic count backward from a number by threes. This additional load was used to encourage the Mimic to interpret the tactile sensations subconsciously. Testing runs with the load applied served as a measure of this trend. The training involved several practice runs, which allowed for learning the rhythms and the intensities of the tactile information. Two training levels were used as test points and were preceded by equal amounts of practice training. All testing and training were preceded by a thorough orientation to the equipment. The dependent variables in the experiment were the mean and standard deviation of the time delay between like heel strikes of the Walker and Mimic.

2. *Apparatus*

The foot load transducers were bonded strain gages mounted in a conventional Wheatstone bridge configuration. The FLTs were located under the toe and heel areas and mounted on foot load plates (Figure 10.3). For these experiments, only the output from one of the FLT toe units was fed over ribbon cable and connected to the appropriate signal processing units.

The FLTs were interfaced with a signal processing circuit. This signal processing circuit (Figure 10.4) is replicated two times, one system for an FLT at the toe of each foot. The circuit generates a carrier oscillator frequency at 70 or 100 Hz, which is suppressed when there is no modulating signal from an FLT. With increasing pressure (weight) on an FLT, there is decreasing suppression of the carrier oscillator output so that increased signal is delivered to the vibrocutaneous interface consisting of four channels. Each signal processing circuit has a gain that is variable between 1 and 10. Each circuit also has a separate balance control for adjusting null offset voltage from each amplifier. The signal processing circuit provided outputs to the two channels of the VCI taped to the chest of the Mimic.

The sound of a metronome was recorded onto the right channel of a cassette tape using a cassette tape recorder and a standard monaural microphone. The Mimic's tape contained simulated white noise recorded using a stereo cassette recorder and played on another personal stereo. The control box (which contained the four signal processing units) was placed on a laboratory cart, which was pushed by the Mimic throughout the experiment.

Data collection and analysis was performed using a video camera, a video recorder, and a color monitor. The video recorder provided an internal clock displaying time increments of one-hundredth of a second onto the recording. It also had four video heads and allowed for frame-by-frame slow motion capability.

3. *Subjects*

Six subjects participated as Mimics during the experiment. The subjects were all adult males from the Wright State University community. None of the Mimics were impaired in terms of walking performance and all were quite capable of complete participation in the experimental procedures.

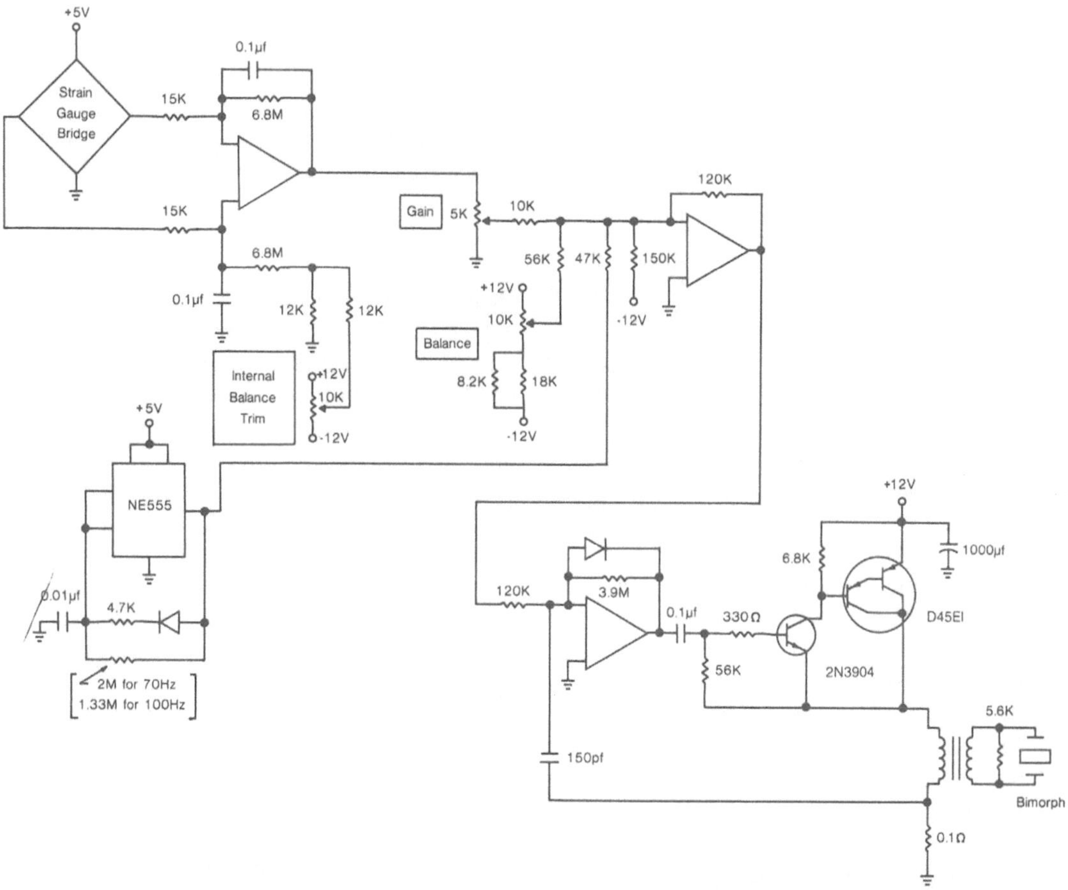

FIGURE 10.4. Schematic diagram of signal processing circuit; all operational amplifiers are LM 353s.

To expedite the conduct of the experiment, non-paraplegic subjects were selected, since they would not have to be outfitted with the FES–orthosis system currently utilized in our laboratory (Phillips, 1989b).

4. Procedure

Each of the subjects was first briefed on the background, purpose, and expected benefits of the experiment. The Walker donned the FLTs while the VCI was taped to the upper chest of the subject (Mimic). Both Walker and Mimic were connected to the control box. One channel of the VCI corresponding to the left toe of the Walker was placed on the right side of the Mimic's chest. Conversely, the other channel of the VCI representing the Walker's right toe was located on the left side of the Mimic's chest. The proper vibrator intensity was adjusted, and the Mimic was familiarized with the tactile sensations of each channel.

The Mimic was instructed that the Walker would be walking at a steady pace behind the Mimic, although the Walker's pace might vary between runs. Furthermore, the Mimic was briefed that when the vibratory sensation occurred on one side of his chest, he should step off with that same foot. The entire ensemble (Walker behind Mimic and Mimic behind laboratory cart) then proceeded down a hallway for completion of training and data collection runs.

The first experimental run was a data collection run. The Mimic (who needed to be isolated from all the tactile feedback provided by the Walker) was blindfolded, and ambient noise was masked using the recorded white noise. In order to keep a steady pace, the Walker donned the headset of the stereo playing the metronome at 100 beats per minute.

TABLE 10.9. Experimental trials.

Trial number	Procedure
1	Experimental data
2	Experimental data
3	Training trial
4	Training trial
5	Practice trial
6	Practice trial
7	Experimental data
8	Experimental data
9	Training trial
10	Training trial
11	Practice trial
12	Practice trial
13	Experimental data
14	Experimental data

Half of the Mimics were instructed to count backward by threes from 100, 75, or 50, while the other half did not (but would do so in their second run). The run began with the Mimic standing still and the Walker walking in place or behind the Mimic. The Walker then tapped the Mimic on the shoulder and both began walking down the hallway. The Mimic pushed the cart, which was also guided by an additional person (to keep the blindfolded Mimic in the center of the hallway).

The camera operator followed the two down the hallway, collecting data by focusing the camera on their feet. For data analysis, the first three or so steps were disallowed so that the Mimic could get in step. Thereafter, the next 20 steps were used. This resulted in 10 data points for the time difference between right heel strikes. The second experimental run was identical to the first run in every way except that the cognitive load was or was not applied, depending on the first run.

The third and fourth runs were training runs. The Mimic was not blindfolded or sound-masked. The Walker used the other side of the tape, which provided a stimulus of 60 beats/min. These runs allowed the Mimic to see the footfalls as he felt the vibratory stimulus. The fifth and sixth runs were also practice runs, although the Mimic was then blindfolded and sound-masked. Again, the Walker walked at 60 beats/min. No data were collected during these four runs (third through sixth).

The seventh and eighth runs were data collection runs identical to experimental runs 1 and 2 except for a reversal in the occurrence of the counting task. The next four runs (9–12) were practice runs, identical to the third through sixth. The last two runs (13 and 14) were data collection runs identical to the first and second, respectively. The 14 experimental trials are summarized in Table 10.9.

C. Results

Quantitative analysis to evaluate the effect of training under different levels of cognitive load utilized the mean difference and variance (in seconds) between right heel strikes of the Walker and the Mimic. An analysis of variance (ANOVA), using the mean difference in heel strikes as the dependent variable, disclosed a significant difference in subject performance between training levels to the 0.0104 level. Based on $\alpha = 0.05$ minimum, both the level of cognitive load and also the interaction of training and cognitive load were found not to significantly impact the effect of training on subject performance.

A Bonferroni (Dunn) *t*-test was used to compare the mean differences between the three levels of training. A significant difference at the 0.05 level was found between subject performance of the no-training level and at the training levels for phases 1 and 2. The effect of training is illustrated by the bar graph in Figure 10.5.

D. Discussion

Various conclusions can be drawn from this study. First, the significant effect of training supports the dual processing code theory (Wickens, 1984) in that training allowed the cognitive feedback stimuli to become preattentive or automatic. With training, less attention was required to process the stimuli, and performance improved. In a practical application, a well-developed orientation and training program could allow persons using an FES–orthosis with a tactile feedback system to reduce the present need of visual feedback, thus permitting users to interact with their environment more effectively.

Significant effects were shown in walking performance as a result of training. This satisfies the original hypothesis that useful information can indeed be transferred using the apparatus described previously (Phillips and Petrofsky, 1986). This information, which is predicted to be sufficient

FIGURE 10.5. Time difference in heel strikes as a function of training level (with and without a cognitive load).

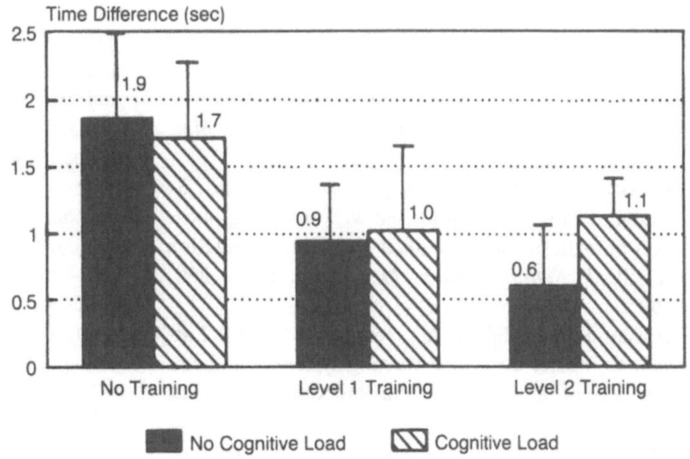

(when used in conjunction with a thorough training program) to interpret the footfalls of a Walker by a Mimic, could be used in a practical sense by a paraplegic individual to interpret his own footsteps.

Finally, these findings indicate that, through continued use and training, it is likely that this information usage could become subconscious and automatic. At this level, the cognitive burden on the visual channel could be minimized or removed to the point that the impaired individual could utilize a standing and walking FES–orthosis system on a busy sidewalk with street crossings, shop in a supermarket, and perform many other activities more independently.

No conclusions can be made at this time with respect to the number, placement, or intensity of the channels of tactile stimuli on the individual, as these possible variables were held constant throughout the experiment. Consideration of these additional variables must remain the subject for future experimental analysis.

References

Brindley, G.S., Polkey, C.E., and Rushton, D.N. Electrical splinting of the knee in paraplegia. *Paraplegia*, 16:428–435, 1979.

Buhrman, J.R., and Phillips, C.A. A mathematical model to analyze control of upright posture and balance. *Automedica* (London) 11:111–122, 1989.

Coxe, W., and Grubb, R. Trauma to the central nervous system. In *Neurological Pathology*, S. Eliasson, A. Prensky, and W. Hardin, Jr., Eds. Oxford University Press, New York, 1974.

Douglas, R., Larson, P.F., D'Ambrosia, R.E., and McCall, R.E. The LSU reciprocating gait orthosis. *Orthopedics*, 6:834–839, 1983.

Hendershot, D.M., and Phillips, C.A. Improvement of efficiency in a quadriplegic individual using an FES-RGO system. *Proc. IEEE/EMBS Tenth Int. Conf.* 10:1577–1578, 1988.

Hendershot, D.M., Moore, M.L., and Phillips, C.A. Cardiopulmonary conditioning when walking with and without FES in the paralyzed. *Fed. Proc.* 44:680, 1987.

Holle, J., Frey, M., Gruber, H., Kern, H., Stohr, H., and Thoma, H. Functional electrical stimulation of paraplegics. *Orthopedics*, 7:1145, 1984.

IPAT Staff. *The 16PF Questionnaire*. Institute for Personality and Ability Testing, Inc., Champaign, IL, 1986.

Jaeger, R.J. Design and simulation of closed-loop electrical stimulation orthoses for restoration of quiet standing in paraplegia. *J. Biomech.* 19:825, 1986.

Kralj, A. Electrical stimulation of lower extremities in spinal cord injury. In *Spinal Cord Injury Medical Engineering*, D.N. Ghista, and J.L. Frankel, Eds. Charles C. Thomas, Springfield, IL, p. 439, 1986.

Kralj, A., Bajd, T., and Turk, R. Electrical stimulation providing functional use of paraplegic patient muscles. *Med. Prog. Technol.* 7:3, 1980.

Marsolais, E.B., and Edwards, B.G. Energy costs of walking and standing with functional neuromuscular stimulation and long-leg braces. *Arch. Phys. Med. Rehabil.* 69:243–249, 1988.

Marsolais, E.B., and Kobetic, R. Functional walking in paralyzed patients by means of electrical stimulation. *Clin. Orthop. Relat. Res.*, 175:30, 1983.

McCall, R.E., Douglas, R., and Rightor, N. Surgical treatment in patients with myelodysplasia before using the LSU reciprocating system. *Orthopedics*, 6:843–848, 1983.

Petrofsky, J.S., and Phillips, C.A. Closed-loop control of movement of skeletal muscle. *CRC Crit. Rev. Bioeng.* 13:35–96, 1985.

Petrofsky, J.S., Phillips, C.A., and Heaton, H.H. Feedback control system for walking in man. *Comput. Biol. Med.* 14:135, 1984.

Petrofsky, J.S., Phillips, C.A., Larson, P., and R. Douglas. Computer synthesized walking: An application of orthosis and functional electrical stimulation (FES). *J. Neurol. Orthop. Med. Surg.* 6:219–230, 1985.

Petrofsky, J.S., Phillips, C.A., Douglas, R., and P. Larson. Computer-controlled walking system: The combination of an orthosis with functional electrical stimulation. *J. Clin. Eng.* 11:121–133, 1986.

Phillips, C.A. Sensory feedback control of upper and lower extremity motor prostheses. *CRC Crit. Rev. Biomed. Eng.* 16:105–140, 1988.

Phillips, C.A. An interactive system of electronic stimulators and gait orthosis for walking in the spinal cord injured. *Automedica* (London) 11:247–261, 1989a.

Phillips, C.A. Electrical muscle stimulators in combination with a reciprocating gait orthosis for ambulation by paraplegics. *J. Biomed. Eng.* 11:338–344, 1989b.

Phillips, C.A. Functional electrical stimulation and lower extremity bracing for ambulation exercise of the SCI individual: A medically prescribed system. *Phys. Ther.* 69:842–849, 1989c.

Phillips, C.A. Physician directed ambulation for rehabilitation of spinal cord injury. *Surg. Rounds Orthop.* 4:41–47, 1990.

Phillips, C.A., and Petrofsky, J.S. Cognitive feedback as a sensory adjunct to functional electrical stimulation (FES) neural prosthesis. *J. Neurol. Orthop. Med. Surg.* 6:231–238, 1985a.

Phillips, C.A., and Petrofsky, J.S. Preliminary report on balance/stance in paraplegia with and without a cognitive feedback system. *Physiologist*, 28:302, 1985b.

Phillips, C.A., and Petrofsky, J.S. A total neural prosthesis for spinal cord injury rehabilitation: The cognitive feedback system with a functional electric stimulation (FES) orthosis. *J. Neurol. Orthop. Med. Surg.* 7:225–234, 1986.

Phillips, C.A., Petrofsky, J.S., Hendershot, D.M., and Stafford, D. Functional electrical exercise: A comprehensive approach for physical conditioning of the spinal cord injured individual. *Orthopedics*, 7:1112–1123, 1984.

Phillips, C.A., Danopulos, D., Kezdi, P., and Hendershot, D. Muscular, respiratory and cardiovascular responses of quadriplegics to an FES bicycle ergometer conditioning program. *Int. J. Rehabil. Res.* 12: 147–157, 1989.

Phillips, C.A., Heard, D.C.Y., Koubek, R.J., and Hendershot, D.M. A sensory feedback system for constant velocity control during EMS-RGO ambulation. *Proc. 12th Annu. Conf. IEEE Eng. Med. Biol. Soc.* 12: 2260–2261, 1990.

Pierrynowski, M.R., Winter, D.A., and Norman, R.W. Metabolic measures to ascertain the optimal load to be carried by man. *Ergonomics*, 24:393–399, 1981.

Runyon, R.P. *Nonparametric Statistics: A Contemporary Approach*. Addison-Wesley, Reading, MA, 1977.

Solomonow, M., Baratta, R., Shoji, H., et al. FES powered locomotion of paraplegics fitted with the LSU-RGO. *Proc. 10th Annu. Conf. IEEE Eng. Med. Biol. Soc.* 10:1672, 1988.

Stallard, J., Major, R., and Patrick, J. A review of the fundamental design problems of providing ambulation for paraplegic patients. *Paraplegia*, 27:70–75, 1989.

Thomas, D.P., and Whiney, R.J. Postural movements during normal standing in man. *J. Anat.* 93:524–539, 1959.

Vander, A.J., Sherman, J.H., and Luciano, D.S. *Human Physiology—Mechanisms of Body Function*, 2nd ed. McGraw-Hill, New York, pp. 549–550, 1975.

Vodovnik, L., Bajd, T., Kralj, A., Gracanin, F., and Strojnik, P. Functional electrical stimulation for control of locomotor systems. *CRC Crit. Rev. Bioeng.* 6:63, 1981.

Wickens, C.D. *Engineering Psychology and Human Performance*. Charles E. Merrill, Columbus, OH, pp. 123–126, 133, 292–296; 1984.

Yngve, D.A., Douglas, R., and Roberts, J.M. The reciprocation gait orthosis and myelomeningocele. *J. Pediatr. Orthop.* 4:304–310, 1984.

Part 4
Prescription of Functional Electrical Rehabilitation

11
The Medical Criteria: Physician Guidelines for Patient Participation

I. Active Physical Therapy*

A. Introduction

Active physical therapy (APT) has been defined as a system in which functional electrical stimulation is applied to the paralyzed extremities of spinal-cord-injured individuals. The electrical stimulation "activates" the paralyzed muscles so that a specific motion is created and the muscle works against a specific load at a specific rate. Both the rate and the load are characteristic of the type of exercise training desired, as discussed below. In APT the actively contracting muscles act upon their environment to produce the exercise effect.

APT is synonymous with functional electrical rehabilitation. However, APT may be contrasted to conventional physical therapy (no FES) in which the environment acts upon the muscle to produce the exercise effect (Phillips et al., 1984). The currently accepted care and treatment of SCI individuals are well summarized in standard texts (Rossier, 1964; Guttman, 1976).

B. The APT System (FES Exercise Modalities)

APT was first described by Petrofsky and Phillips (1983a) to consist of an isokinetic leg trainer and an FES bicycle ergometer. At that time a system for functional ambulation of the SCI individual was also described (Petrofsky and Phillips, 1983b) but separately from APT. The essential feature of the leg trainer and bicycle ergometer is closed-loop (feedback) control of the electrical stimulation in order to obtain isovelocity muscle contractions. These systems deliver FES under microprocessor control and are based upon a system that produced isovelocity contractions in the hindlimb muscles of the cat (Petrofsky and Phillips, 1979). Both the open-loop (no feedback) approach to skeletal muscle stimulation (Vodovnik et al., 1981) and closed-loop (feedback) control of skeletal muscle movement (Petrofsky and Phillips, 1985) have been reviewed.

APT consists of four modalities capable of producing FES exercise and described as follows.

1. Isokinetic Leg Trainer

The leg trainer system was developed to provide a form of isokinetic exercise for paralyzed muscles (Figure 11.1). In actuality, the quadriceps muscle group of one leg is stimulated and various weights are connected to the foreleg via an ankle harness. Repetitive isovelocity contractions are produced at the knee (at an angular velocity of 15° per second). The result is repetitive cycles of foreleg extension and flexion (through 45° of motion) and a 3-second rest interval recurring about six times per minute.

The patient generally exercises for a total of 30 minutes (15 minutes for each leg). Load at the ankle is adjusted daily so that the quadriceps muscle exercises for at least 10 minutes, but fatigues

*From C.A. Phillips, The medical criteria for active physical therapy: Physician guidelines for patient participation in a program of functional electrical rehabilitation. Reprinted with permission from *American Journal of Physical Medicine*, 66:269–286, 1987. Copyright 1987, Williams & Wilkins, Baltimore.

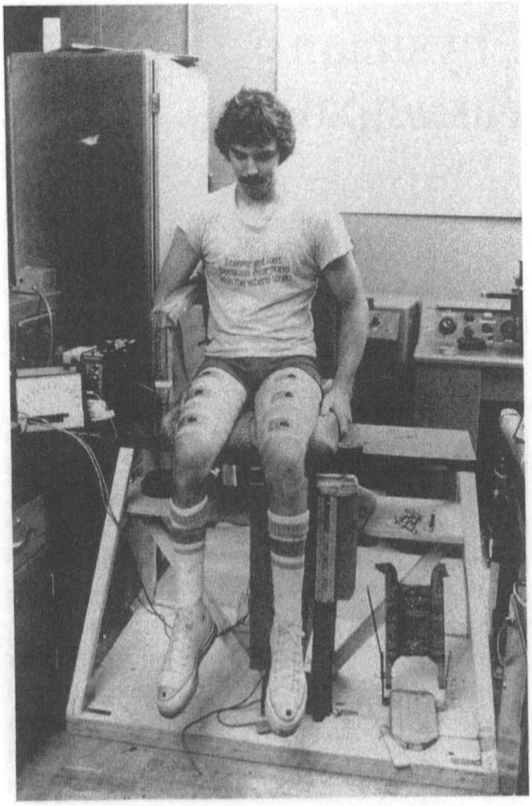

FIGURE 11.1. The isokinetic leg trainer.

within the 15-minute period. The fatigue end point occurs when the maximal stimulation levels are unable to move the foreleg (at the knee) through at least 30°. The isokinetic leg trainer system involves closed-loop computer directed stimulation and has been described in detail (Petrofsky et al., 1984a).

2. *Stationary Bicycle Ergometer*

The bicycle ergometer system was developed to provide a form of aerobic exercise for paralyzed muscles (Figure 11.2). In this system, six channels of stimulation are provided, one each to the quadriceps, hamstring, and gluteal muscle group of each leg. Stimulation is temporally and spatially distributed so that the patient pedals the bicycle at a speed of 50 revolutions per minute. The patient exercises for a total of 30 minutes. Load is adjusted by a friction belt attached to a flywheel so that the subject pedals for at least 20 minutes, but fatigues within 30 minutes. The fatigue end point occurs when the maximal stimulation levels are unable to maintain a pedaling velocity of at least 35 rpm. The stationary bicycle ergometer involves closed-loop, computer-directed stimulation and has been described in detail elsewhere (Petrofsky et al., 1984b).

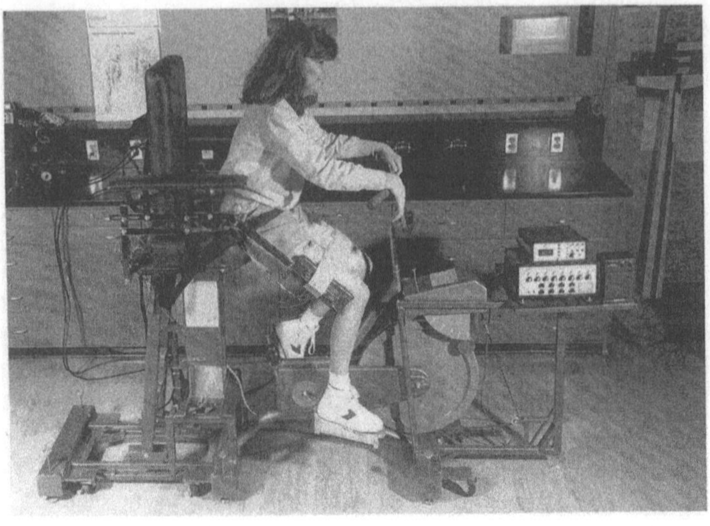

FIGURE 11.2. The FES bicycle ergometer.

FIGURE 11.3. The outdoor exercise bicycle.

3. *Outdoor Exercise Bicycle*

An exercise bicycle adapted for outdoor use has been developed as previously described (Petrofsky et al., 1983) and is shown in Figure 11.3. Its basic design has been patterned after the stationary bicycle ergometer described above.

4. *FES-Orthosis for Ambulation*

Recently, functional electrical stimulation has been combined with a lightweight reciprocal gait bracing system (orthosis) as previously described (Petrofsky et al., 1985c) and shown in Figure 11.4. This FES-orthosis allows functional walking in SCI patients. Weight is shifted to the ipsilateral foot and stimulation applied to the hamstring and gluteal groups of the ipsilateral (weight-bearing) leg. This results in ipsilateral hip extension that is cross-corrected (via cabling) to produce hip flexion of the contralateral leg. Since weight has been off-loaded on this (contralateral) leg, the contralateral leg moves forward. Weight is then shifted to the now forward positioned (contralateral) leg and the process repeated (as above), with the weight now shifted off the ipsilateral foot.

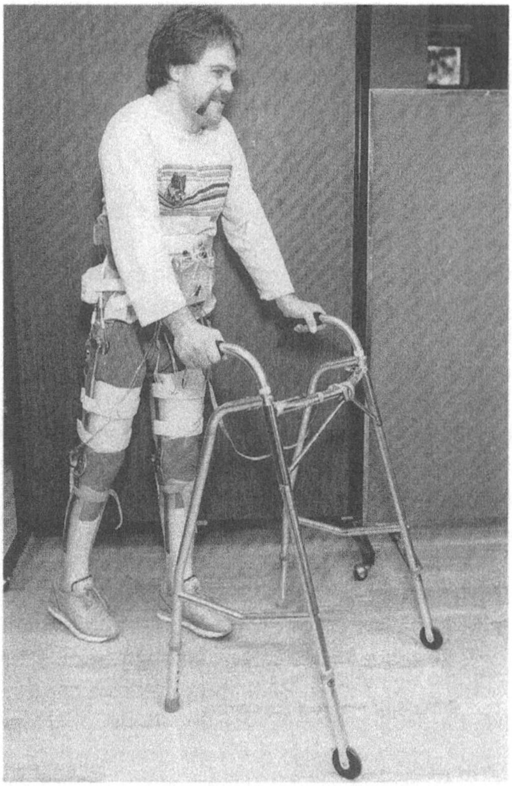

FIGURE 11.4. FES-orthosis for ambulation.

Subjects utilize the FES-orthosis for APT by walking in this manner for one hour daily. Muscle fatigue is minimized, since the orthosis maintains the upright posture for the stance phase of gait. FES is employed only transiently for the swing phase of the gait cycle. Subjects exercise to fatigue by undergoing progressive velocity walking in which the speed of walking increases incrementally once the subject is able to walk continuously for half an hour at the baseline speed. The incremented velocity then becomes the new baseline speed. Currently, our walking subjects also undergo a 30-minute session of FES bicycle ergometry (at zero load) twice weekly.

C. Physiological Response to APT

The four FES exercise modalities (*vide supra*) have been evaluated to varying degrees with respect to the physiological response elicited. Data have been reported on the muscular, cardiovascular, and respiratory systems. Broadly, these reports focus on either the acute (stress) effects or the long-term (training) effects. This section reviews our current knowledge of the physiological responses in terms of the four FES exercise modalities.

1. *Isokinetic Leg Trainer*

With acute isokinetic leg exercise, the heart rate does not change significantly in quadriplegics, paraplegics, or normal controls. The blood pressure does increase significantly in quadriplegics, but there is no significant increase in paraplegics or controls. This blood pressure response is a function of the rest interval between exercise bouts, and the greater response occurs with the shorter rest intervals (Hendershot et al., 1985). A chronic (training) effect of isokinetic leg exercise has also been documented. Petrofsky and Phillips (1983a) reported an approximate doubling of quadriceps muscle strength over a 3-week training period, while Collins et al. (1984) reported a sevenfold increase in quadriceps muscle strength over a 12-week training period. Both groups utilized progressive intensity protocols. Finally, Robinson et al. (1986) reported results of a leg trainer study in which the type of exercise appeared more "aerobic." Specifically, all 20 paraplegics and quadriplegics had an increase in muscle endurance, but only about half experienced an increase in muscle strength.

2. *FES Bicycle Ergometer*

With acute FES bicycle ergometry, the blood pressure and heart rate responses of quadriplegics and paraplegics are significantly greater than those of normal controls at matched workloads (Petrofsky et al., 1985b). A chronic (training) effect of FES bicycle ergometry has also been reported. Petrofsky et al. (1985b) demonstrated a tenfold increase in lower extremity pedaling endurance (of both paraplegics and quadriplegics) that averaged about 2 minutes (at zero kp workload) initially and increased to an average of 15 minutes (at 1/8 kp workload) after 6 weeks. Phillips et al. (1984) demonstrated that mean blood pressure in four quadriplegics at rest was significantly elevated posttraining over the hypotensive levels measured pretraining. His group also demonstrated that the mean blood pressure in the four quadriplegics during exercise was somewhat reduced over the higher mean blood pressures during exercise observed prior to the training period.

Additional chronic (training) effects of FES bicycle ergometry include a small but significant rise in the systolic blood pressure, resting heart rate, and cardiac index of quadriplegic individuals after a 12-week exercise period (Danopulos et al., 1986). Finally, it has been reported that eight quadriplegics who trained over periods varying from 12 to 39 weeks had a 200% increase in maximal workload and a 10% increase in minute ventilation over previous levels (Moore et al., 1987).

3. *Outdoor Exercise Bicycle*

No reports appear in the literature with respect to either acute or chronic (training) effects of the outdoor exercise bicycle.

4. *FES–Orthosis for Ambulation*

The acute effect of FES–orthosis walking for a paraplegic subject is an oxygen consumption that is 1.5–2 times the oxygen consumption associated with normal control walking (Petrofsky et al., 1985c). The chronic (training) effect of walking with the FES–orthosis has recently been reported for one paraplegic and one quadriplegic subject. Walking one hour daily (five times per week) over a 3-month period resulted in energy requirements

(O_2/kg·m) that were decreased by an average 18% posttraining. It was also observed that average walking speeds more than doubled over the initial (pretraining) walking speeds (Hendershot et al., 1987).

It should be noted that the thermal response (body heat load) to arm crank ergometry exercise in paraplegics, quadriplegics, and normal controls has been evaluated. Even with a marginal increase in workload (such as 25 W of arm crank ergometry) in a warm environment, the quadriplegics exhibited poor thermoregulation. It is concluded that quadriplegics must be extremely careful with any type of exercise in non-air-conditioned rooms (Petrofsky and Phillips, 1984).

In conclusion, the various reports collectively indicate selected therapeutic benefits with a prescribed program of APT. The next section will describe the medical criteria by which patients may participate in such a program.

II. The Medical Criteria for Functional Electrical Rehabilitation

A. The Patient Evaluation

The goal of the patient evaluation is optimal prescribing of a program of APT. Consequently, the evaluation results in a "prescription" from which the patient will receive maximal benefit—that is, the best therapeutic outcome compatible with minimal risk to the patient. Overall, the patient evaluation is divided into two general categories: the initial evaluation and the interim evaluation.

1. The Initial Evaluation of the Patient Candidate

The initial evaluation of the candidate patient involves history, physical examination, laboratory tests, and report summary specifically tailored to result in an APT prescription.

History

A brief (but pertinent) history is obtained as noted on the upper half of Figure 11.5.

Physical Examination

A systematic evaluation of pertinent physiological systems is obtained as indicated on the lower half of Figure 11.5.

Laboratory Tests

The following tests are administered to each potential subject prior to participation in an FES program: urinalysis, complete blood count, chemical profile, electrolytes, creatine phosphokinase (CPK), chest X-ray (not necessary if taken within the past 3 months and normal), lower extremity radiographs, electrocardiogram, and pulmonary function testing.

Report Summary

Laboratory tests, X-rays, and recommendations are summarized ad libitum (Figure 11.6), and physical therapy referral (when necessary) is presented in accordance with Figure 11.7.

The FES Prescription (Level of Clearance)

The results of the patient evaluation (as above) are interpreted with respect to the medical criteria (see below). The result is a level of clearance as indicated by the APT prescription (Figure 11.8). Specific recommendations may also be placed upon the exercise protocols (e.g., frequency, duration, upper workload limit). These recommendations may also include physiological parameters (e.g., heart rate, blood pressure, body temperature). Conceptually, these recommendations would be programmed onto a "prescription cartridge" (Petrofsky et al., 1985d) that would interact with the exercise system (Petrofsky et al., 1985a). While such prototype systems do exist, their application has been severely restricted. This is due to the lack of reliable, repeatable, noninvasive physiological monitoring devices for interfacing the subject to the system.

2. The Interim Evaluation of the Patient Participant

The interim evaluation of a patient already participating in a program of APT involves two general issues: continuation/termination of his or her

Patient Name ____________________

Physical Evaluation
Active Physical Therapy Program

Date of Exam: ____________________

Date of Birth: ____________________

Age ________ Race ________ Sex ________ Marital Status ________

Personal Physician ____________________

Date of Accident/Illness ____________________

Nature of Same ____________________

Other pertinent past history:

Allergies:

Pertinent family history:

Present medications:

Physical examination:

Ht ____ Wgt ____ BP ____ P ____ R ____

General

HEENT

Lungs

Cardiovascular

Abdomen

Genitalia (?Catheter)

Neuromuscular/Skeletal

FIGURE 11.5. Physical evaluation form for the active physical therapy program.

Name ____________________

Physical evaluation (cont.)

X-ray/Lab results:

Recommendations:

FIGURE 11.6. Page 2 of the physical evaluation form for the active physical therapy program.

Physical Therapy Referral

NAME ______________________________

PHYSICAL THERAPY AS INDICATED:

________ EVALUATION

________ PASSIVE RANGE OF MOTION

________ ACTIVE RANGE OF MOTION

________ RESISTIVE EXERCISES

________ TILT TABLE

________ ULTRASOUND

________ GAIT TRAINING

________ OTHER (specify) ______________________________

PRECAUTIONS:

FIGURE 11.7. Sample physical therapy referral form.

A.P.T. Prescription

Subject Name ______________________________

Date of Exam ______________________________

With the noted restriction, this person appears fit for exercise under the following protocols:

Quadriceps Trainer ________

Hamstring Trainer ________

Indoor Bike ________

Outdoor Bike ________

Standing w/R.G.O. ________

Walking w/R.G.O. ________

Standing/Sitting w/F.E.S.-R.G.O. ________

Walking w/F.E.S.-R.G.O. ________

Other Protocol:

____

____

____

____

Restrictions:

____

____

____

Signature ______________________________

Rec'd by ______________________________ Date ______________________________

FIGURE 11.8. Sample active physical therapy prescription form.

present level of clearance, and progression beyond that current level of clearance. It is assumed that the patient has successfully completed an initial evaluation and has satisfied the general medical criteria (see below).

Interval Evaluations for Participant Continuation/Termination

It is recommended that interval evaluations be performed when any intercurrent condition occurs that would affect continuation in a program of APT. Some of these are listed in Section II. B-1, below. The physician who performs the initial evaluation may elect to temporarily discontinue a patient's participation in APT. Appropriate specialty referral might then be obtained for that patient. Program reentry would depend on the specialist's recommendation. As an alternative, the physician who performed the primary evaluation may proceed to treat the intercurrent condition. In that case, the same physician would make the decision regarding program reentry. Interval evaluations for participant continuation/termination occur because of some intercurrent condition. These evaluations are necessarily abbreviated and more focused versions of the general evaluation initially performed on the patient.

Interval Evaluations for Patient Progression

Satisfactory accomplishment of a program of APT requires progression through the various FES exercise modalities: isokinetic leg trainer, exercise bicycle ergometer, and FES–orthosis ambulation. These FES modalities are designated in the columns of Table 11.1.

Interval medical evaluation of a patient involves all eight of the physiological systems that were reviewed during the initial evaluation of the patient. These organ systems are indicated in the rows of Table 11.1.

Specific medical criteria need to be satisfied for progression to each FES exercise system (Table 11.1). These criteria are based on our current FES modalities (see above) and on our experience with those modalities (as described in Section I). The criteria for progression through FES exercise modalities are subject to change as more experience is acquired.

B. The Medical Criteria

The purpose of the specific medical criteria (as with the patient evaluation) is the optimal prescribing of a program of APT. The physician's own medical experience and judgment should decide which patients are likely to have the best therapeutic outcome. The following medical criteria will assist toward that objective. These criteria have also been formulated to define those patients who will be placed at minimal risk by FES exercise. The medical criteria are subdivided into two categories: long-term participation (selection, continuation, promotion, termination) and immediate (short-term) monitoring.

1. The Medical Criteria for Patient Participation

The medical criteria for long-term patient participation are functionally organized around eight organ systems as follows.

Nervous System

Level of Neurological Injury. The upper limit of spinal cord injury, for participation in active physical therapy, is currently defined at C4/C5. However, the upper limit for FES–orthosis ambulation is somewhat lower (see below). The C4/C5 limit is established to allow sufficient diaphragmatic drive (C3-C4-C5) via the phrenic nerve.

Successful progression from isokinetic exercise (leg exerciser) to aerobic exercise (bicycle ergometer) requires adequate ventilatory drive. This is more precisely determined by pulmonary function testing than by defining a specific level of neurological injury.

The lower level of neurological injury for participation in APT is currently defined at T11/T12. At this high level of injury, the primary damage is to the interneurons of the cord (with peripheral motor neurons below the level of injury remaining functional). Lower level injuries result in an equina-type injury in which peripheral (alpha-motor neurons) are damaged. It is well established that FES is not successful with such lower motor neuron lesions (Dubowitz and Brooke, 1973; Guttmann, 1976).

Degree of Cutaneous Sensation. Our experience indicates that electrical muscle stimulation sufficient to cause a functional muscle contraction does

TABLE 11.1. FES exercise progression criteria.

System	Leg exerciser[a]	Bicycle ergometer[b]	FES-orthosis[c]
Nervous			
Neurological level			C6/C7[d]
Cutaneous sensation			
Other modes			Adequate vision Adequate hearing
Muscular			Adequate aerobic exercise[e]
Skeletal			
Bone			
Joint			
Hip			Adequate extension[f]
Knee		Adequate flexion[g] Medial-lateral stability[i]	Adequate extension[h]
Ankle		Minimal dorsiflexion[j]	Adequate dorsiflexion[k]
Cardiovascular			
Respiratory			Adequate aerobic exercise[e]
Urogenital			
Cutaneous			Adequate foot care[l]
Psychological			

[a]In our current programs, this is the entry-level FES exercise modality.
[b]In the current commercial system, this is the entry-level FES exercise modality.
[c]See note (*e*) for bicycle exercise prerequisite.
[d]Neurological function sufficient for extension of the forearm and adequate stabilization of the wrist.
[e]Muscular, cardiovascular, and respiratory system function sufficient for 30 minutes of continuous bicycling at minimal (0 kp) load.
[f]Degree of extension at hip sufficient for the patient to assume an upright posture (i.e., absence of pelvic tilt).
[g]Minimum of 75° of foreleg flexion at the knee (preferable 90°).
[h]Sufficient extension of foreleg at the knee to allow upright posture (i.e., knee at zero degrees hyperextension).
[i]Absence of medial-lateral knee instability; when present (± 5–10°), must be correctable with a knee brace orthosis.
[j]Sufficient dorsiflexion of foot at the ankle so that bicycle pedaling occurs at not greater than 5–10° of plantar flexion.
[k]Sufficient dorsiflexion of foot at the ankle so that normal standing occurs at zero degrees plantar flexion (i.e., absence of *pes equina*).
[l]A high level of foot hygiene (skin and nails) is necessary. No paronychia, abrasions, or blisters are permitted. Cushioning of vulnerable areas is advised.

require power levels sufficient to result in an uncomfortable cutaneous sensation being noted by some individuals. The two interacting variables appear to be absolute degree of cutaneous sensation and tolerance level to uncomfortable stimuli. Both are highly idiosyncratic.

In any case, operation of the electrical muscle stimulation power control(s) should be performed slowly, since a rapid increase in the power level may have the effect of startling a patient and may cause an unpleasant sensation or a secondary accident. An individual who (for any reason) is hypersensitive to the electrical muscle stimulation is not a suitable candidate for APT.

Other Sensory Modalities. The patient may participate in APT (involving the leg exerciser and the bicycle ergometer) even with severely impaired audition or vision. Adequate precautions for the patient's safety must, of course, be taken.

APT involving the FES-orthosis, however, requires significant voluntary interaction with the system and with the environment. For this final phase of APT, a moderate-to-severe loss of hearing or vision would be a relative contraindication. Sensory substitution (e.g., hearing aids, physical guidance by the therapist) could be utilized to mitigate this deficit.

Muscular System

Muscular spasms must be infrequent, or if frequent, controlled by medication. This criterion is necessary because the engineering design of the

APT system has a "spasm" override. It is potentially hazardous to stimulate movement in a prescribed pattern (i.e., exercise chair or bicycle ergometer) during spastic muscle activity (which would oppose or accelerate that prescribed pattern). Consequently, upon detection of a spasm, the computer will shut down; after passage of the spastic activity, it must be manually reactivated.

Skeletal System

The skeletal criteria are functionally divided into bone and joint criteria.

Bone Criteria. A major criterion for prescription of APT is the degree of disuse osteoporosis (due to the relative immobility of the SCI patient). The inability of conventional X-rays to quantify the degree of osteoporosis is acknowledged (Meema and Meema, 1981). New emerging quantitative techniques will provide a more definitive answer (Hangartner and Overton, 1982; Hangartner et al., 1985).

Conventional radiographs evaluate the thickness of the cortical bone (and its trabecular pattern). This can usually be semiquantified on a five-level scale: normal, mild, moderate, moderate–severe, or severe osteoporosis. The radiologist should be requested to use this five-level scale when patients are being evaluated as APT candidates.

Regardless of the radiographic method of analysis, it is recommended that only cases in the moderate osteoporosis range (or better) be candidates for APT. The primary predisposing factor to fractures in the SCI individual is disuse osteoporosis (Comarr et al., 1962; Freehafer and Mast, 1965). Fractures have been the primary risk in our program of APT (Phillips and Petrofsky, 1985).

Joint Criteria. Hip, knee, and ankle joints must also be evaluated, usually by plane radiographs. Of specific interest are radiographic criteria of degenerative joint disease (DJD) or heterotrophic ossification (HO) of joints. No or minimal DJD or joint HO is acceptable for a program of APT. Moderate or severe DJD or joint HO, as revealed by radiographic examination, would contraindicate a program of APT.

At present, it is not known whether APT will promote the development of a Charcot joint (degenerative changes secondary to stresses placed on an anesthetic joint). Until more evidence is available, common medical wisdom would dictate that APT not be prescribed when there is moderate to severe DJD. Final resolution of this important problem may ultimately require intermittent knee arthroscopy on a series of APT patients.

It is also not known at present whether APT will either promote or retard heterotrophic ossification. Again, common medical wisdom would dictate that APT be prescribed only when there is no or minimal heterotrophic ossification.

A history of whether there has been hip or knee disarticulation should be obtained. While not an absolute contraindication, the joint involved should be adequately braced (or otherwise protected) so that a recurrence does not occur during APT.

Cardiovascular System

The role of exercise in patients with angina and coronary artery disease in general remains controversial. In such SCI patients, APT should be prescribed only in close consultation with the attending cardiologist. Uncontrolled hypertension is an absolute contraindication.

The effect of electrical stimulation on pacemaker patients is largely unknown. Until further investigations have been conducted, FES should be avoided in anyone who concurrently utilizes an implanted pacemaker.

The peripheral vascular circulatory system must be uncompromised. The presence of intercurrent lower extremity thrombophlebitis or phlebothrombosis is an absolute contraindication.

Respiratory System

Compromised ventilatory response to APT will significantly affect the patient's outcome. This is a consequence of the aerobic nature of APT (exercise bicycle ergometer). The results of pulmonary function testing will help determine the patient's functional status. Evidence of chronic obstructive pulmonary disease (COPD) requires that patient participation in APT be coordinated with a pulmonary physician. During periods of intercurrent respiratory infections, APT is contraindicated.

Urogenital System and Reproductive Systems

Evidence of chronic renal disease requires that patient participation in APT be closely coordinated with a urologist. During episodes of intercurrent urinary tract infections, APT is contraindicated. There is (as yet) no evidence that APT will either exacerbate or attenuate a chronic renal condition. However, an acute urinary tract infection does represent an intercurrent infection that (overall) lowers a patient's stamina.

The safety of electrical stimulation during pregnancy and delivery has not been established. Until further information is available, FES should be avoided in the pregnant woman.

Cutaneous System and Subcutaneous Fat

Evidence of any chronic skin disease should require that patient participation in APT be closely coordinated with a dermatologist. Any acute skin disease must be evaluated by a physician before proceeding with APT. The presence of any cutaneous disorder over an area of electrode application is a contraindication to the continuation of electrical stimulation over that area.

Subcutaneous fat acts as an electrical insulator and impedes the flow of electrical current between the skin surface and the target muscle. Therefore, excessive obesity will contraindicate the application of APT. In general, patients should not exceed a 25% increase of their ideal weight (based on their frame and measured height). At 50% above ideal weight, the obesity is excessive and would contraindicate APT.

Psychological System

Any psychological evaluation or testing should be referred to a psychiatrist or clinical psychologist. Those SCI subjects who indicate a positive history of drug or alcohol abuse will usually prove either unreliable or manipulative during their program of APT. Such a history is a relative contraindication to APT.

Minimal psychological testing should include a Minnesota Multiphasic Personality Inventory (MMPI). Insufficient data are currently available regarding optimal "personality profiles" for APT. Conversely, there are no data regarding a contraindicatory "personality profile." However, subjects who consistently score high on the antisocial (sociopathic) index of the MMPI have historically been unreliable, dishonest, and manipulative when pursuing a program of APT. While not an absolute, a high score in this area is a relative contraindication to a program of APT.

2. *The Medical Criteria for Patient Monitoring*

The medical criteria for immediate (short-term) patient monitoring refer to the actual exercise session. These criteria are basically divided into heart rate and blood pressure responses, direct observation of the patient, and direct observation of the exercising system.

Cardiovascular Monitoring

Standard guidelines for blood pressure (BP) and heart rate (HR) monitoring are useful in the application of APT. Overall, systolic BP should not exceed 180 mmHg in a person aged 20–40 years. The diastolic BP should not exceed 120 mmHg. HR should not exceed 160 beats/min. Finding BPs or HRs above these levels during acute exercise necessitates termination of the exercise.

Systolic BP should not fall below 80 mmHg, and diastolic BP should not fall below 50 mmHg. Some quadriplegic BPs may be near these levels, but should not fall below them during APT. HR should not fall below 50 beats/min, even in a quadriplegic subject.

Objective Patient Monitoring

Objective patient monitoring involves both observing and talking to the patient. Level of consciousness is a good index of cerebral perfusion. An earlier index is visual acuity and visual fields. It is most helpful to frequently talk to the patient (e.g., "How do you feel?") and then carefully listening to the response.

Additional objective information may be acquired by holding the subject's hand (conveniently done when palpating the radial pulse). The levels of tension (grip), diaphoresis (moistness), and tremor (anticipation/fatigue) can be readily detected by this maneuver.

It is important to be prepared for untoward reactions and to remain in constant communication with and observation of the patient. As experience accumulates in objective monitoring of APT patients, the physician will develop individual criteria.

Exercise System Monitoring

Although the prescribing physician may seldom observe the actual exercise system operating on a routine basis, he or she should instruct assistants to monitor the "machine" operation (during exercise) in addition to the patient. APT utilizes computer-controlled electromechanical systems to which a disabled human subject is interfaced. An operational checklist of "system performance" should be available to the operator (and provided by the manufacturer). Such a checklist should be referred to as often as necessary during the exercise session.

C. Conclusion

The mature physician respects the judgment of his patient. The mature patient respects the judgment of his physician. This is the traditional basis by which informed consent is obtained from the patient by the physician. Active physical therapy is an emerging treatment modality: its ultimate benefits are as unknown as its ultimate risks.

This section has reviewed the extant literature, which overall supports the premise that functional electrical exercise can result in improved physical conditioning of the spinal-cord-injured patient. Adherence to the medical guidelines presented here should result in the selection of patients that are at minimal risk as participants. With the acquisition of a broader patient data base, these medical guidelines probably will be modified. So also, will our appreciation of the therapeutic outcome.

III. A Computer Program for the Prescription of Functional Electrical Rehabilitation*

A. Introduction

Phillips and Petrofsky (1983) described a method of rehabilitation called active physical therapy. Electrical stimulation of muscles, which had been used for 30 years for rehabilitation, involved only the application of two electrodes above the muscle, followed by stimulation with different patterns of current. This allowed a small increase in strength of the muscle. These systems did not provide either controlled resistance exercise to increase strength or aerobic exercise to develop endurance. Such principles are well known to physical fitness trainers. Based on these principles, Petrofsky and Phillips (1983a) developed the method of APT for rehabilitation. This method involves using functional electrical stimulation to stimulate paralyzed muscles in the limbs of individuals with spinal cord injuries (Phillips et al., 1984). While conventional physical therapy allows the environment to act on the muscle to exercise, APT allows the muscle to act against the environment.

APT is generally divided into four categories. Ideally, a patient progresses from one category to the next. The first category is the isokinetic leg trainer (Petrofsky et al., 1984a). Various weights are attached to the foreleg of the patient through an ankle harness. Then FES is used to develop repetitive contractions at the knee. The load against which the patient works can be adjusted daily to achieve maximum training. This exercise develops strength in the muscles of the leg (e.g., Collins et al., 1984).

The bicycle ergometer is the second state of the training program (Petrofsky et al., 1984b). The bicycle ergometer allows for stimulation of the quadriceps, hamstring, and gluteal muscles of the leg. Once again the load can be adjusted to the individual's needs. This is aerobic exercise. It benefits the cardiovascular system and conditions for endurance (e.g., Danopulos et al., 1986). It is important that the muscle be able to work at the most efficient level.

The third part of the APT program is an outdoor bicycle (Petrofsky et al., 1983). This is a device similar to the ergometer, adapted for outdoor use.

The final phase of the program involves combining FES with a reciprocating gait orthosis (Petrofsky et al., 1985c). This system allows walking in individuals with spinal cord injuries. To avoid injury to the patient, the strength and endurance required for this type of exercise must be developed by the prior exercises before FES-RGO is attempted (Phillips and Petrofsky, 1985).

Section II presented the guidelines for participation in an APT program. The medical criteria, originally set forth by Phillips (1987), are based on his experience with FES exercising systems. These

*With C. Brunsman of Wright State University.

FIGURE 11.9. Main menu flow chart.

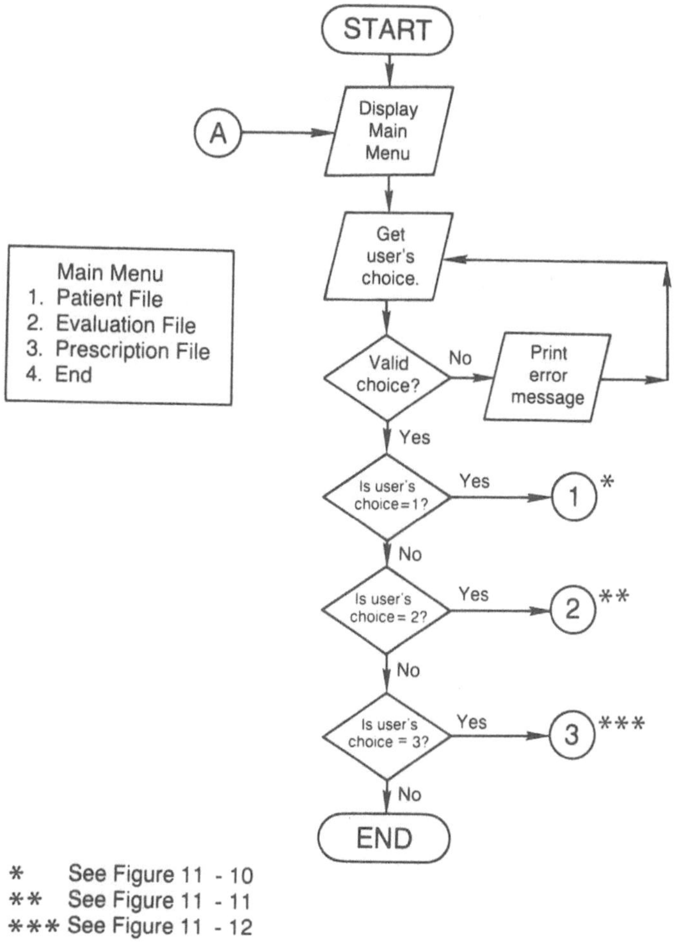

criteria identified the patients who would be at minimum risk for such a program. This does not mean that other patients may not participate, but they may require closer monitoring, etc. The eight organ systems that must be examined to determine suitability for an APT program are briefly summarized as follows.

1. *Level of neurologic injury.* The upper level of injury is set at C4/C5 to allow for sufficient diaphragmatic drive with the phrenic nerve. The lower level is T11/T12. This is to ensure that there is no peripheral nerve damage.
2. *Muscular system.* Muscle spasms must be infrequent or controlled by medication. Electrical stimulation of muscles during spastic activity is dangerous. During spasms, the computer system controlling the exercise system shuts down.
3. A. *Skeletal system, bone criteria.* The major consideration here is disuse osteoporosis. Presently, five levels of osteoporosis are identified: normal, mild, moderate, moderate–severe, and severe. The APT should be in the moderate range or better.
3. B. *Skeletal system, joint criteria.* Hip, knee, and ankle joints must be examined for degenerative joint disease. There should be no or little degenerative joint disease present as detected by plane radiographs.
4. *Cardiovascular system.* A person with uncontrolled hypertension should not participate in APT. The effect of APT on coronary artery disease and angina is not yet clear. Such patients

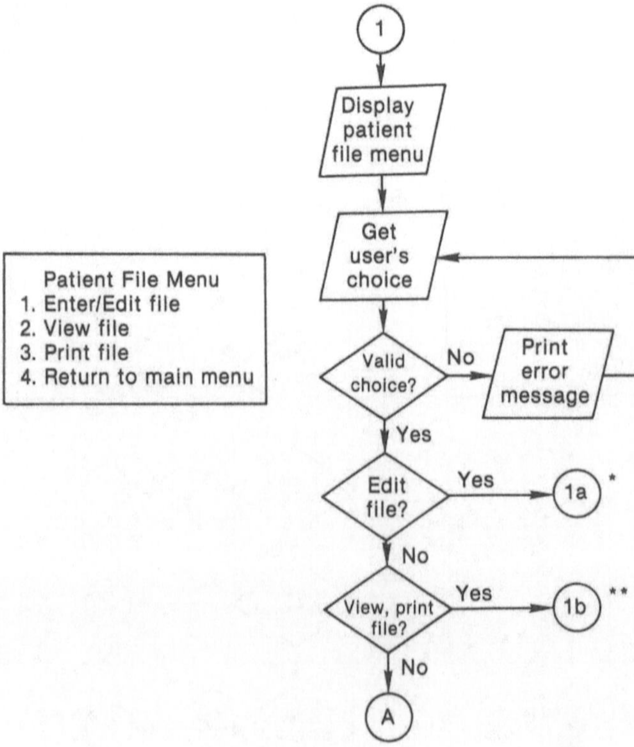

FIGURE 11.10. Patient file menu.

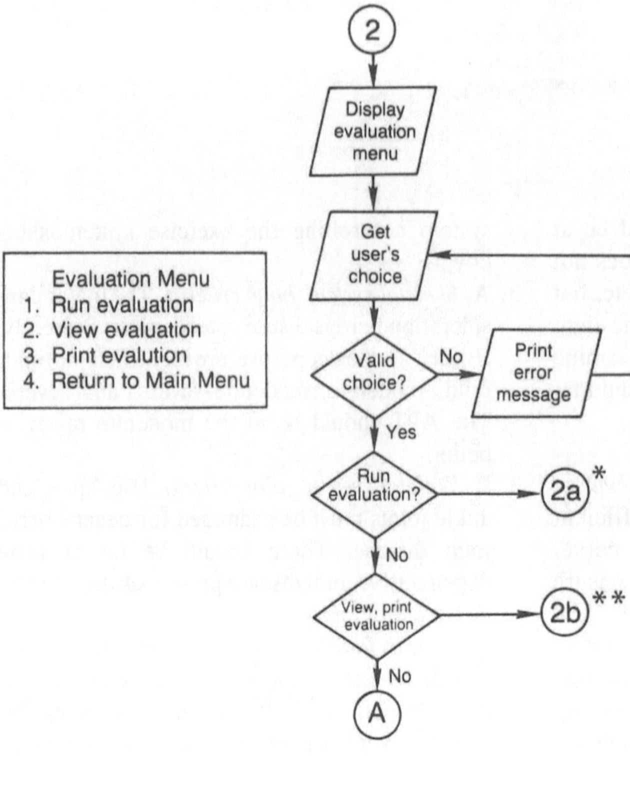

FIGURE 11.11. Evaluation menu.

FIGURE 11.12. Prescription menu.

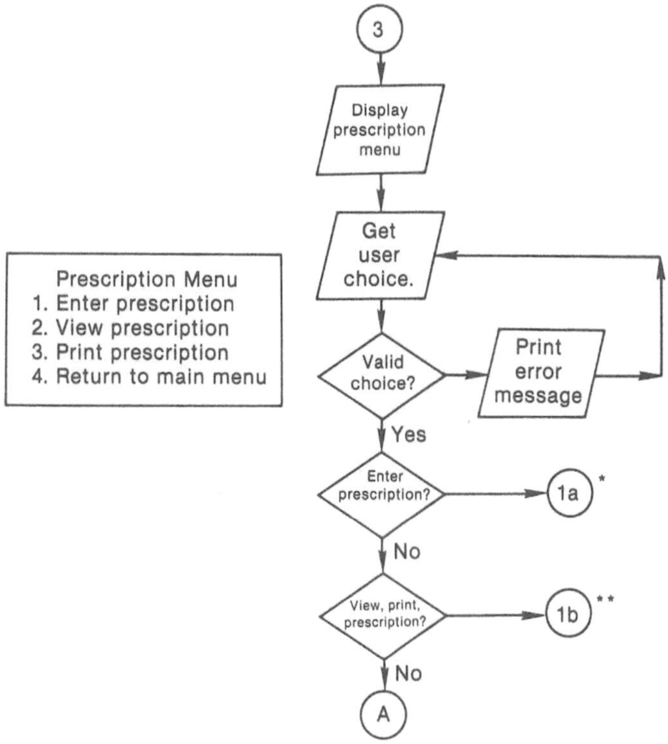

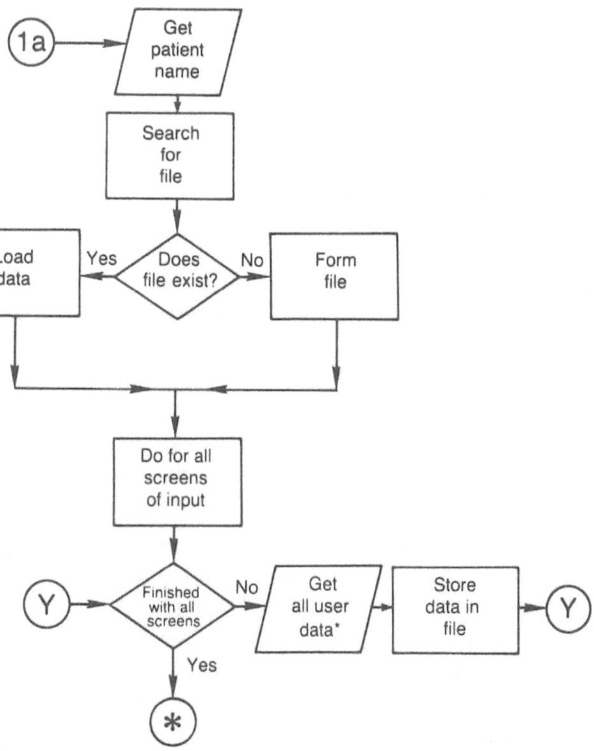

FIGURE 11.13. Enter/edit patient data or enter/edit prescription data.

* Go to 1 if Patient file (Fig. 11 - 10)
* Go to 3 if Prescription File (Fig. 11 - 12)

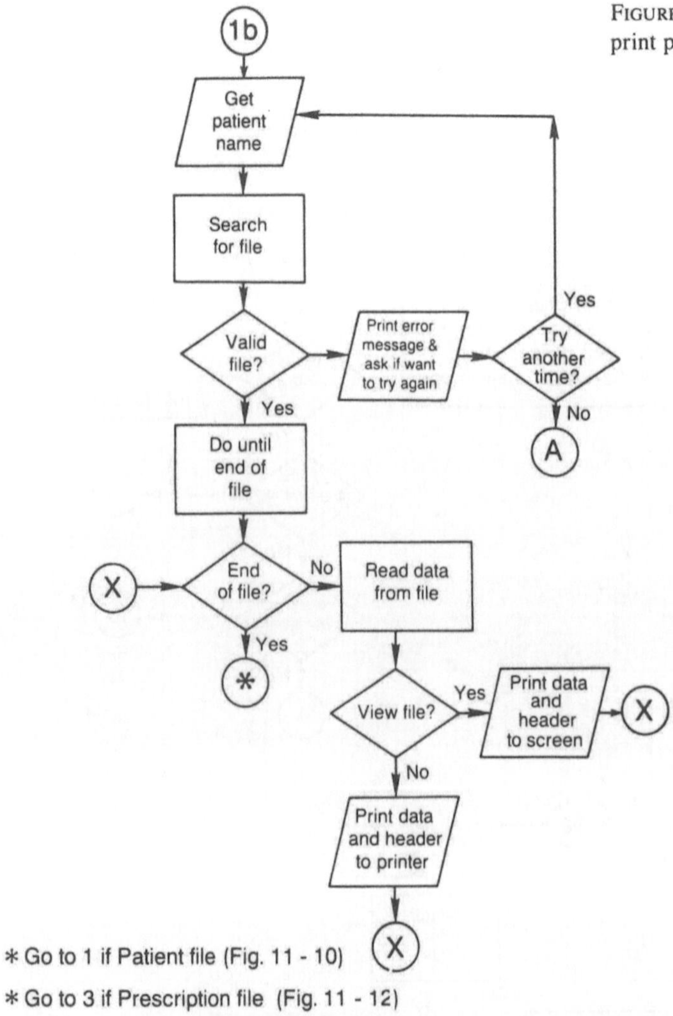

FIGURE 11.14. View/print patient file or view/print prescription file.

may be admitted with direct supervision of a cardiologist.

5. *Respiratory system*. Weak ventilatory response will limit progress in APT. Evidence of COPD requires the participation of a pulmonary physician. Participation in APT during periods of respiratory infection is completely restricted.
6. *Urogenital system*. Chronic renal disease requires the coordination of APT with a urologist. Because of lower stamina and other possible effects, the patient must not participate during intercurrent urinary tract infections.
7. *Cutaneous system*. Chronic skin disease requires coordination with a dermatologist. Any acute skin disease in the area of electrode placement prevents the continuance of electrical stimulation in that area.
8. *Psychological system*. An MMPI should be given to each patient. A history of drug abuse or antisocial behavior seems to indicate that a patient may be unreliable or manipulative. This is not an absolute constraint.

The purpose of this section is to describe a computer program that organizes information to help a physician evaluate a patient. The program compares the patient's data to the medical criteria listed above and allows the physician to write a prescription for APT, if appropriate.

B. Development

The target users of this program are physicians in the rehabilitation field. Minimum requirements

FIGURE 11.15. Run evaluation subroutine.

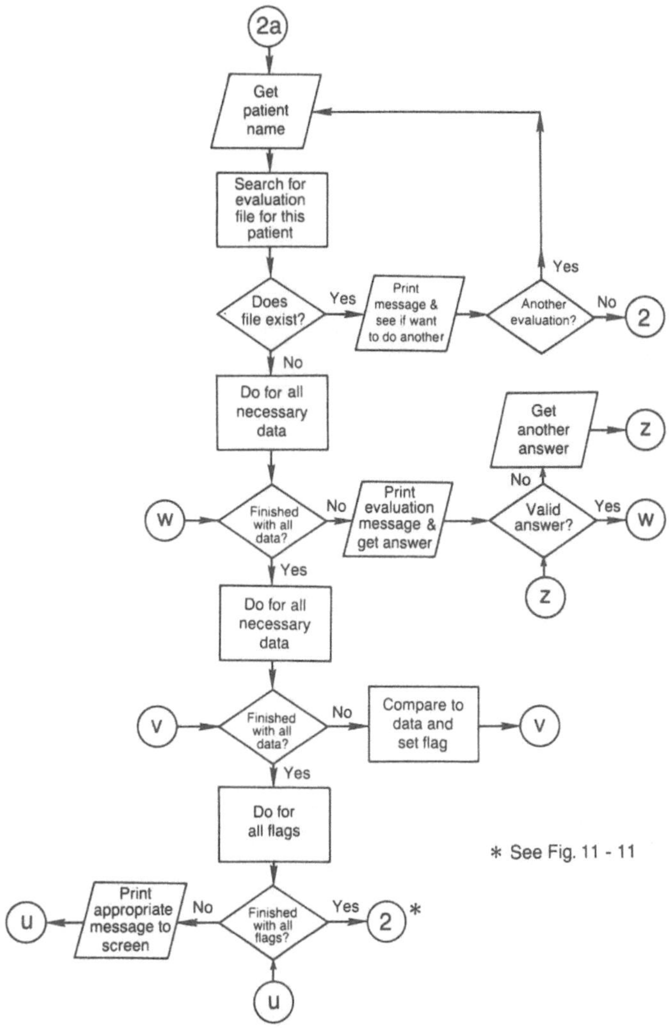

for the use of this program include an IBM PC or clone with a monitor, one floppy drive or a hard drive, 128 kilobytes of memory, and a dot-matrix printer. Since the program is written in BASIC, the user must also have BASIC capabilities on the computer. This language is provided with many systems as sold. The data can be stored on the floppy drive and the program can be run from the hard drive (or the other floppy drive). The interactive menu-driven program requires very little experience to use. Even someone who is fairly unfamiliar with computers should be able to respond to the prompts by entering the necessary data. However, the physician does need to use the training received during his medical education in interpreting the data presented by the program. Training in the actual use of the program should take very little time, no more than an hour or two. The major cost of the program is in the hardware requirements. However, many doctors' offices are already equipped with personal computers. The program should save the physician time in evaluating SCI cases.

The program allows the physician to enter or edit patient data, and then view or print the information that has been entered. The patient file will contain a brief patient history, data on physical examination and laboratory tests, a report summary, and a place for comments, following the format of the sample patient evaluation form given by Phillips (1987). After patient data have been entered, the doctor can run an evaluation and print or view it. The physician will also be able to enter,

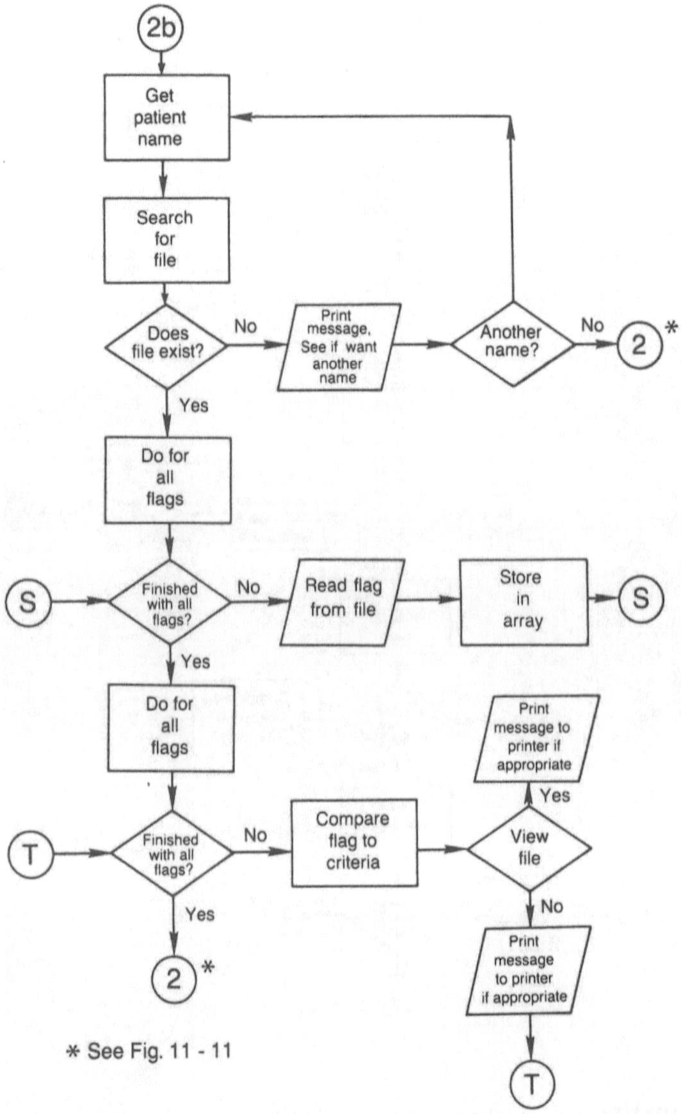

FIGURE 11.16. View/print evaluation subroutine.

edit, print, or view a prescription. Flow charts for this program are given below.

C. Program Operation

From the flow charts (Figures 11.9 through 11.16), a program was written that allows a physician to have a patient's physical evaluation, evaluation for the appropriateness of APT, and APT prescription stored for further reference on a floppy disk. Since many doctors' offices also require the use of paper backups, a copy of these three files may be printed for the patient's records, as well.

Generally, the physician would first enter the patient's physical examination data. This includes all the information from a normal physical examination as well as a physical therapy referral. The information included in such a form is shown on Figures 11.5, 11.6, and 11.7. A copy of typical output is also included at the end of the paper. The first two screens, recommendations, and the final screen of information are stored in one file. The neuromuscular and skeletal data include sensory functions, motor functions, range of motion, skin hygiene, and reflexes. Because of the large amount of information a physician may want to enter here, this information is contained in a separate file. The X-ray and laboratory results are also stored in a separate file because a physician may want to enter a large amount of information.

After data entry, the physician would run an evaluation by responding to several on-screen questions that relate to the patient's fitness for an APT program. The computer program then informs the physician of any absolute constraints for the patient's participation in APT, any special considerations that should be studied before prescribing APT, any special cautions, and possible problems the patient might experience in an APT program. From these data, the physician can make a well-informed decision on the patient's suitability for APT. The responses to the questions are then used to set a flag variable to 1 or 0. If the response indicates a constraint, concern, etc. for APT prescription, the flag is set to 0. Otherwise, the flag is set to 1. It is simple binary data of this type that are stored in the evaluation file. When the file is to be viewed or printed, the program simply loads the flags and prints the appropriate messages.

If the physician decides to prescribe APT, there is an option in the program that will allow him or her to write the prescription. All the prescription information is stored in one file.

The program allows a physician to simply organize all the information needed to start a patient in an APT program. By pointing out any possible constraints or problems that a patient may encounter in the course of APT, the program allows a doctor to make informed decisions about prescribing APT.

D. Discussion

An interactive computer program would assist in the prescribing of active physical therapy programs by allowing the user (usually a physician) to enter data, which may be called up for later viewing. By pointing out any discrepancies between the patient's data and the medical criteria for APT, the program will help the physician make a knowledgeable decision about prescribing an active physical therapy program.

This program will be of assistance to any physician treating patients with spinal cord injuries, especially those who are not familiar enough with APT to make a good decision about a patient's suitability for this type of therapy. The program could also be useful in training physicians who were interested in working in rehabilitative medicine. Eventually, this type of program could be expanded to help a physician decide when a patient is ready to move from one APT stage to the next. In addition, a patient's entire APT history can be stored in one convenient location.

E. Summary

Active physical therapy involves the use of continuous resistance exercise to increase the muscular strength of spinal-cord-injured patients and aerobic exercise to develop physical endurance. This program utilizes functional electrical stimulation to produce work by the muscle of the patient on the environment. APT can generally be divided into four categories: the isokinetic leg trainer, the bicycle ergometer, the outdoor bicycle, and the use of FES and a reciprocating gait orthosis to produce walking.

Before entering such a program, a patient must meet certain medical criteria, which include specific standards for the neurological, muscular, skeletal, cardiovascular, respiratory, urogenital, cutaneous, and psychological systems.

A program was written to allow physicians to organize information about patients. An informed decision about prescribing APT can be made on the basis of how well a patient meets the medical criteria for entering an APT program. This program can be run on an IBM or compatible system with BASIC. A program user's guide is presented in Chapter 12.

References

Collins, S.R., Glaser, R.M., and Feinberg, S.D. Progressive training of paralyzed leg muscles with electrical stimulation. *Fed. Proc.*, abstract 4262, 1984.

Comarr, A.E., Hutchinson, R.H., and Bors, E. Extremity fractures of patients with spinal cord injuries. *Am. J. Surg.* 103:732–739, 1962.

Danopulos, D., Kezdi, P., Stanley, E.L., et al. Changes in cardiovascular circulatory dynamics after a twelve week active bicycle rehabilitation in young tetraplegics. *J. Neurol. Orthop. Surg.* 7:179–184, 1986.

Dubowitz, V., and Brooke, M.H. *Muscle Biopsy: A Modern Approach*, Saunders, Philadelphia, 1973.

Freehafer, A.A., and Mast, W.A. Lower extremity fractures in patients with spinal cord injuries. *J. Bone J. Surg.* 47:683–694, 1965.

Guttmann, L. *Spinal Cord Injuries: Comprehensive Management and Research*, 2nd ed., Blackwell Scientific Publications, Oxford, 1976.

Hangartner, T.N., and Overton, T.R. Quantitative measurement of bone density using gamma-ray computed tomography. *J. Comput. Assist. Tomogr.* 6:1156–1162, 1982.

Hangartner, T.N., Overton, T.R., Harley, C.H., et al. Skeletal challenge: An experimental study of pharmacologically induced changes in bone density in the distal radius, using gamma-ray computed tomography. *Calcif. Tissue Int.* 37:19–24, 1985.

Hendershot, D.M., Petrofsky, J.S., Phillips, C.A., and Moore, M. Blood pressure and heart responses in paralyzed and nonparalyzed during isokinetic leg training. *J. Neurol. Orthop. Surg.* 6:259–264, 1985.

Hendershot, D.M., Moore, M.L., and Phillips, C.A. Cardiopulmonary conditioning when walking with and without FES in the paralyzed. *Fed. Proc.* 46:680, 1987.

Meema, H.E., and Meema, S. Radiogrammetry, in *Noninvasive Measurements of Bone Mass and Their Clinical Application*, S.H. Cohn, Ed., CRC Press, Boca Raton, FL, pp. 5–50, 1981.

Moore, M.L., Hendershot, D.M., and Phillips, C.A. Cardiopulmonary conditioning by FES induced bicycle ergometry. *Fed. Proc.* 46:679, 1987.

Petrofsky, J.S., and Phillips, C.A. Constant velocity contractions in skeletal muscle of the cat. *Med. Biol. Eng. Comput.* 17:583–592, 1979.

Petrofsky, J.S., and Phillips, C.A. Active physical therapy—A modern approach for rehabilitation of the disabled. *J. Neurol. Orthop. Surg.* 4:165–173, 1983a.

Petrofsky, J.S., and Phillips, C.A. Computer controlled walking in the neurological paralyzed individual. *J. Neurol. Orthop. Surg.* 4:153–164, 1983b.

Petrofsky, J.S., and Phillips, C.A. The use of functional electrical stimulation for rehabilitation of spinal cord injured patients. *CNS Trauma J.* 1:57–74, 1984.

Petrofsky, J.S., and Phillips, C.A. Closed-loop control of movement of skeletal muscle. *CRC Crit. Rev. Biomed. Eng.* 13:35–96, 1985.

Petrofsky, J.S., Heaton, H.H., and Phillips, C.A. Outdoor bicycle for exercise in paraplegics and quadriplegics. *J. Biomed. Eng.* 5:296–297, 1983.

Petrofsky, J.S., Heaton, H.H., and Phillips, C.A. Leg exerciser for training of paralyzed muscle by closed-loop control. *Med. Biol. Eng. Comput.* 22:298–303, 1984a.

Petrofsky, J.S., Phillips, C.A., Heaton, H.H., and Glaser, R.M. Bicycle ergometer for paralyzed muscle. *J. Clin. Eng.* 9:13–19, 1984b.

Petrofsky, J.S., Phillips, C.A., Almeyda, J., et al. Aerobic trainer with physiological monitoring for exercise in paraplegic and quadriplegic subjects. *J. Clin. Eng.* 10:307–316, 1985a.

Petrofsky, J.S., Phillips, C.A., and Hendershot, D. The cardiorespiratory stresses which occur during dynamic exercise in paraplegics and quadriplegics. *J. Neurol. Orthop. Surg.* 6:252–258, 1985b.

Petrofsky, J.S., Phillips, C.A., Larson, P., and Douglas, R. Computer synthesized walking: An application of orthosis and functional electrical stimulation (FES). *J. Neurol. Orthop. Surg.* 6:219–230, 1985c.

Petrofsky, J.S., Phillips, C.A., and Petrofsky, S.H. Electronic physicians prescription system for FES patients. *J. Neurol. Orthop. Surg.* 6:239–246, 1985d.

Phillips, C.A. The medical criteria for active physical therapy: Physician guidelines for patient participation in a program of functional electrical rehabilitation. *Am. J. Phys. Med.* 66:269–286, 1987.

Phillips, C.A., and Petrofsky, J.S. Computer controlled movement of paralyzed muscle: The medical perspective. *Proceedings of the First Vienna International Workshop on Functional Electrostimulation*, paper 5.2, 1983.

Phillips, C.A., and Petrofsky, J.S.: A review of fracture cases in spinal cord injured individuals participating in closed-loop functional electric stimulation experiments. *J. Neurol. Orthop. Surg.* 6:247–251, 1985.

Phillips, C.A., Petrofsky, J.S., Hendershot, D.M., and Stafford, D. Functional electrical exercise: A comprehensive approach for physical conditioning of the spinal cord injured individual. *Orthopedics*, 7:1112–1123, 1984.

Robinson, C.J., Bolam, J.M., Chinoy, M., et al. Response to surface electrical stimulation of the quadriceps in individuals with spinal cord injury. *Proc. RESNA*, 6: 282–284, 1986.

Rossier, A.B. *Rehabilitation of the Spinal Cord Injury Patient*. CIBA-GEIGY, Summit, NJ, 1964.

Vodovnik, L., Bajd, T., Kralj, A., et al. Functional electrical stimulation for control of locomotor systems. *CRC Crit. Rev. Bioeng.* 6:63, 1981.

12
The Functional Electrical Rehabilitation Patient Evaluation and Prescription Program*

This program is designed for physicians in the rehabilitation field. It allows them to store on disk the files that pertain to the prescription of active physical therapy for patients with spinal cord injuries. These files can then be viewed or printed as needed. The program also aids the physician in evaluating a patient's fitness for a functional electrical rehabilitation program by comparing pertinent patient data to the medical criteria of Chapter 11, Section II. This chapter presents detailed instructions on the use of the program.

I. Getting Started with the Functional Electrical Rehabilitation Program

A. System Requirements

System requirements for the Functional Electrical Rehabilitation Patient Evaluation/Prescription Program are as follows.

IBM PC or compatible.
MS-DOS.
GW-BASIC or compatible interpreter
At least 128 kilobytes of RAM.
One disk drive or a hard drive.
One program disk (provided upon written request to: Chandler A. Phillips, M.D., Department of Biomedical and Human Factors Engineering, Wright State University, Dayton, OH 45435).

*With C. Brunsman of Wright State University.

B. Typographical Conventions

Keystrokes that the user is to type are designated by boldface type, as shown below.

A command shown in **boldface** represents a user entry.

When a special key is to be pressed (e.g., the enter key), it is shown in angle brackets: <Enter>

C. Setting up the Program

The program may be run from one floppy disk drive or from a hard drive and one floppy disk drive. All the patient data are stored on a disk placed in the floppy drive, but the program may be run from a hard drive or a floppy drive. If you have a single-floppy system, you may skip to Section II.C.2, "Starting the Program."

1. Installation onto a Hard Drive

Start your computer:	Place the DOS disk in the drive and turn on the computer. If DOS is on your hard disk, you may just turn on the computer. If the computer prompts you for the date and time, type the date (e.g., 03-14-88) and press the enter key. Then enter the time. The computer uses a 24-hour clock, so to indicate 2:00 P.M. you would type 14:00 and press the enter key.
Form a directory on your hard disk for the program:	If you loaded DOS from the A drive, hence see the A > prompt on your screen, change the drive designation to the hard drive by typing:

C: <Enter>

If you already have a C> prompt you need not type the drive change command. Now, with the C> prompt on your screen, type the following commands:

md \ apt <Enter>

cd \ apt <Enter>

The prompt should now look like:

C: \ APT>

Copy the program to your hard drive: Now that you have the prompt shown above, insert the program disk into drive A and type the following from the DOS prompt.

C: \ > **copy a:apt.bas c:apt.bas** <Enter>

2. *Starting the Program*

First, you must load DOS by inserting the DOS disk in drive A and turning on the computer, or by simply turning on the computer if DOS is on your hard drive. After answering the prompts from DOS, you will get a DOS prompt, that is:

A> or C>

Assuming that you have installed the program onto your hard drive, at the C> prompt you must type:

cd/apt <Enter>

Now, since the program is written in BASIC, you must load BASIC from the DOS prompt. This can be done by typing at the appropriate prompt:

BASICA/S:1300 <Enter>

or

BASIC/S:1300 <Enter>

The command you use depends on your system. The "quick reference" part of your system documentation should tell you which is the proper command. Notice that the BASIC command is followed by "**/S:1300**." This is necessary to save enough memory space for your files. Do not omit this part of the command.

Once you have loaded BASIC, type the following to remove the menu line at the bottom of the screen:

key off <Enter>

Now, if you are using a single-floppy system, run the program by typing:

run "A:APT" <Enter>

or, if you have installed the program on your hard drive:

run "APT" <Enter>

An introductory screen will appear as follows:

ACTIVE PHYSICAL THERAPY
PRESCRIPTION PATIENT EVALUATION
Carol S. Brunsman

When this screen appears, place a formatted disk into drive A. The disk must be formatted before this program is run. If you are unsure about how to format a disk, refer to your DOS user's manual.

After a short pause, a "main menu" screen will appear, as below. From this menu you may choose to enter into any of the three sections of the program (described later) or to exit to BASIC.

ACTIVE PHYSICAL THERAPY
PRESCRIPTION PATIENT EVALUATION

MAIN MENU

1. Patient File
2. Evaluation File
3. Prescription File
4. Exit to BASIC

Please enter your selection (1, 2, 3, or 4).

Now, the following line appears:

Please enter your selection (1, 2, 3, or 4).

This is a prompt, asking you to choose one of the selections. At this point you may refer to the later sections of this chapter to see how to choose one of these selections.

3. *Saving Data and Exiting the Program*

You may exit the program at any point by (a) choosing selection "4" from any of the submenus to return to the main menu and (b) then selecting "4" from the main menu to return to BASIC. It is important not to exit the program any other way; if you depart from the procedure just given, you will lose the files you have created.

II. The Patient File Option

To choose the patient file option, from the main menu type:

1

Notice: you do not type <Enter> after typing the "1."

> PATIENT FILE MENU
>
> 1. Enter or Edit Data
> 2. View File
> 3. Print File
> 4. Return to Main Menu
>
> Please enter your selection (1, 2, 3, or 4).

Once again, the cursor prompts you to enter your selection. Option 1 is to enter or edit patient data. You must do this before you can either view or print patient data (options 2 and 3). When you have created a patient data file, you may view or print it. To choose option 1, type:

1

Notice once again that you do not type <Enter> after you type the command. After you type "1," the program will ask if this is the first entry for this file. At this prompt you must type:

y or **Y** (if you have never entered patient data)
<Enter>

or

n or **N** (if you have previously entered patient data.)
<Enter>

Only the first time you enter patient data onto the current disk you are using for files will the correct response to this question be "**y**." If the disk you are using already contains patient files created with this program, you must respond with an "**n**."

Next, you will be prompted for the patient's last name. If you have more than one patient with the same last name, enter some other identifying letter or letters before the last name. For example, enter:

JaneSmith <Enter>

if you also have a patient named John Smith. The patient's records are stored under the name you enter here, so it is important that each patient has an identifying name. It is also important that you remember the name you type here, so that you can later access the patient's data.

If you have never entered data for the patient whose name you have just typed, the following screen will appear:

> PATIENT EVALUATION
> Name:
> Exam Date:
> Birthdate:
> Age:
> Race:
> Sex (M or F):
> Marital Status:
> Personal Physician:
> Date of Accident/Illness:
> Nature of Accident/Illness:
> Other Pertinent History:
>
> Allergies:
>
> Pertinent Family History:
>
> Present Medication:

If you have ever entered data for this patient, the data you entered will appear following the "?" prompt.

The cursor first appears at the top line, to prompt you for the patient's name. Type the patient's name and then press the <Enter> key. For example, type:

JaneSmith <Enter>

To enter data for each line, simply type what you wish to enter, and press <Enter> after you have entered that piece of data. You may continue to do this for the entire screen of data.

If you had entered data previously, which now appear following the prompt, you have two choices. First, you may accept the current data shown by simply typing:

<Enter>

Your second option is to change data you had entered previously. To do this, simply type your new response at the cursor and press the <Enter> key. For example, to change "Jane Smith" to "John Smith" at the prompt:

Name: ? JaneSmith

you simply type:

JohnSmith <Enter>

You may continue in this fashion for the entire screen of data.

One important thing to know is the size of the field (i.e., the number of characters) that can be entered for each prompt. If you enter more than the allowed number of characters, the data will not be saved to the file and you will change the appearance of the screen so you cannot see the next prompt. The characters [each capital or lowercase letter, number, symbol, punctuation mark, and space counts as a character] allowed for each prompt on the first screen are as follows.

Data	Length
Name	30
Exam date	8
Birth date	8
Age	2
Race	20
Sex	1
Marital status	10
Personal physician	30
Accident date	8
Nature of injury	30
Patient history	100
Allergies	25
Family history	50
Medications	100

When you have entered all the data from the first screen, the second screen will appear. Notice that the patient's name you entered at the beginning will appear at the top. Enter data from the second screen as you did for the first screen. The maximum character lengths are as follows.

Data	Length
Height	3
Weight	3
Blood pressure	7
Pulse	25
Respirations	15
General (body physique)	25
HEENT	50
Lungs	50
Cardiovascular	100
Abdomen	50
Genitalia	25

Once again, when you have finished entering the data for this screen, the screen will clear and a new prompt will appear. This prompt is for neuromuscular and skeletal data. It is suggested that such data as motor and sensory function, range of motion, skin hygiene, and reflexes be entered here. As can be seen from the illustration that follows, this screen is different from the preceding screens. After the main prompt, a "?" will appear, and you may enter up to 200 characters. When you have typed in those 200 characters, press <Enter> and another "?" will appear. This is the prompt for 200 more characters of data. In all, you will get five of these "?" prompts in this screen for a total of 1000 characters. [It is not necessary for you to count characters as you type; if you should enter more than 200 characters in response to any prompt, the screen will read " " and you can edit the entry.]

Neuromuscular/Skeletal:
[e.g., Sensory and Motor Functions, Range of Motion, Skin Hygiene, Reflexes]

? [The user may enter up to 200 characters at this prompt.]
? [Second "?" indicates another prompt.]
? [There will be five prompts, as indicated, allowing the user to enter up to 1000 characters of data at this screen.]
?
?

When you have pressed the <Enter> key after the last prompt from the neuromuscular/skeletal screen, the screen will clear and a prompt for X-ray and lab results will appear. This screen works the same as the last one. You will be given five prompts, each with a maximum of 200 characters each. This space can be used as the physician needs and is not further subdivided for this purpose.

X-ray/Lab Results:

? [As in the Neuromuscular/Skeletal screen, the user receives five prompts. The maximum here is *200* characters per prompt, for a total of 1000 characters.
?
?
?
?

When you have pressed <Enter> after the last prompt for the X-ray and lab results, a prompt for recommendations will appear. This screen allows the user to enter up to 200 characters of recommendations. When finished, press <Enter> and the final screen will appear. This is the physical therapy referral screen. Data are entered in the same manner as the data for the first two screens. When the last piece of data has been entered, the screen will display the patient file menu again. A sample screen is as follows.

```
PHYSICAL THERAPY REFERRAL
Name:                Cathy Peters
Physical Therapy as indicated:

  Evaluation:                    ?
  Passive Range of Motion:
  Active Range of Motion:
  Resistive Exercise:
  Tilt Table:
  Ultrasound:
  Gait Training:
  Other:

Precautions:

Physician:
```

The maximum character lengths for each piece of data on this screen are:

Data	Length
Evaluation	10
Passive range of motion	10
Active range of motion	10
Resistive exercise	10
Tilt table	10
Ultrasound	10
Gait training	10
Other	100
Precautions	100
Physician	30

When the screen is displaying the patient file menu, you may choose to enter or edit more data, view a file, print a file, or return to the main menu. To view a file, choose option "2" by typing:

2

Do not type <Enter> after you type "2." The program will then prompt you for the patient's last name. Enter the name exactly as you did in the patient evaluation menu. For example, type:

JaneSmith <Enter>

If a file exists for this patient, the first screen of data will appear on the screen. If a file does not exist, you will be asked if you wish to try another name. If so, enter

Y or **y** <Enter>

You will once again be prompted for a name. If you do not wish to try another name, enter

N or **n** <Enter>

You will then return to the patient file menu.

If the file for the patient exists, the first screen of data will appear on the screen. You may view this as long as you like. When you want to continue, just press any key, as the prompt asks you to do. You may do this for all screens of the file. After you have finished viewing the final screen and have pressed any key, the patient file menu will again appear on the screen.

If you wish to print the patient's file, make sure the computer is connected to a printer, which is plugged in, turned on, and on-line. If you are unsure about this, refer to the owner's manual for the printer. Then from the patient file menu, choose option "3" by typing:

3

Once again, do not press the <Enter> key after you type "3."

The computer will then prompt you for the patient's last name, and you must enter it exactly as you did when creating the file. As in the view file menu, if the file exists, the file will begin printing; otherwise you will be asked if you would like to try another name, and the procedure outlined above applies. When the printer has finished, the patient file menu will again appear on the screen.

Finally, to return to the main menu, choose option "4" by typing:

4

Once again, do not type <Enter> after you type "4."

III. The Evaluation File Option

To choose this option, at the main menu type:

2

Do not type <Enter> following the "2."

The evaluation file menu will then appear as follows.

EVALUATION FILE MENU
1. Run Evaluation
2. View Evaluation
3. Print Evaluation
4. Return to Main Menu
Please enter your selection (1, 2, 3, or 4).

Before you may view or print an evaluation file, you must run an evaluation. Once you have run an evaluation, you can view or print it using options 2 and 3. To run an evaluation, choose option "1" from the evaluation file menu by typing:

1

As before, do not type <Enter> after typing the "1."

The program will now prompt you for the patient's name. Type the name and press <Enter>. If you have already evaluated this patient, the program will not allow a reevaluation. Instead, it will ask if you wish to try another name. If so, type:

Y or **y** <Enter>

The program will then prompt you for another name. If you do not wish to try another name, type:

N or **n** <Enter>

The program will then return you to the evaluation file menu.

If an evaluation file does not already exist for the patient, you will be prompted by a series of questions. Here are some samples of those that you answer with a "y," "Y," "N," or "n" and then the <Enter> key.

DEGENERATIVE JOINT DISEASE CLASSIFICATION

Is there degenerative joint disease in the hip? (Enter Y or N) ? **n**
Is there DJD in the knee? (Enter Y or N) ? **y**
Is the DJD present mild? (Enter Y or N) ? **n**
Is there DJD in the ankle? (Enter Y or N) ? **y**
Does the patient have hypertension? (Enter Y or N) ? **n**
Does the patient have coronary artery disease? (Enter Y or N) ? **n**
Does the patient have angina? (Enter Y or N) ? **n**
Does the patient have a weak ventilatory response? (Enter Y or N) ? **n**
Is COPD present? (Enter Y or N) ? **n**
Does the patient suffer from frequent respiratory infections? (Enter Y or N) ? **n**
Does the patient have chronic renal disease? (Enter Y or N) ? **n**
Does the patient suffer from frequent urinary tract infections? (Enter Y or N) ? **y**
Does the patient have chronic skin disease? (Enter Y or N) ? **n**
Is there acute skin disease in the area of electrode placement? (Enter Y or N) ? **n**
Does the patient have a history of drug use? (Enter Y or N) ? **n**
Does the patient have a history of anti-social behavior? (Enter Y or N) ? **y**

Questions that cannot be answered by yes or no include:

Area of Spinal Cord Injured. This calls for a 1, 2, 3, or 4. Simply type one of these numbers. Do not type <Enter> after the number.

Level of Injury. Enter the number of the vertebra that was injured. If no vertebra exists corresponding to the level you entered in the first question, you will be told that no such vertebra exists and asked to enter another number. Here the program will accept only numeric data. Do not enter any letters. For example, if you said that the area of the spinal cord injured was cervical, and then said that the "**12th**" vertebra was injured, an error message would appear and you would have to reenter the number of the vertebra that was injured. You must type <Enter> after you have entered the number.

Osteoporosis. This prompt calls for a 1, 2, 3, 4, or 5. Once again, you must enter only numeric data. If you enter a number other than 1, 2, 3, 4, or 5, an error message will appear and you will be prompted to enter a number 1 through 5. You must type <Enter> after the number.

Some sample screens for area and level of injury, and osteoporosis, are presented next.

EVALUATION QUESTIONS
Enter the area of spine that is injured.
1. Cervical
2. Thoracic
3. Lumbar
4. Sacral
Please enter your selection (1, 2, 3, or 4).
Enter the number of the vertebra that is injured?
Does the patient have frequent muscle spasms? (Enter Y or N) ? **y**
Are the spasms controlled by medication?

OSTEOPOROSIS CLASSIFICATION
1. Normal
2. Mild
3. Moderate
4. Moderate–Severe
5. Severe

Enter the level of osteoporosis present (1, 2, 3, 4, or 5) ?

Once you have entered all the evaluation data, the first evaluation screen will appear. When you have finished viewing it, you can type any key to continue, per the prompt. Then the next evaluation screen will appear. Up to four such screens will appear, including a screen for:

Constraints prohibiting participation in APT
Cautions to be considered in prescribing APT
Specialists that must participate in APT with patient
Chronic illnesses during which the patient must not participate

When you have finished the final screen and have pressed any key, the evaluation file menu will appear on the screen.

If a patient has no restrictions as noted above, the message that there are no restrictions for this patient will appear briefly on the screen and then the program will return to the evaluation menu.

After an evaluation has been run, it may be viewed or printed. To view an evaluation file, type:

2

at the evaluation menu. *Do not type <Enter>*. You will then be prompted for the name of the patient. If the patient has not been evaluated, you will be asked if you wish to try another name. If so, respond with:

Y or **y** and then <Enter>

You will be prompted for another name. If you do not wish to try another name, type:

N or **n** and then <Enter>

This will return you to the evaluation file menu.

If the patient has been evaluated, the first evaluation screen will appear on the screen. It will remain until you press any key. After you have viewed the final evaluation screen and pressed a key, to ask the program to continue, you will be returned to the evaluation menu. If there are no restrictions, the "No restrictions" message will appear briefly, and you will be returned to the evaluation file menu.

If you wish to print the evaluation, type:

3

from the evaluation file menu. *Do not type <Enter> after you type "3."* You will then be prompted for the patient's name as before. If this patient has not been evaluated, you will be asked if you wish to try another name, and you should respond as described above. If the file exists, the printer will start printing the evaluation. When finished, the program will return to the evaluation menu. Remember to make sure that the printer is connected and ready to print before choosing the third option.

Finally, you may return to the main menu by typing:

4

Do not type *<Enter> after you type "4."*

IV. The Prescription File Menu

If you decide to prescribe the APT for a patient, you may write a prescription by typing:

3

from the main menu. *Once again, do not press <Enter> after pressing the "3."*

The program will now display the prescription file menu. Each of these parts works exactly like the patient file menu, except that there is only one screen of data, like the sample that follows.

APT PRESCRIPTION	
Name:	Doug Weber
Exam Date:	3-9-91
With the noted restrictions, this person appears fit for exercise under the following protocols.	
Quadriceps Trainer:	ok
Hamstrings Trainer:	
Indoor Bike:	ok
Outdoor Bike:	ok
Standing with RGO:	
Walking with RGO:	
Stand/Sit with FES-RGO:	
Walking with FES-RGO:	
Other Protocol:	none
Restrictions:	none

The character lengths available for each piece of prescription data are:

Data	Length
Name	30
Exam date	8
Quadriceps trainer	10
Hamstring trainer	10
Indoor bicycle	10
Outdoor bicycle	10
Sitting with RGO	10
Walking with RGO	10
Sit/stand with FES/RGO	10
Walk with FES/RGO	10
Other protocol	100
Restrictions	200

To save your data and exit the program, choose selection "4" (return to main menu) from the prescription file menu and then select "4" (exit to BASIC) from the main menu, without pressing <Enter> in either case. To exit BASIC, type "system" at the flashing prompt, then press <Enter>. Now, remove the floppy disk from the disk drive and store it in a safe place. This concludes your interaction with the APT program.

Epilogue

In *The Plague*, Albert Camus vividly narrates an outbreak of bubonic plague in Europe during the medical "dark age." He poignantly describes the various ways in which each of his characters copes with the hopelessness of escaping the inexorable scourge. The physicians of the era could offer no comfort, no cure, and no hope!

Today, bubonic plague is a medical rarity (in most parts of the world) thanks to the development of vaccines and antibiotics. Physicians have learned that the history of medicine is one of progress, and today they can indeed give their patients comfort, cure, and hope.

However, even in today's medically "enlightened" age, there are many diseases the physician cannot cure. Some are chronic, some are terminal. But today, the physician will tell the patient suffering from terminal cancer or AIDS, "Don't give up hope!" From the history of medical progress, the physician has learned that he or she can at least give the patient comfort, hope, and medical support, while scientists and researchers around the world pursue a cure.

My rendering of the phrase shown in Figure 1 is: "I will support and practice (my) assistance upon (those) afflicted according to my ability and judgment." These "afflictions" take many forms, including trials and tribulations as well as disease and disabilities. It is the latter two, however, for which the physician is most often consulted.

Ever since beginning my endeavors in the realm of functional electrical rehabilitation (as chronicled in this book), I have been struck by the dichotomy of some physicians' "judgment" regarding disability as compared to disease.

What, then, of the patient with a chronic disability? Here we often see the dichotomy in some physicians' judgment. Some physicians (the same physicians even as above?) will tell the patient suffering from chronic spinal cord injury or other chronic neuromuscular paralysis, "Accept your disability! Adjust yourself and your environment around it. There is no hope of anything better!"

Regardless of (or oblivious to) the tremendous advances of modern technology, such physicians will even go so far as to say that these technological

ΟΡΚΟΣ

***διαιτήμασί τε χῥήσομαι ἐπ' ὠφελείῃ καμνόντων
κατὰ δύναμιν καὶ κρίσιν ἐμήν,***

FIGURE 1. Phrase from ΟΡΚΟΣ (the oath) of Hippocrates.

FIGURE 2. Hope comes into the world.

achievements are "false hopes." It is better that their patients be realistic (i.e., have no hope!). To these physicians, is chronic paralysis the bubonic plague of their own medical "dark age"? I leave it to the reader of this book to decide. I also leave the reader with this concluding story.

Pandora's Box*

In a country, far away, there once lived a little boy who had no father nor mother. He lived alone in a little house. A strange man had left him there. He had no little boy or girl to play with him.

One day he was made very happy, when the same strange man came and left a little girl.

He grew very fond of his little playmate. She had a strange name; she was called Pandora.

What happy times the children had, playing all day long in the garden. They were without any care whatever. Pandora did not have to cook any food, for they ate the fruit that grew in the garden. They had oranges, grapes, and all kinds of fruit.

One day, Pandora came across a box, made of wood, in one of the rooms of the house. It was a very strange looking box. It had a handsome face on the lid. Around the face were pretty flowers.

She wanted to know what was in the box, but she did not dare to open it. Day after day she looked at the box, and the more she looked at it the more she wanted to know what was in it.

At last she said to the boy, "Do tell me what is in that box."

"I cannot tell you," said the boy. "Do not let us talk about it; let us run into the garden and eat some of the fruit."

Pandora went into the garden, but all the time she was thinking of the box.

"What can be in that strange box?" she said to herself. "Can there be a new dress for me? Why does he not want to tell me about it?"

Day after day she talked to the boy about the box. At last the boy said, "I am sick of hearing you talk about the box."

"If you want me to stop talking about it, tell me what is in it," said Pandora.

"I do not know what is in it," said the boy; "a strange man came here one day and left it in the house."

"What kind of a man was he?" asked Pandora.

"He was a very odd looking man; he had wings on his cap, and wings on his feet."

"Oh, I know that man," said Pandora; "he is the man that brought me here. The box is for me; he left it for me. Come, my dear boy, let us open it at once."

*From A.J. Demarest and W.M. Van Sickle, *New Education Readers, Book III*, American Book Company, New York, pp. 66–70, 1901.

"We must not do that," said the boy; "when the man comes back, he will open it for us. That will be the time for us to see what is in it."

This made Pandora very cross. She would not play that morning, and the boy felt very sad and unhappy.

"If Pandora will not play with me, I shall go into the garden and play alone," said he to himself.

When Pandora was alone in the house she went and stood by the box and looked at it for a long time.

"What a pretty box it is!" she thought. "If I open it and look in, no one will know of it."

The little boy did not care for any of his games. He wanted Pandora to play with him; he was very unhappy without her.

He knew that Pandora liked flowers, so he thought he would pick some for her. Then he went into the house to give the flowers to his playmate.

Pandora did not hear him coming. At that time she was not thinking about the little boy.

He thought, "I will stand here and see what Pandora is doing." When he saw that Pandora was about to open the box, he said to himself, "I will stay here and see what is in it, too."

What do you think they saw coming out of that strange box? It was a swarm of bees. They flew around the little boy and made him cry out, "O Pandora, I have been stung."

"So have I," said Pandora. "The room is full of bees!"

The bees stung everything in the room. Then they flew out and stung everybody and everything they met. They stung the pretty flowers, and made them fade. They stung the fruit; they stung the grass, and made it turn yellow. They made everything look as if a curse had come over the world.

"Why did I open the box?" cried Pandora. "Why did I let those bees come and sting us?"

By and by a little tap was heard on the inside of the box. "Let me out," said someone from within.

"I will not open the box," said Pandora.

"If you will let me out I will make the world bright once more."

"If I were you I would open the box," said the little boy.

Pandora lifted the lid. What do you think came out this time? No, it was not a swarm of bees; it was a very pretty thing with wings (Fig. 2). It flew round the room; and Pandora and the boy did not feel the stings any more.

"Who are you?" said Pandora.

"I am Hope. I have come to fill the world with mirth and sunshine. Without me, life would not be worth living. I give hope to everybody."

"Do not go away from us; do not leave this world," said Pandora.

"Yes, stay with us," said the little boy. "You are needed in this world. We cannot do without you. If you go away, joy will go with you. Life will not be worth living."

"Sweet Hope, do stay with us and brighten this world," said Pandora.

And from that day, sweet Hope never went away; she is in the world today. Did you ever meet her?

Index*

*With V. Bauer of Wright State University

MIX
Papier aus verantwortungsvollen Quellen
Paper from responsible sources
FSC® C105338

If you have any concerns about our products,
you can contact us on
ProductSafety@springernature.com

In case Publisher is established outside the EU,
the EU authorized representative is:
Springer Nature Customer Service Center GmbH
Europaplatz 3, 69115 Heidelberg, Germany

Printed by Libri Plureos GmbH
in Hamburg, Germany